The Polish Elite and Language Sciences

Tomasz Zarycki

The Polish Elite and Language Sciences

A Perspective of Global Historical Sociology

palgrave
macmillan

Tomasz Zarycki
Robert Zajonc Institute for Social Studies
University of Warsaw
Warsaw, Poland

ISBN 978-3-031-07344-1 ISBN 978-3-031-07345-8 (eBook)
https://doi.org/10.1007/978-3-031-07345-8

Cover illustration: Allan Baxter/Getty Images (The Old Library Building of the University of Warsaw)

This Palgrave Macmillan imprint is published by the registered company Springer Nature Switzerland AG
The registered company address is: Gewerbestrasse 11, 6330 Cham, Switzerland

Preface

Language sciences, linguistics, and philology in particular are concerned with creating meaning. Their exponents might claim to be doing no more than describing the meanings that various forms of language and literature carry, but I contend that they themselves are deeply implicated in the creation of such meanings. In particular, they attempt to regulate the ways in which linguistic forms take on social and political meanings, and lay down guidelines for the interpretation of literary works. This book is an attempt to show how such social sciences engagements can be reconstructed from the perspective of historical sociology by drawing on the example of one European nation. I do not consider the Polish case, analyzed here, typical or even representative from any particular standpoint. Rather, its usefulness lies in the challenges it poses for many of the commonly used social science models. This is because Poland, with its complex history and non-obvious status among European states, does not seem to fit well into any of the classical types of states or societies. In other words, Poland defies categorization. It is undeniably European, but it is not part of the Western core. Nor is it a typical post-colonial nation. Post-colonial countries are a common object of modern Western critical theory, including the theory of Pierre Bourdieu, who was clearly sensitive to colonial and post-colonial issues, as is evident in e.g. his studies of Algeria. For most of its history, however, Poland has been sui generis, not only in the sense just described, but also for having been an object of, and an influence on, Western European (principally German) and Russian rivalry for centuries. At the same time, Poland has been both a multi-ethnic and multi-confessional empire and (since WWII) a homogeneous nation-state.

Poland was stripped of independent statehood and split between three European empires in the late eighteenth century. This state of affairs endured until the end of WWI. Austria and Russia were classical empires, and it was these two countries that controlled most of the Polish lands and most strongly influenced the formation of the modern Polish nation as well as its social sciences. After 1945, Poland became part of the wider Soviet Empire, which also played a significant role in the development of modern Polish society and its social sciences. Therefore, this book can be primarily understood as an attempt to inscribe the case of Poland in the developing paradigm of the global history of empires. At the same time, it is an attempt to link it to research on the role of the social sciences and humanities in the creation of modern nations, as well as the role of linguistics in the construction of social meanings. As I see it, this combination of different theoretical, temporal, and thematic perspectives on Poland will illustrate the potential of their mutual integration.

Linguistics and literary studies are two closely related academic fields which started to develop rapidly and extensively in the second part of the nineteenth century, as they had clear global dynamics as well as critical roles in state-building efforts for most European countries, as well as stateless ethno-religious groups. They are therefore a fascinating realm of the intersection of global and regional political forces trying to use that critical sector of the social sciences and humanities. For that reason, this study shows the international and intra-Polish political and social context of the development of language sciences, which does not normally attract a great deal of attention from historians of Poland.

This study of Polish linguistics and literary studies is an attempt to demonstrate the way(s) in which field theory (and other elements of Bourdieu's broader theoretical apparatus) can be used to reconstruct the history of a selected social sciences field on the internal periphery of Europe. In the process, it demonstrates the enormous potential of the theoretical approach proposed here, especially for research outside Western Europe. At the same time, it points to possible tools that can help to adapt Bourdieu's approach to other historical, cultural, and geographical contexts. In particular, I refer to the works of Immanuel Wallerstein and Stein Rokkan, as well as Bob Jessop and George Steinmetz, as their approaches for contextualizing social fields research outside Western nation-states evince a great deal of potential. At the same time, I argue that the theoretical models of these and other authors can (and should) be used flexibly, and that there is no need to adhere to orthodox interpretations. I propose contextualizing them with each other, as this can reveal additional variables that influence the dynamics of social fields under specific historical and geographical conditions.

At the same time, this book aspires to contribute to the development of global historical sociology of knowledge. In particular, it attempts to reconstruct a rather complex system of dependencies, influences and conflicts at scales varying from global to local; a system that has created and sustained the contemporary social sciences and humanities. These fields have always had parallel ambitions to achieve universality and become key components in the development of national cultures and ideologies.

The book shows how ideas from the global centers of these disciplines were used during the formation of the field of these sciences in the periphery, the role they played in peripheral state building, and the way(s) in which they were used by both the communist regime and local social and political elites, especially the intelligentsia. More importantly, this study shows the often overlooked role that empires, particularly the Austrian and Russian empires (and the later regional hegemony of the Soviet Union), played in the development of social sciences and humanities, in particular Central European linguistics.

Therefore, while this book is specifically about Poland, its ambitions are much broader. They include developing an innovative view of what can be called a peripheral state, i.e. a state that is much weaker, usually younger, and in many ways more dependent on external forces and actors than a typical Western state. The book reflects on how social elites relate to such a weak and dependent state and how they function within it. It is also shown how such a weak, peripheral state can produce its own intellectual and academic fields and elites and their own sub-fields, and how these are related to external actors, including other states and global academic hierarchies.

One of the critical arguments of the book is that a relatively underused concept proposed by Pierre Bourdieu, viz. the field of power, is needed to make these processes comprehensible. For example, modern Poland statehood, formally established in 1918, was preceded by the emergence of the field of power in the late nineteenth century. The role of the national field of power is also crucial for understanding the dynamics of Polish society, especially its elites, and its intellectual field during the German and Soviet occupations, as well as during the communist period, where an assertive, at times even authoritarian, state was not able to control the entire social elite of the country, as was the case in the Soviet Union.

Although this book primarily concerns Poland and its history, it differs considerably from most books on Poland. The key difference lies not only in the reliance on Bourdieu's theoretical framework, but also in the international contextualization of the dynamics of Polish elites. Most studies devoted to the history of Poland, and to the history of its elites and social sciences, are strongly Polish-centric. While external influences are considered, they are usually perceived as detrimental to what is seen as a healthy, historically distinct, and autonomous dynamic of national development. In this book, however, the modern Polish field of power is seen as a product of the empires that governed Poland in the nineteenth century, in particular the Russian Empire. Other periods of Poland's development are also analyzed in a broad global perspective, with respect to both the position of the Polish state in international hierarchies and to the place of its economy, culture and sciences in the global network of material and intellectual dependencies. All of these realms are seen here as fields in Bourdieusian terms, i.e. as spaces of tensions, conflicts, and alliances, with sectors differing in the degree of their external dependence.

This book can therefore be seen as an attempt to rewrite the modern history of Poland from the perspective of its social sciences. It aims to make this case study a model for the application of Bourdieu's approach to the historical analysis of non-core Western societies. This makes it a reflexive study of the application of Bourdieu's social theory. At the same time, it also critically studies the application of Western social theory in Poland, which is largely seen as a peripheral country. The study of the Polish social sciences, with a special emphasis on linguistics and literary studies, points to the peculiar dynamics of peripheral intellectual and academic fields and their external dependencies. These insights are intended to offer a critical extension of Bourdieu's theory of state and social elites outside the Western core. It is also shown how these theories can be used to reinterpret and expand post-colonial theory, global history, and comparative studies on post-communism.

The book consists of four chapters. The first outlines the theoretical context. This includes selected elements of Pierre Bourdieu's sociology, led by the concepts of field of power and homology, which I use to show how social meanings are created. I then draw on world-systems theory, political cleavages theory, and global sociology of knowledge and culture to show how the specificity of peripheral social fields can be accounted for. I additionally show that theoretical considerations based on empirical analyses of peripheral countries' social cleavages can contribute to our understanding of the operation of the global knowledge production system. In the second chapter, I analyze the history of Poland and science on Polish territory from around the mid-nineteenth century to the present. I attempt a relational analysis of the history of the country with a strong emphasis on the structure of the field of power and international dependencies. The third chapter is devoted to the field of Polish linguistics and literary studies, whose history is also analyzed in relational terms and in reference to the contexts outlined in the first part. My analysis aims at laying bare the non-obvious entanglements of a given field of Polish studies and makes it possible to understand the significance of the international context for its development. The fourth chapter is the general conclusion.

This book would not have been written without institutional support. I first wish to thank the National Science Center of Poland (NCN) for awarding me a grant under the title "Sociological theory of meaning: the field of power and social sciences in process of semiosis" (NCN project no. 2015/17/B/HS6/04161). The project was carried out at the Robert Zajonc Institute for Social Studies (ISS) at the University of Warsaw. I am grateful to my colleagues at the institute for their support, in particular, the director of the institute Anna Domaradzka, as well as Alicja Newecka and Aneta Maciągowska. The bulk of this book was written during a fellowship at the Swedish Collegium for Advanced Studies (SCAS) in Uppsala as a Johan Peter Falck Fellow funded by Riksbankens Jubileumsfond during the 2020/2021 academic year. I am grateful to all my colleagues at SCAS for inspiring talks and comments on my project presentation. I would also like to thank the staff of the Collegium and its Principal Christina Garsten. Hanna Dębska and Tomasz Warczok contributed greatly to this project. We were also supported by Andrzej Turkowski. I am grateful to them for their inspiring and committed assistance, in which, apart from my study of Polish linguistics and linguistics presented in this book, research on the Polish field of philosophy of law was conducted by Hanna Dębska and on the field of Polish sociology by Tomasz Warczok. These studies were further complemented by research on the Polish field of Eastern policy experts, conducted at the ISS by Andrzej Turkowski, and on the Polish field of historical sciences at the turn of the communist and post-communist periods, conducted by Valentin Behr. The study presented here has benefited greatly from the opportunity to collaborate with these projects. I also had the opportunity to present the research discussed in this book during my stay as a visiting professor at École des hautes études en sciences sociales (EHESS) in early 2020, and I would especially like to thank Morgane Labbé and Gisèle Sapiro of EHESS for their help in organizing this trip and their hospitality. I would also like to thank the two anonymous reviewers of this book for their valuable comments and suggestions for improving the manuscript. Its final form is, of course, the sole responsibility of the author. Last but not least I am very grateful to Stephen Canty for his dedicated work on the language editing of this book.

Warsaw, Poland — Tomasz Zarycki

Contents

1

Global Context of Knowledge Production in the Peripheries

1.1 Meaning Production in a Sociological Perspective

1.1.1 Theoretical Starting Points: Bourdieu and Jessop

This study draws extensively on social field analysis, a sociological approach developed primarily by Pierre Bourdieu (Bourdieu and Wacquant 1992). In their optics, social systems are imagined or modeled as social fields, and social actions as activities that take place in the context thereof, i.e. in abstract, often multidimensional, social arenas, which Bourdieu likened to the spheres of social games. Their dimensions are defined by specific assets theorized by Bourdieu as types of capital possessed by social actors participating in games in specific fields. These fields are typically formed around their own types or sub-types of capital, which become the key assets of their inner world and form the basis of its autonomy. This chapter proposes an extension of social field theory; one which attempts to combine Bourdieu's concept of a field with selected theoretical concepts from other spheres, especially those

T. Zarycki, *The Polish Elite and Language Sciences*,
https://doi.org/10.1007/978-3-031-07345-8_1

pertaining to the theory of meaning, in a novel way, in order to correlate field theory with the theory of meaning creation and semiosis. Bourdieu's field theory is therefore compared to several structuralist, and some non-structuralist, approaches in order to construct a theoretical framework that will enable semiosis to be studied within the broader context of its sociological underpinnings, in particular, its political, historical, and other related aspects, i.e. as part of the social production of meaning. The proposed combination of theoretical tools is then used to analyze the processes by which meanings were created in the Polish social sciences during the twentieth century.

My own approach to semiosis takes the "Cultural Political Economy" model proposed by Ngai-Ling Sum and Bob Jessop (Sum and Jessop 2013) as a point of reference. Sum and Jessop draw on several grand theories, including Antonio Gramsci's version of historical materialism, and historical semantics—especially as developed by the German semantic conceptual history school with a special emphasis on the role of Reinhart Koselleck and Niklas Luhmann and his followers. Sum and Jessop are also influenced by the co-evolution of semantics and social structure, and by Foucault's archaeology of discourse and discursive formations. Their ambition is similar to Ferdinand de Saussure's original, but never realized, project to develop semiology as the study of language in both its linguistic and extra-linguistic contexts. Sum and Jessop distinguish two basic processes, or even realms, viz. semiosis and structuration, which they see as the context in which any social action occurs, as well as aspects of it. In their view, semiosis is a mechanism of sense and meaning-making, while structuration denotes the non-symbolic aspects of material worlds and delimits the "compossible combinations of social relations" (Sum and Jessop 2013: 168). They see both semiosis and structuration as "modes of complexity reduction", supposedly transforming relatively meaningless and unstructured "complexity" into relatively meaningful and structured complexity. They therefore assert that complexity reduction mechanisms play a crucial role in any human action. This assumption, borrowed from Niklas Luhmann and his theory of social systems, is one with which I do not unreservedly concur (Luhmann 2012). I would rather perceive complexity, or use Luhmann's more precise notion, "external", or "extra-system", complexity as a social construct that serves to label spheres not fully explored or not fully controlled by specific actors or systems. I propose to view human activities as involving complexity building, or to use Luhmann's terminology, to perceive a system's internal complexity as the only "real" form of complexity, i.e. the one that is projected onto the outside world. I am following here the assumption of classic structuralist thought, which perceives human exploration of the world as beginning with rudimentary symbolic oppositions which are later multiplied and developed into complex grand systems in the process of creating actual images of the social world. At the same time, I see the actual complexity of the natural world as impossible to even estimate, especially given our limited understanding of both the micro-biological world and the cosmos. Some thinkers argue that these realms are actually governed by a few relatively simple "general laws" or mechanisms, which are beyond our comprehension at this stage, and this possibility cannot be excluded. That being the case, the possibility that our subjective perception of things as "complex" might attest more to the inability of certain individuals to comprehend or model a given reality than to their objective complexity. Therefore, I am not convinced by Sum and Jessop's contention that making sense of reality involves reducing complexity for actors (and observers) by directing attention to, and focusing action on, only some of its countless aspects. I would rather see complexity as a human concept, a social construct, related to the fact that increasing the complexity of the perceived world is also a basic human game. In particular, we have to recognize that accusations of failing to see or fully understand the supposed complexity of "things" are a classic rhetorical move. This is nowhere more apparent than in political discourse, where disputants invariably accuse each other of presenting overly simplistic visions of the social world and proposing simplistic solutions to its perceived ills. This is also a common accusation in academic discourse, where the most elaborate (and sometimes book-length) arguments are frequently reduced to their basic components and deprecated as "simplistic". I therefore remain unconvinced by Sum and Jessop's contention that social actors are under constant pressure to reduce the complexity of the social world they are engaged in, in order to make sense of it. As they suggest, "such reductions are never wholly 'innocent': in construing the world, sense-

and meaning-making frame lived experience, limit perceived courses of action, and shape forms of social contestation, alliance-building, and domination" (Sum and Jessop 2013: 149). I would rather talk about limiting the number of possible dimensions to be constructed and (obviously) not only by purely symbolic means. This limitation does not come from "reality" or "nature" itself. Rather, it is conditioned by the resources of the actors and the characteristics of what Luhmann would call a specific social system, and what we could also see as a society or social group.

I nevertheless subscribe to some of the assumptions underlying the Sum and Jessop model. In particular, I strongly support their contention that language should not constitute the primary model for studying social processes (and especially not semiosis), social practices, or their results. Sum and Jessop criticize what they call "discourse-imperialist exorbitation of language", which is meant using language as a model to understand the world beyond language. This approach is shared by e.g. Claude Lévi-Strauss, Julia Kristeva, Ernesto Laclau, and Chantal Mouffe. Sum and Jessop draw on critiques of this language-based school of thought by the likes of Perry Anderson (2016), Geoff Boucher (2009), Michel Pêcheux (1982), and Gareth Jones (1996). Boucher in particular criticized Laclau and Mouffe for mistakenly inflating ideological discourse into the social substance, thereby negating the specificity of the ideological aspect and rendering the effectiveness of economics and politics invisible. It should be noted that Bourdieu himself criticized linguistics for the very same reasons, albeit not from such an avowedly Marxist standpoint. He made this explicit in his *An Invitation to Reflexive Sociology* (1992). This work is highly critical of the "semiological vision of the world", or "textism"—a vision in which social reality is constituted as a text (Bourdieu and Wacquant 1992). I subscribe to the view that the "discursive", or "symbolic", aspect is only one component of the social world and social processes, and that it may be understood not only as a global philosophy of social analysis, but also as a mode of criticism of the dominant branches of linguistics, especially discourse analysis. A notable critic of the one-sided language-centered approach in linguistics is Michiel Leezenberg (Leezenberg 2002), who convincingly identifies and explains the power-related but usually naturalized aspects of any symbolic—not merely linguistic—activity. However, Leezenberg's critique is not structural, i.e. it does not propose any methodology to study communication. Thus, it does not offer an alternative model of imagining relations between language, or culture, on the one hand, and politics, institutional setup, or economy, within the broad meanings of those terms, on the other. The focus is on criticizing speech act theory as lacking a power dimension, thereby presenting the performative aspects of language as acts of individual will. This problem also affects pragmatics, usually defined as that branch of semiotics, along with syntactics and semantics, focused on "performance". Its dominant variants do not view acts of effective language use as exercises of power that merely alter the symbolic world, but ones which also change the material world and whose effectiveness requires material resources.

Here I would like to point out that the abovementioned reductionism of linguists, as well as that of several social theorists who have made language the essence of social reality in their philosophies, is only one of many possible reductionisms in social science. In the same way, we can point to the economic-centrist approach of several branches of economics, but primarily of orthodox Marxism, which reduces all spheres of social activity to economic mechanisms. We can also point to narrow politically-centrist perspectives of certain doctrines in the political sciences, e.g. geopolitics. It is worth noting that geopolitics can also be found in the geographical doctrine responsible for narrow environmental determinism (e.g. the works of Friedrich Ratzel). Another form of reductionism is represented by narrow institutionalism. All these reductionisms seem inevitable to some extent. This is because they are implicit functions of specific social science disciplines. These functions include decontextualizing specific social processes and presenting them as "issues" defined primarily by one specific aspect, be it linguistic, political, economic, or any other. This kind of decontextualization provided by specific disciplines, in particular by the theoretical models and tools of their "applied" variants, legitimizes specific actions, e.g. political, language, or economic reforms. This book aspires to avoid these reductionisms while drawing on Bourdieu's model of the social world with its basic division into the realms of economics, culture, and politics (within the broad meaning of the term). It should be borne in mind that, in Bourdieu's system, these correspond to economic, cultural, and social capital, with political capital as an institutionalized form of social capital (Bourdieu 1986). The logics of these capitals define their respective fields. Obviously, they can all be divided into subtypes depending on specific contexts.

1.1.2 The Role of the State and Other Social Spheres in Production of Meaning

One major type of political field, which is of importance to this work, is the state field. The autonomy of its logic from the realms of culture, and especially economics, is emphasized in the early works of Bob Jessop, in particular those inspired by the writings of Nicos Poulantzas (Jessop 1985). As Jessop reminds us, Poulantzas insisted on the autonomy of the state, i.e. on its institutional separation as a heterogeneous structure distinct from the structure of the economic system. He held that the state is an institutional ensemble rather than a unitary political entity. It is shot through with contradictions and has no political power of its own. Importantly, according to Poulantzas, the state should be seen as a social relation. State power is a form-determined condensation of the balance of forces in the political and politically relevant struggle. Thus, Poulantzas sets himself firmly against any crude instrumentalist account of the state (Jessop 1985: 339). Adopting this perspective and translating it into Bourdieusian optics implies that the state can be perceived as a relatively autonomous field, composed of several sub-fields. Its operation, as Poulantzas implies, is not based on the simple realization of the interests of the economic elite, but rather on the synchronization i.e. relative homology, of its actions with the entire configuration of forces (especially economic forces), acting in a given society. It can therefore be argued that the relationship of the modern state to capital is ambiguous, as Bourdieu has also suggested. The capitalist state simultaneously serves and opposes capital. It sustains the cultural system, which opposes capital, even while legitimizing it.

More generally, Sum and Jessop argue that one-sided, language-centered interpretations of the social world neglect the crucial role of what they call "structuration". It could be said that they see it as "the other side of complexity reduction" (Sum and Jessop 2013: 153), although in line with the assumptions presented above, I prefer to see it as a mode of complexity *production.* However that may be, they argue, that the opposition between semiosis and structuration is a classic agency vs. structure scheme. In their view, this may be seen as the equivalent of the relations between linguistic (symbolic) behavior and material (or extra-linguistic) context. This seems to be an overgeneralization. First, what Sum and Jessop label structuration involves two major spheres, viz. the abovementioned realms of economy and politics, enframed in the modern state. Second, any combination of social structure and action involves all the components of the social world (symbolic, institutional, economical, etc.). Sum and Jessop express this as follows: "All social practices have sense and meaning for their agents: without this, they could not be designated as social. But no social practice is reducible to its semiotic moments. It also has what, for now, we will call 'substantive' aspects, including social forms, institutional contexts, social embedding and social effects" (Sum and Jessop 2013: 154). What distinguishes specific actions and structures are the relative proportions and relations between these (and possibly more) of their essential components. Given the above, the strong opposition between semiosis and structuration, as presented by Sum and Jessop, could justifiably be described as essentialist. Since any social action simultaneously involves several aspects (including economic and political aspects) of meaning-making and structure-making, while being confined within the dominant symbolic, economic, and political structures, it can be assumed that whether that action is perceived as primarily symbolic or material is chiefly contextual. In other words, specific actions or structures in specific contexts are perceived as primarily—or even uniquely—as symbolic (linguistic), economic, or institutional (political). They may well have stronger symbolic or other components, but the extent to which their secondary aspects are concealed partly determines their overall effectiveness. Thus, while e.g. economic reforms are invariably presented (and construed) as attempts at economic rationalization, they always have important symbolic (cultural) aspects and political implications that are seldom mentioned, let alone debated. In the same way, symbolic and cultural reforms are rarely presented as being related to the realms of politics and economics, even though they always are. To conclude, the distinction between what Sum and Jessop call semiosis and structuration seems to be highly contextual in practice.

The question of how relations between these spheres emerge is of crucial importance. Sum and Jessop introduce the notion of sedimentation, which covers all forms of routinization that eventually lead to the contested origins of discourses, practices, processes, and structures being forgotten. However, this "forgetting" seems to be not so much the natural process of waning memory but rather the result of structural transformation. The view that semiosis is essentially the relatively successful synchronization of the dimensions of the organization of several socially produced realms or fields is central to my own approach. Such an act of synchronization alters the connections between two given spheres. If one of them is mostly symbolic, certain concepts are given new meanings and rationalizations, and their earlier ones are "forgotten". This may involve synchronization between purely symbolic oppositions, especially those on which language is based, on the one hand, and apparent "material" oppositions, especially those from the economic or political realm, on the other. One example from the post-communist context could refer to the ways in which economics and politics were interconnected during and after the communist system. This change has impacted the way(s) in which the opposition between entities and individuals endowed with capital and those deprived of it, as well as that between administrative units operating pursuant to different statutes, are labeled and perceived. Under communism, the poor were seen as "victims" of social inequalities in need of state assistance, while since its collapse they have been perceived as "losers" who only have themselves to blame for their plight, and who may need the cultural system to educate them on how to survive in free-market conditions (Danilova 2014). These changes can be theorized as synchronizations between the economic, political, and symbolic realms, which can be envisaged as fields in Bourdieu's terminology. It has to be emphasized, however, that the moment of such synchronization does not occur in the commission of a simple act of an individual or collective will, or in the result of the coincidence, or as Sum and Jessop call it, the routinization, although all those aspects must be present.

Such a synchronization, which to some extent has cultural, political, and economic consequences and ramifications, must be approved by those controlling the relevant realms and often involves considerable energy, or to use Bourdieu's language, capital investment. Obviously, some of these acts are merely local. As such, they require minimal investment (e.g. the establishment of conventions to label specific objects). Moreover, none of them are wholly successful. Their forms may be consensual or confrontational to a greater or lesser extent. However, what is often presented as the imposition of certain symbolic forms, e.g. in the form of new oppositions, is usually a form of symbolic control, although it is never complete. Moreover, what may seem to be a one-way process of imposition is always interactive in nature and involves reciprocal impact and mutual constitution between e.g. symbolic fields and the fields and sub-fields of economics or politics. In this context, Bourdieu writes about the "... dialectical relationship between the objective structures and the cognitive and motivating structures which they produce and which tend to reproduce them" (Bourdieu 1977: 83).

1.1.3 Homology as Mechanism of Meaning Production

I now present the crucial element of my own approach. I posit that what I call (relative) synchronization between realms may be seen as equivalent to the mechanism approximated by several other theoretical models and their diverse labels: by what Bourdieu terms "homology"; by what Althusser terms "interpellation"; and also by the concepts of "resonance" and "synergy". The concept of "isomorphism", especially as it is theorized in the new institutionalism (Wang 2016), might be deemed worthy of mention. Homology, however, differs from isomorphism, a term widely used in organizational sociology. Isomorphism denotes a simple diffusion of institutional patterns rather than an interactional process, whereas homology involves mutual influence and complex negotiations between agents of different levels. However that may be, the mechanism of homology, or occurrence of similarity of structures, is theorized in many similar ways. Another analogy worth noting is general systems theory. The relations between the field analysis and system approaches are discussed in more detail below, particularly in the context of the world-systems theory on which the present work draws heavily. Here, however, I would like to digress a little and discuss an important analogy between the two models. I refer here to the approach of the Polish scholar Marian Mazur and his *Cybernetyczna teoria układów samodzielnych* (A Cybernetic Theory of Autonomous Systems, Mazur 1966). In Mazur's variant of the system theory, the two basic circuits of any system are those of energy and information. Homeostasis is the crucial mechanism that ensures the stability of the system. This is provided by the homeostat, which actually regulates by synchronizing the relations between the energy and information (control) circuits of the system. This role of the homeostat may be compared to the role of the field of power in Bourdieu's field theory. Both are sites where different circuits or interests meet and influence each other, and the entire system is regulated through the establishment of particular point of reference standards. Interestingly, in another of his works, *Cybernetyka i charakter* (Cybernetics and Character), Mazur mentions the possibility of seeing the characteristics of the homeostat as defining features of character when analyzing a human individual as an "autonomous system" (Mazur 1976). This suggests the analogical concept of summarizing the characteristics of society through the prism of the nature of its field of power.

It should be borne in mind that Bourdieu presented homology as a crucial mechanism in his vision of the functioning of the social world in the main chapter of his *Outline of a Theory of Practice* (1977), even though his theory of social fields had yet to be fully developed at the time it was written (the French original was published in 1972). He saw homology as the unifying principle of practices among domains and as a "relation of transformation objectively established between them" (Bourdieu 1977: 83). Bourdieu was writing about relations between domains rather than fields at the time. This was to change later. He mentions connections based on overall resemblance, calling them a "mode of apprehension never explicitly or systematically limiting itself to any one aspect of the terms it links, but taking each one, each time, as a whole, exploiting to the full the fact that two 'data' are never entirely alike in all respects but are always alike in some respect, at least indirectly (i.e., through the mediation of some common term)" (Bourdieu 1977: 111). As mentioned above, homology plays a crucial role in the relationship between objective social structures and cognitive structures according to Bourdieu. He subscribed to Durkheim's hypothesis concerning the genetic relationship between mental structures and social structures. He later expressed it as follows: "Indeed there exists a correspondence between social structures and mental structures, between the objective divisions of the social world – especially the division into dominant and dominated in the different fields – the principles of vision and division that agents apply to them" (Bourdieu 1996: 1). At the same time, he was interested in the embodiment of these structures, and their unconscious internalization while being socialized. The relationship between structure and dispositions was also the founding mechanism of the concept of habitus, which was to become central to Bourdieu's theory (Bourdieu 1977: 84). As he argued, "every successfully socialized agent possesses, in their incorporated state, the instruments of an ordering of the world, a system of classifying schemes which organizes all practices, and of which the linguistic schemes (to which the neo-Kantian tradition – and the ethnomethodological school nowadays – attribute unjustified autonomy and importance) are only one aspect" (Bourdieu 1977: 123–124). Several recent studies have convincingly illustrated these mechanisms. For example, Vincent Pouliot's work on the diplomatic field (Pouliot 2010) captures the connection between individual habitus and international relations especially well by offering a striking illustration of how individuals acting as representatives of their countries on the international scene come to "embody" the structural properties thereof. This embodiment may also be related to the widest imaginable spectrum of the dimensions of material cultures. Bourdieu writes about the relations between "corporal space" and "cosmic space", which enables him to describe the difference between genders as they emerge from the study of the house. Here, it is worth calling to mind some of the fundamental oppositions he reconstructed by means of his anthropological field study of Kabyle tribes: "The house is organized according to a set of homologous oppositions – fire:water:: cooked:raw:: high:low:: light:shade:: day:night:: male:female … the same oppositions are established between the house as a whole and the rest of the universe …"

(Bourdieu 1977: 90). By using the vocabulary of social fields which Bourdieu later adopted, it can be argued that each social field is homological in one way or another to the global social space, or it can be seen as a refraction of its structure to varying degrees. However that might be, the relations between fields are a fundamental part of the structural study of social space in Bourdieu's system. Among the formal methods of its study, the (multiple) correspondence analysis (MCA) stands out as Bourdieu's favorite tool. In his view, this was an essentially relational technique whose philosophy corresponded to exactly what social reality was (Lebaron and Le Roux 2018).

This perspective may be related to the classic structuralist argument of Ferdinand de Saussure that the negative, oppositional differences between signs are of crucial importance to the emergence of meaning. As he pointed out, the key relationships in the structuralist analysis, e.g. nature/culture, life/death, etc. are binary. These are analogous to the symbolic oppositions reconstructed by Bourdieu. Saussure argued that "concepts are defined not positively, in terms of their content, but negatively by contrast with other items in the same system. What characterizes each most exactly is being whatever the others are not" (Saussure 1983: 115). Saussure also noted that, in writing, "the values of the letter are purely negative and differential" and that we need only be able to distinguish one letter from another (Saussure 1983: 118). As for his emphasis on negative differences, Saussure argued that although both what he called the signified and the signifier are differential and negative when considered separately, the sign in which they are combined should be seen as a positive term. He added that "the moment we compare one sign with another as positive combinations, the term difference should be dropped (…). Two signs (…) are not different from each other but only distinct. They are simply in opposition to each other. The entire mechanism of language (…) is based on oppositions of this kind and upon the phonic and conceptual differences, they involve" (Saussure 1983: 119). Thus, in Saussure's model, arbitrarily constituted negative differences establish a structure of meaning, which can be theorized as the smallest level of social field. This could justifiably be termed a "semantic field", if this label had not already been taken by theorists who adopt a considerably different approach; one that examines families of words or concepts which share some basic nature; and in particular, by Anna Wierzbicka in her theory of "semantic primitives" (Wierzbicka 1992). However, I would now like to move to the models of social reality and social processes inspired by de Saussure. As mentioned above, they usually assume that performative self-reference reproduces its tautological effectiveness. These approaches assume a self-sustaining symbolic sphere in which hegemonic ideologies introduce certain social practices, which in turn are reiterated, thereby legitimizing their ideological justification. As mentioned above, such an approach is best represented by the writings of Louis Althusser, Ernesto Laclau, and Chantal Mouffe, who argue for the crucial agency of ideological performativity in achieving interpellation effectiveness of what Althusser (1994) called "ideological state apparatuses". As mentioned above, these approaches perceive the performative effect as operating in the purely symbolic, ideological, or linguistic dimension, whereas I would argue that they cannot be considered purely symbolic actions, as they necessarily involve the use of political and economic resources, which are usually rendered invisible when used by any ideology. I would even go so far as to argue that the performative effect is assumed at the moment the homology is established. In this context, the question of arbitrariness would seem to be of importance. Culture- or langue-centered approaches tend to assume a rather contingent and arbitrary nature of symbolic actions, which may even be seen as expressions of the free will of individuals. I would, however, stress the embedded nature of any symbolic action. On the one hand, it is restricted by pre-defined, historically established structures, and above all, by an accumulated resource of actors who undertake seemingly effortless symbolic actions. On the other, however, there is the issue of possible pre-existing non-historical structures conditioning cultural expression. On the one hand, biological (e.g., innate cognitive structures), and on the other, transcendent sources of culture to which some authors directly or indirectly allude. For instance, even if de Saussure argues that structures of meaning are arbitrary and their precarious position is not secured by any transcendental instance, he does refer to the idea that meaning is somehow derived from an ultimate guarantor, a transcendental locus, even if he is trying to falsify that notion. Nevertheless, he admits that there is something that exists at a fundamental level, perhaps in the biological makeup of the human brain, which structures the functions of the human mind, as in later forms of structuralism. On the whole, the structuralist concern with the scientifically objective, and the unambiguously definable, can be seen to produce moralism, which in the wider perspective can be perceived as an attempt to replace God with Structure.

1.1.4 The Role of Homology's Restricted Nature in Semiosis

I now lay out my own understanding of semiosis. This should not be seen as merely a facet of the semiosis-structuration mechanisms, as in the Sum-Jessop approach, but as a component of any social action and structure. More specifically, semiosis can be understood as a function of homology. In this perspective, meaning would emerge from, or be the result of, synchronizing specific fields or their dimensions. That includes those that are mostly symbolic, those that are mostly economic or political (institutional), and those that are considered "natural", but controlled by some social systems. Such homologies can be systemic, even global, e.g. when oppositions between empires or systems such as "capitalist" and "communist" emerge, and their binaries homologically shape multiple social fields down to the local level. But these homologies can also be very local, even individual. Their most important feature is that they are never complete, and the remaining differences between the structures of specific fields become the locus of the meaning.

How Bourdieu saw the relations between the individual and the social world would seem to be highly pertinent here, in particular his argument that "the homology of world-views implies the systematic differences which separate singular world-views, adopted from singular but concerted standpoints. Since the history of the individual is never anything other than a certain specification of the collective history of his group or class, each individual system of dispositions may be seen as a structural variant of all the other group or class habitus, expressing the difference between trajectories and positions inside or outside the class (…) Personal style, the particular stamp marking all the products of the same habitus, whether practices or works, is never more than a deviation in relation to the style of a period or class so that it relates back to the common style not only by its conformity" (Bourdieu 1977: 86). Of crucial importance here is the notion of deviation, which can be defined as the degree of incompleteness of a given homology. In this case, the homology in question is between individual style, or personal social trajectory, and the wider field of social classes. It creates what might have been called an individual semantic field had the concept of "semantic field" not already been taken by theorists such as Anna Wierzbicka, as it has been mentioned above. We can therefore distinguish between local/individual homologies and structural homologies. The mechanism of interpellation, as defined by Louis Althusser, can also be seen as a form of homology between structural and individual fields. Here, it should be kept in mind that interpellation, as introduced by Althusser, is the imaginary mechanism of mutual recognition by which ideology operates to constitute concrete individuals as human subjects. In his view, subjects of experience are ideologically produced and thereby equipped to perform their allocated roles in the social division of labor (Althusser 1994).

It bears repeating that homology, which can be identified with the synchronization of the structures of fields, or the coordination between their dimensions, is always partial in my approach. Meaning is therefore not produced by the exact reproduction of the dimensions of structure of fields, but by the (usually small) deviations between them. This requires a degree of synchronization, which may be close to a full linkage, but never complete or strong enough to make the relation between any two fields noticeable. Complete autonomy of fields implies a lack of any relationship between them, which in turn implies the impossibility of semiosis. Any message based on such full synchronization will appear as redundant. This seems to accord with the argument put forward in the literary studies of Jan Mukařovský, one of the founders of the Prague Linguistic Circle ("Prague School"), which is discussed in more detail below. Mukařovský, unlike the Marxist critics, for whom the value of a literary work lies in the ideological correctness of its representation of the social environment, judged the value of a literary work on its innovativeness in departing from literary tradition and conventions, and its ability to challenge the way society is perceived. He insisted on the dialectical model of a "semiological fact" double function (viz. autonomous and communicative) of a work of art (Gandelman 1988). This can be related to the fact that any act of establishing a homology requires both a degree of autonomy of a specific field and a degree of synchronization. Famous literary works, which were the object of Mukařovský's interest, constitute a specific object in which both individual and broad social agency is activated. But, as mentioned above, very individual and very structural homologies can be identified.

1.1.5 Bourdieu's Insights into Workings of Homology

I now return to the question of homology in Bourdieu's approach; this time in reference to a later stage in the development of his conceptual framework, when his theory of social fields, construed as the scaffolding of the social world, had already been developed. Olivier Roueff points out that although homology was central to Bourdieu, even at that stage, it remained under-theorized as a concept, and its actual workings were something of a mystery (Roueff 2013). In particular, Bourdieu applied the notion of homology when attempting to explain the structures of the field of cultural production (Bourdieu 1983). He mentioned homologies between on the one hand the fundamental opposition that gives the field of cultural production its structure and on the other hand the oppositions that comprise the structure of the fields of power and class relations (Bourdieu 1983: 325). His contention that demand for cultural production is not the product of a conscious arrangement between producers and consumers is particularly important. He rather saw that "It results from the correspondence between the space of the producers, and therefore of the products offered, and the space of the consumers, which is brought about, on the basis of the homology between the two spaces ..." (Bourdieu 1983: 326). Bourdieu also argued that "the encounter between a work and its audience (which may be an absence of immediate audience) is, strictly speaking, a coincidence which is not explained either by conscious, even cynical adjustment (though there are exceptions) or by the constraints of commission and demand" (Bourdieu 1983: 326). He explained this in terms of the homology between positions in the space of production and positions in the space of consumption, including in the field of power. Bourdieu's paradigmatic example of homology of structures is the opposition between the dominant and the dominated fractions in the field of power and the field of class relations.

At the same time, Roueff argues that casual relations are difficult to trace as far as the mechanism of homology is concerned. Roueff, following Bourdieu, argues that a battle to control the conversion rates between capitals is being waged at the heart of the process of homology. Moreover, this battle is related to the power struggles between the elites of each field, and is therefore also a struggle to establish a hierarchy between the capitals of each field, i.e. a struggle for a (symbolic) meta-capital. Roueff proposes to talk about the "social magic" of the "coincidence" between supply and demand, which is so universal that it has "all the appearances of a pre-established harmony", as Bourdieu puts it. Roueff calls the intermediaries working between fields the "magicians" of structural homology. However, homologies are necessary but not sufficient conditions for enabling these intermediaries to adjust supply and demand (or convert capital). Although structural homologies between different fields are widely observed by scholars in their empirical studies, their causal status concerning the relations between fields or between supply and demand is not entirely elucidated. Like Bourdieu, Roueff is most interested in the cultural sphere and the relations between the fields of cultural production and cultural consumption, and believes that cultural intermediaries, such as agents, gallery owners, programmers, critics, and financiers, play a crucial role. They try to reduce uncertainty of demand for cultural production and to ensure optimal conditions for the products they invest in. This involves anticipating the categories of reception and involving them at different stages of the production process and the distribution of its products, so as to maximize the likelihood of obtaining the expected sanctions from the consumers. In short, they try to control the conditions of reception, both objective and subjective, by attempting to shape the representations and materializations of the expected categories of demand during the production process. This includes controlling supply and its effect on scarcity, through the simple selection of contending producers, by reducing the range of possible products, or by advising producers as to how to orient their production. Control also includes legitimate modes of appreciation by critics, broadcasters, schools, or cultural education institutions.

Recent studies provide several instructive examples that confirm the logic reconstructed by Bourdieu. A study on the homology between definitions of "ecological" (organic) products in the field of sales of consumer goods and other fields in a particular field of consumption and lifestyle choices affected by new strongly moralized categories is a case in point (Suckert 2018). Roueff argues, following Bourdieu, that intermediaries translate structural homologies into conscious strategies for controlling reception, or much more frequently, into series of trial-and-error events obscure to themselves. This may call to mind the Sum and Jessop (2013) semiosis scheme, in which the emergence of a new symbolic frame passes through the stages of variation, selection, retention, and sedimentation. This model is compatible with a previous proposal of mine to view human activity as generators of complexity, predominantly by producing diverse visions, models, or narratives, rather than efforts to reduce external complexity. The multiplicity of narratives generated during the initial stage of the emergence of a new paradigm creates what Sum and Jessop call "variation". This should not be confused with the actual complexity of the physical world or what Luhmann would call "external complexity". In this context, we can also remind Jessop's concept of "collibration" defined as "the rebalancing of different forms of governance within and beyond the state in the shadow of hierarchy. 'Collibration' is more than a technical, problem-solving fix: it always involves specific objects, techniques, and subjects of governance, and it is tied to the management of a wider 'unstable equilibrium of compromise'" (Jessop 2012: 76). Jessop's notion of collibration seems therefore to be strongly related to Bourdieu's idea of homology. Jessop emphasized the need of identifying and exploring the extra-semiotic conditions that both enable meaning-making and make it more or less effective, not only in terms of comprehension, but also in terms of practical action.

The relations between cultural production and cultural consumption only constitute one example of many interesting pairs of homologically coupled fields. Moreover, it is a specific type of example; one in which one of a pair of fields is "producing" while the other "receiving", as the names given by Bourdieu attest. Even in this case, however, the asymmetry is not complete, and there are clear mutual influences. In the language of Bourdieu's economic metaphor, "demand" does not simply adapt to "supply", but also shapes it. What is crucial here is the production of meaning or semiosis as a result of these sorts of homologies. This works both ways, as on one hand linking production with consumption shapes the meanings of cultural products, which become associated with certain class positions or other identities of consumers of these products. On the other hand, the same homology influences the meanings of specific sectors of field power and class positions. As a result, to some extent, class positions and power positions may be expressed through specific cultural forms identified with certain consumption patterns.

1.1.6 The Political Dimension Meaning Production

Relations between the political field, in particular the field of institutionalized politics, with the crucial role of parties, and other actors, such as political leaders, on the one hand, and other fields on the other, also provide a lot of illustrative examples on the workings of homology. To some extent, relations between the political field and the field of political preferences may call to mind the relations between the fields of cultural production and the cultural consumption discussed above. There is also a clear sector of intermediaries whenever there is an interface between institutionalized politics and voting behavior. These intermediaries are mostly impersonalized by journalists, intellectuals, the media, and political campaign experts. They all work hard on establishing specific homologies between the "supply" of the political field and the "demand" of the voters. The semiosis created by the homology of the two fields also works both ways: parties acquire identities created by the social characteristics of their dominant electorate, while sectors of the fields of power and class positions may become strongly identified with the parties they tend to vote for. In other words, specific social classes or other groups, including regional ones, may develop allegiances to specific parties, which appear to be stronger predictors of their voting behavior than any other of their characteristics (Campbell et al. 1960). However, several other factors and actors simultaneously influence party strategies and voter behavior. These include historical heritages of cultural and institutional structures, economic interests, state infrastructure, the legal system, international entanglements, and many others. Their fields often interact with the fields of party politics and voter behavior, thereby establishing various kinds of homologies that create or impact the meanings generated by their respective fields. It should be emphasized that any interaction between fields is always mutual, even in the case of clear asymmetry, which may give the impression that one field is completely subordinated to another.

At the same time, the sociology of politics provides several interesting examples of the workings of homologies between large social fields. First, several debates on the relationship between voters (demand) and political parties (supply) are of relevance here and the studies on possible homologies are instructive. Given that the political game has several dimensions in mature democracies, especially European ones, and that these relations are far from simple, the number of possible combinations between them, and consequently the number of possible homologies, is quite large. Of interest in this context is the notion of political and social cleavages, as these may be seen as an equivalent of a dimension of the fields of politics and related fields. Studies on political cleavages often focus on their relationship to social cleavages. This places homology in the center of the study of political behavior and electoral politics. The establishment of clear homologies between the fields of politics and field of class positions is a common goal of political strategists. However, they may also want to avoid such clear homologies, e.g. when attempting to form catch-all parties. Moreover, the strong homology between the political field and other important social fields is not normally considered a desirable outcome from the standpoint of democracy theory. It may produce a deep social fracture, considerably complicating the functioning of society. This kind of polarization is sometimes called the state of "duality of society", which is sometimes diagnosed in Central and Eastern Europe (Éber 2011). However, even in cases like this, these sorts of diagnoses can be read rather as warnings or predictions, as full homology does not seem to be an attainable state. Similarly, as Szymon Wróbel has pointed out, the effect of Althusserian interpellation must always be incomplete, unfinished (Wróbel 2012). Civil society theorists usually warn against this kind of scenario and promote systems that cross-cut political and social cleavages as the basis of a "healthy" society. Political communication and trust are much easier to maintain in these sorts of partially synchronized systems, as people usually find themselves on opposite sides of conflicts on specific political or social issues and the fact that these are not linked makes opposing positions difficult to predict. More generally, such a configuration of cleavages makes politics (and other realms) autonomous, thereby producing a sphere whose positions are not easily translatable to positions in other fields.

At the same time, it should be noted that political sociology has interesting measures of what can be interpreted as homology. One such is the index of class voting (Alford 1962). This is defined as the share of voters representing a given social class who vote for a party claiming to represent that class. Analogical indexes can be created for other social divisions, such as religion, education, or ethnicity. These will measure the degree of homology. Absolute homology can probably only be produced in an institutional way, where the system clearly differentiates between specific sectors of society, and which often leads to the fragmentation of the country. This was certainly the case in the former Yugoslavia. By contrast, the erstwhile pillarization (religious segregation) of Dutch society did not bring about the breakup of the country. Of relevance to this discussion is the theory of political cleavages in post-war Western Europe developed by Seymour Martin Lipset and Stein Rokkan (Lipset and Rokkan 1967)—not so much its Parsons-inspired functionalist introductory sections, but more its historical component. In that later part of their work Lipset and Rokkan argued that the main cleavages in post-war Western Europe were a product of French (political) and British (economic) revolutions. They saw the former as producing what they called the center vs. the periphery cleavage and the state vs. the (Catholic) church cleavage. The British revolution was seen as the source of the economic left vs. right (employees vs. employers) cleavage and the urban vs. rural (industry vs. agriculture) interests cleavage. Of crucial importance in their scheme was the famous hypothesis of the "freezing" of these cleavages, which began to emerge in the nineteenth century and were fully institutionalized at the beginning of the twentieth. Lipset and

Rokkan argued that, in many cases, the original conflicts that produced the specific cleavages have ceased to have any bearing on them. One of the principal reasons for this was the marginalization of one side of a given conflict (e.g. rural or Church interests). The structural feature of the organization of the party system, however, remained, as did the semantic (discursive) structure of the political field. The role of these deeply entrenched divisions (both in voters' identities and the organizational system) as symbolic organizing principles of the political system continued. In some cases, it may still continue, if only through inertia or reproduction. This thesis provoked an interesting debate on the extent to which political cleavages in post-war Western Europe were "real" or "symbolic". What was considered "real" was, of course, mostly understood in terms of more or less tangible interests and deeply ingrained identities. It is not normally possible to completely distinguish identities from interests, as they are thoroughly and inextricably intertwined in the same way as what Sum and Jessop call "semiosis" and "structuration". However, these distinctions, as stated above, continue to be routinely made, and they can be seen as performative, i.e. identifying a specific cleavage or interest as "material", "political" or "symbolic" is not an objective statement, but rather a position whose effectiveness is conditioned by the resources of those who adopt it. Lipset and Rokkan, for example, separate identity-based from interest-based cleavages, even if the expression of any interest also involves assuming an identity of some kind. Cleavages considered "symbolic" or cultural are usually divisions which can be considered more "historical" than others, i.e. as being legitimized by past events. It can be argued that they are erroneously identified as "real" because of a time lag, but non-obvious "real functions" of what appear to be merely cultural cleavages can always be identified. For instance, the cultural capital around which symbolic cleavages are built enables power relations to be stabilized and is often a legitimatization, if not a proxy, of economic or political interests. Another intriguing aspect of political cleavages is that they involve altercations over definitions and meanings of key concepts. These may appear to be purely linguistic in nature, as is the case with e.g. disputes over the meaning of axial notions, such as "liberal", "left", "right", "traditional", "modern", or "progressive". In the context of Central and Eastern Europe, these arguments can be related to debates on e.g. "the real left/right" and "the meaning of being socialist/communist/Soviet or post-socialist/post-communist/post-Soviet". These seemingly symbolic conflicts can have far-reaching implications, as the logic of political cleavages is often coordinated with several other crucial fields. They thus define the logic of friend and foe.

Another intriguing aspect of political sociology is the geographical dimension of politics. The spatial aspects of social processes are studied as part of several disciplines, and they are addressed in several chapters of this book. My own studies on electoral geography, in particular, the electoral geography of Poland (Zarycki and Nowak 2000), provide several telling insights into the working of homologies in the political sphere (within the broad meaning of that term). First, in the late 90s, using factor analysis, I identified a two-dimensional geographical structure—or two clear and very stable dimensions along which the Polish political map was organized. First, there was the opposition of conservative-religious regions vs. secular-former-state-farm-dominated cleavage, and second, there was an urban vs rural cleavage. These appeared to be strongly homological to structures of the political scene or political field when reconstructed in different ways, especially in terms of ecological, i.e. spatially defined, data, such as voting results by county, as shown in my geographical analysis. Poland's political system was characterized by a highly unstable party system in the 1990s and 2000s. Political parties mushroomed, fractured, merged, and vanished with an impressive intensity. Despite this, a very clear and stable two-dimensional structure could be discerned in analyses of the political scene and voting behavior, almost irrespectively of the type of elections. This demonstrates that there was, and still is, a highly homological relationship between the two fields, even if there is a rather weak relationship between parties and their electorates, primarily for the simple reason that parties are inherently unstable in Poland.

1.1.7 Revolutions as Re-adjustments of Homologies

As mentioned above, these sorts of homologies, however unexpected, do not arise by pure happenstance, but are the results of the systematic and deliberate efforts of specific actors. It could even be said that there is a system of "working to establish homology", which involves efforts intended to synchronize diverse fields. Social change, especially revolution, requires the synchronization of fields and the creation of new common meanings. These are necessary for both an effective political mobilization and concentration of resources. The "making of revolution" thus requires working on homologies, and synchronizing conflicts in different fields. Interestingly, a revolutionary situation often involves a dispute over what qualifies as "real". For instance, during the fall of communism, political capital was devalued, while the cultural capital of opposition intellectuals was re-evaluated. But as mentioned above, it can be argued that any field and any cleavage is partly "symbolic" and partly "real", but above all, economic or political. This is because it is not possible for a field to exist without some non-symbolic resources, and no "material" field can exist without some symbolic components. The notion of symbolic capital introduced by Bourdieu is the interface between the symbolic and material aspects of assets. Symbolic capital makes material (or more generally non-symbolic) resources seem symbolic, and enables symbolic work which is erroneously deemed to be as purely symbolic.

What are usually considered "historic events" can be interpreted as moments of "re-synchronization" of specific crucial dimensions, most often through a re-arrangement of crucial cleavages in the field of power (Dobry 1992). There is no shortage of several "field-based" examples of revolutionary moments or interpretations that are easily interpretable in terms of homology and the logic of fields. Bourdieu, for instance, explained the crisis of May 1968 in terms of the "synchronization of crises latent in different fields" in his *Homo Academicus* (Bourdieu 1988: 173). The book devotes an entire section to the issue of synchronization, mostly understood as the emergence of coincidence "between the dispositions and the interests of agents occupying homologous positions in different fields". The same chapter additionally discusses a model transforming a "regional crisis", i.e. one confined to a sector of social space, into a "general crisis, a historical event". Bourdieu argues that such synchronization is also largely dependent on the time scales of specific fields, or their sectors, as it also requires that cycles of change be synchronized. As George Steinmetz notes, this interpretation of May 1968 made by Bourdieu, constitutes an alternative view of that historic moment, which is usually seen as a radical move from tradition to modernity (Steinmetz 2011). Another classic reinterpretation of this type comes from Immanuel Wallerstein (1989), who challenged the prevailing interpretation of the French Revolution as a bourgeois revolution. He proposed to view it as a moment when what can be called (in Marxist terms) the "ideological-political superstructure" caught up, i.e. when it was synchronized with a considerably transformed "economic base" and altered relationships of production and ownership. Wallerstein argued that the Revolution was less important for France than is generally thought, although it had vital global implications, especially in the symbolic sphere, as it became an important model, and even a point of reference for reforms, in many other countries. Wallerstein argued in particular that the Revolution defined the stable frames of reference for the three "classic" political currents of the modern era, viz. conservatism, liberalism, and radicalism. This argument also accords with observations that Bourdieu made later in his *On the State* (Bourdieu 2014: 346). Here, he argued that there is no justification for making the French Revolution the yardstick for measuring the advancement of modernization. He repudiated the prevailing view that the French Revolution was bourgeois and emphasized the persistent role of what he called "state nobility". He argued further that the French Revolution did not change much in the reproduction of that elite, least of all in its access to the monopoly of juridical capital through its control of cultural capital.

Poland provides another interesting case study of synchronizing fields that can be seen as having a revolutionary aspect. For 15 years after the fall of communism, a classic "freezing of cleavage system", to use the term coined by Lipset and Rokkan, could be observed. In particular, there was a divide created during the communist period that pitted the communists, who had ruled the country until 1989, against the anti-communist opposition. As I have argued elsewhere (Zarycki 2002), that cleavage could be seen as a side effect of the Soviet domination of Poland and could be classified as a center vs. periphery cleavage (to use the Lipset-Rokkan typology). The center was located in Moscow, while Poland was one of the many peripheries of what could be considered the Soviet empire. The cleavage was a byproduct of the opposition between the representatives and supporters of the Soviet center and their opponents. It can be argued that the cleavage in Poland was institutionalized in the late 1970s when the anti-communist opposition started to consolidate, especially when they set up the Workers' Defense Committee (KOR), and the anti-communist left-liberals formed a coalition with the Catholic Church. That cleavage was transformed into the "post-communist" cleavage in which the "former communists" were pitted against the "former anti-communists" (Grabowska 2003). Even though what was originally the defining feature of the "communist camp" had ceased to exist, the cleavage continued to shape the meanings of not only political life, but also those in the many other spheres of social life. Many personal identities and social processes have been molded by the "post-communist vs. anti-communist" dichotomy. Even the general process of social change in the 1990s, which continued until at least 2005, was usually understood in terms of that confrontation. For the dominant anti-communist side, social progress was identified with "de-communization", which was meant jettisoning the remnants of "communism", which could be defined in many mutually contradictory ways. While for the former communist faction of the field of power, and a multitude of homological lesser fields, "progress" meant defending the elements of the progressive heritage of the former communist state against anti-modern anti-communists, who were generally perceived as "fundamentalist" by virtue of their free-market orientation and religious conservatism. That "post-communist" cleavage dominated the Polish political scene until 2005, when the former anti-communist camp started to disintegrate and suffered a resounding electoral defeat. After that, there emerged a new cleavage, which gradually become dominant in the political and power fields. This new divide can be interpreted as another center-periphery cleavage; one that resulted from dependence on the Western core in the classic world-systemic context. It could be interpreted in this light as a delayed synchronization effect of Poland's shift from Moscow to the Western zone of domination, which began in 1989. The new context of external dependence resulted in a cleavage that pitted steadfast supporters and beneficiaries of Poland's integration into Western structures, i.e. Euro-enthusiasts who accepted the new dependence, against Euroskeptics opposed to various aspects of it. From the perspective of the geometry of the political and electoral fields, that change could be seen as a realignment of the cleavage system or, more precisely, the collapse of the two former cleavages (anti- vs. former-communist and urban vs. rural) into a single cleavage (Zarycki 2011, 2015). This could be interpreted as a 15-year delay in the structural homological coordination of the main dimensions of the country's social fields. Interestingly, there was a similar realignment in Hungary, only it began in the early 90s and was manifested by a radical and abrupt change in the ideological orientation of the Fidesz party. The party leader, Victor Orbán, decided to change its orientation from liberal to conservative, in effect adapting early to the new geopolitical context, i.e. redefining the party from anti-communist/Soviet to anti-Western or Euroskeptical. This transformation soon redefined the entire cleavage system in Hungary in the same way as in Poland, only 10 years earlier. The prominent role of Orbán in that process led Zsolt Enyedi (Enyedi 2005) to open a discussion on the role of agency in cleavage formation. The question of agency in homology could also be raised in this context. Orbán could be seen as an actor having clear individual influence on the resynchronization of the cleavage system in Hungary.

1.1.8 Geographical Dimensions of Homology

The issue of control over specific territories provides a useful insight into the workings of semiosis. It could be argued that any attempt to map a territory, both in the direct geographical, as well as the metaphorical sense, may be interpreted as an attempt to establish a homology between the fields representing or constituting the given territory and other fields that are more symbolic in nature and which aspire to a new interpretation, i.e. control, of it. These fields, which have a strong symbolic (semantic) function may be seen as ideologies that often compete to interpret and to represent a given territory. Ideologies understood in this way also constitute territories (and more generally other social fields), as territories cannot exist in the social sense without their organizing ideologies, which may also acquire the form of maps. This may be related to what Foucault called the formation of discourses and the genealogy of knowledge. Foucault argued that these "need to be analyzed, not in terms of types of consciousness, modes of perception and forms of ideology, but in terms of tactics and strategies of power. Tactics and strategies deployed through implantations, distributions, demarcations, control of territories and organizations of domains which could well make up a sort of geopolitics where my preoccupations would link up with your methods" (Foucault 1980: 77). I would posit that this kind of strategy can be seen as an attempt to establish homology between the ideological image to be imposed and the actual territory or more precisely, its existing and better legitimized representation.

One such example, discussed below in more detail, is the emergence of the symbolic map of Eastern Europe generated in the late eighteenth century. Its appearance, which can probably also be interpreted in terms of the Sum-Jessop schema of "variation, selection, retention, and sedimentation", has been masterfully reconstructed by Larry Wolff (1994). As he argues, the orientalist frame of perception of Central and Eastern Europe as a whole, with special roles assigned to Poland and Russia, emerged as a fruit of the ideologies of the Enlightenment and was strongly influenced by the writings of French philosophers such as Rousseau and Voltaire. This image of the region, as "not quite" European and in need of Western civilizing support, seems to be synchronized with the economic rearrangement of the European continent at roughly the same time. What I have uppermost in mind is the emergence of the world system, as reconstructed by Immanuel Wallerstein. Wallerstein argued that Central and Eastern Europe played a crucial role in establishing the system, as it became its first periphery; one that provided a critical supply of natural resources and raw materials, especially grain, timber, and furs (Wallerstein 1974–1989). The establishment of a strongly asymmetric system of economic exchange between Western and Eastern Europe, which is in still place, was very clearly synchronized with what can be called the sequence of the selection, retention, and eventual "sedimentation" of the orientalist imagery of Eastern Europe proposed by French intellectuals. These discourses are still pertinent, as Western Europe continues to dominate its Eastern peripheries in the economic and other fields. The late eighteenth century, which coincided with the fall of the Polish-Lithuanian Commonwealth, can therefore be considered the period during which the homology between the symbolic map of Europe and the then emerging European economic system (which can be understood as a field), was firmly established.

Any geographical, especially geopolitical, discourse, including any cartographic one, may be seen as an attempt to build a homology. From a more general perspective, map-making can largely be seen as establishing homology (synchronization) between diverse representations of territories. In the most classical case, this homology would juxtapose the more direct representations of the physical arrangement of objects in space, e.g. topographical maps based on direct measurements, with more "ideological" interpretations of them, including thematic maps or narratives. A special and radical type of ideological map is what can be called the "mythical map". By this is meant maps, most often ancient, that are partly based on suppositions and fantasies framed in transcendent categories such as sacred and profane. Geographers such as Yi-Fi Tuan, and anthropologists of religion such as Mircea Eliade, are among the best known scholars reconstructing these sorts of symbolic maps (Tuan 1977). One interesting aspect of their underlying logic is the central role assigned to sacred places. These most typically take the form of shrines, with examples from Jerusalem, Mecca, Rome, and Constantinople. They all have their inner logic of symbolic geometries, which are usually homological in some fashion to upper-level symbolic geographies of the universe, empires, states, and particular cities. As for cities, especially national capital cities, their central squares, and sometimes their entire central districts, also have symbolic geographies homological to the wider projections of cosmic and geographic spaces. The cities of Brasilia and Washington, DC are prime examples. The homologies between these urban configurations of space and wider symbolic maps remain in force for as long as a specific political system functions. This demonstrates that these sorts of homologies are to be established and mentioned with tangible resources, i.e. that they do not reproduce themselves out of sheer inertia, as is sometimes assumed. Another type of geographical homology is established between the images and the representations of the spatial arrangement of territories which may be seen as dominated, e.g. peripheral regions like the example of Central and Eastern Europe mentioned above. Orientalist imaginaries can generally be considered homological with the status of regions that are peripheral politically, and especially, economically. Electoral geography is a case in point. Here, I would like to return to the Polish political system and its spatial dimension. The new center-periphery cleavage on the Polish political scene (described above) has a clearly discernable East–West dimension when it comes to spatial patterns of electoral behavior. When it emerged as a clear and stable pattern of the political geography of the country, competing ideological interpretations of that electoral geography began to coexist (Zarycki 2007a). These could be related to the two sides of the contemporary Polish field of power. The side related to the pro-Western, liberal camp is of particular interest, as it very strongly orientalizes the Eastern regions of the country, which happen to be the strongholds of the conservative, Euroskeptic camp. This is thus an illustrative example of being synchronized on a pan-European scale: the fields of economic and political power, and those of symbolic geographies, which play the crucial role of the classic oriental stereotype of the East—on the continental, as well as the national and regional scales. The entire discipline of geography can even be seen as work on homologies. What are sometimes called the performative aspects of map-making are actually related to the establishment of the homologies between them as proposed interpretations of territories, and their other, usually less contested, representations, e.g. topographical maps, which also have their own, albeit better naturalized, ideologies. In the same way, the internal structures of maps (Zarycki 1998) can be seen as homological to structures in several other fields, including those of class hierarchies. These sorts of homologies are also the basis for the development of identities of social groups, which often emerge in relation to the areas of their concentration (Wacquant et al. 2014).

1.1.9 Centrality of Field of Power

I now wish to turn briefly to the role of the field of power in the above theoretical framework. Even if the field of power has a crucial role in Bourdieu's theory of fields, it remains largely under-theorized, as does the mechanism of homology in his system. It is also frequently neglected by scholars who subscribe to Bourdieu's theoretical model, even though he was adamant that no social field can be fully understood unless its relationship to the field of power has been identified. Interestingly, Bourdieu came up with the concept relatively late, only really developing it in his *State Nobility*, whose French original was published in 1989. The field of power has been technically defined as a meeting platform for dominant agents from differentiated areas of activity (economy, politics, military, religion, etc.) who struggle over the commanding principle of legitimation (Bourdieu 1989). Of interest in this context of interest are anthropological studies of the field of power, especially those which try to follow actual, physical interactions between figures that are active in it (Bourdieu and Wacquant 1993). It is significant that the field of power, besides its abstract dimension, frequently materializes in the form of physical meetings between its members, which often know each other and interact directly. These meetings include receptions, dinners, and high-level conferences, some of which have already been studied within this theoretical paradigm (Hjellbrekke and Korsnes 2018). Bourdieu argued that the field of power is predominantly the site of negotiating the dominant principle of hierarchization. Even if some dominated and highly autonomous fields can reverse it locally, they always do so in reference to the field of power. Another function of the field of power is to define the exchange rate of field-specific species of capital.

In addition to those crucial functions of the field of power, as they have been defined by Bourdieu, its role may also be defined in terms of the central apparatus that regulates homologies between its own specific dimensions, as well as those of other key fields. This function primarily consists in defining the nature and the geometry of the key cleavage of the field of power. This cleavage in turn becomes a point of reference for the organization of other fields. Bourdieu frequently addressed the ways in which the key cleavage of the French field of power, one based on the opposition of economic and cultural capitals, is homologically related to crucial divisions in other fields. For instance, in his study of France's elite universities, he compared 1967 with 1984–1985. His conclusion was that the structure of the distribution of schools was similar, although he noted some changes related to the symbolic devaluation of the university and intellectual fields (Bourdieu 1996). In his studies of the French state, Bourdieu found a fundamental opposition between the Ministry of Finance and private banks on the one hand and the Ministry of Infrastructures and "all the agencies connected with the development of social housing" (Bourdieu 2005: 114) on the other. This opposition was homological to the main cleavage of the French field of power. Many of his other studies pointed to other homological linkages between the French field of power and other specific fields. In Bourdieu's model, the field of power is strongly related to the state and its monopolizing centralizing functions. However, the present text argues that it is possible to analyze and theorize the field of power without the direct context of a nation-state. George Steinmetz (Steinmetz 2008) suggests that, in imperial conditions, a social field may be defined in relation to the imperial and national (local) fields of power. In conditions external to a strong and clearly developed nation-state, the distinction between the fields of politics and power seems to acquire particular importance. The possible existence of a global field of power, toward which national social fields and some specialized global fields may be oriented, is discussed below. It can be surmised that such a field exists, and that it first materialized as meetings of the global elite in places like Davos, and as international organizations. Alternatively, it can be held that there is no global field of power (just as there is no global government). Thus, the field of the global hegemon of the world system, viz. the United States, is said to play the role of the global field of power. If this is the case, then the hegemon is possessed of an additional power; one which may be even more important than the privilege of controlling the global reserve currency. One of the arguments for this vision of the US field of power as being coextensive with the global field of power is the scale of international lobbying in Washington D.C., which dwarfs the rest of the world combined (Medvetz 2012).

Another interesting issue discussed in this book is the emergence of a new, national autonomous field of power and the complex relations between states, politics, economics, and culture on the periphery of empire and in places of weakened statehood. I would argue that in this context, it is crucially important to distinguish the field of power from the field of politics, and that the case of countries like Poland illustrates this very clearly. Also important in this context is that Bourdieu seemed to assume the basic opposition he found between economic and cultural capital in the field of power of France was somehow universal. As I argue in the present publication, this may not necessarily be the case, especially in the weaker states of what I call peripheries after Wallerstein. Thus, given the importance of the field of power's function of producing and projecting crucial social cleavages, if the field's structure is organized around specific cleavages, it will have a very different role from what it does in France.

In recent years, more and more systematic studies on the structures of fields of power in countries other than France have emerged (e.g. Schmitz et al. 2017). Most concern Western European countries, especially Scandinavian states (e.g. Hjellbrekke et al. 2007). Among the more interesting ones is a study of the Danish field of power, which suggests that it has a different configuration of the main cleavage than does France (Lunding et al. 2021). Specifically, it claims that the opposition between economic and cultural capital is replaced by the opposition between economic and political capital. It goes on to point to crucial role of social positions based on the rules of power delegation (e.g. dominant political functions or positions of union leaders). Thus, in Denmark, cultural capital appears as partially supplanted by delegated forms of capital. The authors suggest that the Danish cultural elite can be better understood as entertainers than as contenders. In any case, "Thus, dynamics of delegation, rather than control over the means of cultural production, complement the most established fractions of the field of power in the division of the labour of domination" (Lunding et al. 2021: 126). This model, however, is not all that different from the opposition identified by Bourdieu in France. After all, the pole of cultural capital in the French field of power is largely based on the system of higher education and research, and on arts institutions, which are largely maintained by the state. Moreover recent studies indicate a systematically weakening strength of the cultural capital pole in the French field of power (Denord et al. 2018). Be that as it may, the present work has a further ambition to insert itself into discussions about the differential structure of the field of power in countries other than France. Specific proposals for such models in relation to Poland and Russia are presented in the following chapters.

The agency of the field of power is another crucial issue. The field of power seems to be both a reflection of the relations of power in the entire social system, which it represents or on which it focuses, and the locus of their enactment. This may at times be mostly a rather passive but nevertheless visible manifestation of the specific configuration of power relations and social hierarchies in a given society. This might especially be the case when the field of power is strongly fractured, and there is relatively little actual interaction between its sectors and specific members. In other situations, the field of power may become an active actor—a place of crucial negotiations on the future configuration of power relations. This seems to be most clearly apparent during negotiated transitions, e.g. the transmission of power that accompanied the end of communism in Poland against the backdrop of a disintegrating Soviet Union. Its main coordinating site was the "round table talks" (Osiatyński 1996). These involved several months of direct discussions between the communist government and leading opposition figures. At their conclusion, an agreement was implemented as the institutional framework of the transition process, despite criticisms, and even protests, on the part of those who felt unjustifiably excluded.

In any case, this book aims to analyze Bourdieu's theoretical model and its applications outside France. Thus, it can also be seen as an attempt to contribute to the wider debate on the universality of Bourdieu's theses (e.g. Lahire 2001; Silva and Warde 2012; Coulangeon and Duval 2013).

1.1.10 Field of Power and Semiosis

Even if the field of power is mostly a manifestation of processes taking place in other social fields, the mechanisms crucial to semiosis can be observed emerging from its dynamics. After all, by defining homologies, the field of power also defines the basic meaning of the structures of other social fields. The study of the workings of the field of power also provides some insight into several structural mechanisms of meaning generation. In view of that particular role of the field of power, at stake here is the geometry of the main axis of its field or the arrangement of the main oppositions, constituted by groupings of representatives of diverse social forces, within it. The coalitions and alliances, often in the most literal sense of these words, between them are the basic mechanism of meaning-making. These coalitions are never exclusively symbolic or purely political or economic; they always involve every type of resource represented in the field of power. They usually involve attempts at grouping forces around programs of challenging a common enemy and putting aside past feuds to form new clear axes of conflict, thereby changing the systems of meaningful oppositions in a given society. The analysis of the geometry of the field of power is thus also closely related to the issue of meaning generation. In this respect, the definition of the equilibrium point of forces, i.e. a break-even point that may define a wider middle zone in any cleavage and which defines the neutral position in any conflict is crucial. The middle point in any of the major cleavages defines the widely accepted neural or "common sense" position. This is usually hegemonic and allows the actors assuming it to legitimately distance themselves from the positions at either end of the spectrum, which are then labeled "radical". Thus, what is usually disputed in debates on the field of power are the "proportions" of its various aspects. After all, one of the key ways of negotiating homologies is to compare the diverse spheres and try to impose the most strategic equilibrium point in a given dimension. A common argument adduced in negotiations of this kind, mostly by the party defending itself against the establishment of unwanted homologies, is that the realms used in the comparison are in fact "incomparable". The issues of the questionable cause and false analogies between "incommensurable quantities" are often raised. The old saw admonishing the comparison of apples and oranges is also commonly heard in these situations. Making comparisons and taking measurements of seemingly incommensurable objects, concepts, or even narratives, and referencing them to the geometries of specific social fields is, however, a common aspect of any process in the field of power. In fact, it is the essence of any meaning-making activity, and such critiques are best read, not as attempts to prevent making these comparisons (i.e. establishing homologies between what I proposed at the beginning of this chapter to call semantic fields in the spirit of de Sassurean oppositions), but rather as raising the stakes in the debates concerning them. Claiming that some realms are "incomparable" is usually tantamount to claiming that a comparison in the given configuration should be a very asymmetric one, and that the degree of homology in a given situation is rather low. However, the homology between two dimensions does not have to be high for meaning to appear from a comparison of any two fields. There are similar debates on the "correct" use of metaphors, especially political metaphors, which lie at the base of public discourse (Lakoff and Johnson 1980). These may also be seen as tools of establishing or communicating homologies. In these contexts, metaphors do not have any performative power per se, and their effects are defined by the contextual dimension of their use, as has been noted by Michiel Leezenberg (Leezenberg 2001). Any debates on the appropriateness of the use of metaphors in public discourse can therefore be interpreted with reference to some kind of high-level conflict over establishing homologies. The metaphors used by prominent members of the field of power are often subject to heated debate, as they may be seen as a direct effort at reinforcing specific homologies.

1.2 Poland as Semi-Periphery: Uses and Readings of Immanuel Wallerstein's Approach

1.2.1 The World-Systems Theory and Its Competitors

This book primarily intends to explain the dynamics of what is generally considered a peripheral field of the social sciences, or a field of the social sciences that world-systems theory relegates to the periphery, or more precisely, semi-periphery. Peripherality in this context is understood in terms of the definition proposed by Immanuel Wallerstein (Wallerstein

1974–1989). Wallerstein made an important distinction between peripheries and semi-peripheries, the latter being defined as a zone located between the core and the peripheries proper. Its importance lies in certain aspects of the Wallersteinian scheme. For example, it is consequential in the context of the cyclical mechanism of the world system. When there is an upturn, the semi-peripheries share several benefits with the core, in particular, the profits from the exploitation of the peripheries. During downturns, however, they become a buffer for the core, both economically and politically. Their assets rapidly dwindle as investors from the core dispose of their holdings in semi-peripheral businesses, and they are expected to shield the core from any political upheaval arising within the limits of the peripheries. However, in some contexts, this distinction, which can also have even more degrees, may not be all that useful, as peripheral and semi-peripheral areas (i.e. non-core regions) have several features in common. Moreover, this typology of peripheries has attracted a fair share of criticism, not least on the grounds of functionality and arbitrariness. It should be noted that this distinction may be relative (or relational), variable, and even debatable in nature, as are most elements of Wallerstein's approach. It should further be noted that, in this context, Wallerstein's approach to theorizing global-scale dependencies and center-periphery structures is only one of several.[1] It needs to be borne in mind that world-systems theory emerged from the earlier and broader stream of dependency theory. The latter originated mostly from Latin American scholars such as Raul Prebisch, Theotônio dos Santos, Anibal Quijano, and Fernando Henrique Cardozo whose works were popularized in the US and Western Europe by Andre Gunder Frank. Wallerstein, along with Giovanni Arrighi and Samir Amin, later worked on modeling the hierarchical global division of labor to analyze world history. One of the original arguments of Wallerstein and the dependency theorists was that core-periphery relations emerged on a global scale prior to formal colonialism. Wallerstein noticed in particular that a relationship resembling colonialism had existed between Poland and Western Europe during the sixteenth century. He made these observations after

[1] For a brief review of similar or related approaches (esp. economic centric ones), see e.g. (Klimczuk and Klimczuk-Kochańska 2019).

working closely with Polish economic historians (Kola 2018). This book focuses on Wallerstein's model, as it is arguably the most sophisticated and best known of the theoretical models of global inequalities. This is not to say that I am not aware of its competitors and critics, or that I consider it superior in every respect, let alone flawless and unconditionally convincing (e.g. Pieterse 1988 or Skocpol 1977). With that in mind, whenever the terms "peripheries" or "peripherality" are used in this book, they most often refer to Wallerstein's scheme. Poland, the main case study examined, will be seen as a peripheral or, more specifically, semi-peripheral, country with a semi-peripheral field of power. This is central to understanding the logic of a country's social fields as developed in Bourdieu's later works (Bourdieu and Wacquant 1993).

This section is therefore primarily devoted to integrating the two major theoretical perspectives introduced above. I begin with social field theory, mainly in the form proposed by Pierre Bourdieu, and world-systems theory, mainly in the form proposed by Immanuel Wallerstein. I then attempt to combine these two approaches. While this is not a novel undertaking, it nevertheless attempts to correlate these two theoretical perspectives in a way that aspires to be more thorough and systematic than most previous works. This necessitates resolving the important issue of how to combine the concepts of system and field. It should be noted that both have been metaphorically adopted from the natural sciences—one from physics (Martin 2003) and the other from biology (Bertalanffy 1973). This may be seen as a part of a broader trend of the social sciences borrowing and adapting concepts from the natural sciences (Marković 1985).

The possible linkages between the two paradigms can be understood in two fundamental ways. They are generally viewed as divergent or contradictory. This approach is manifested in the works of e.g. Julian Go (2008b). Go argues that the field approach is superior to several others, including the four following main frames of theorizing global-scale arrangements (Go 2008b), viz.: (i) the "realist" international relations theory; (ii) the neoinstitutionalist/world society approach; (iii) normative international relations (IR) theory; (with norms created by "norm entrepreneurs"); and (iv) what he terms "hegemony/world-systems theory", which as he argues, defines the system in terms of a strict economic core-periphery hierarchy. I largely share Go's critical view of the approaches he discusses with the exception of those inspired by Bourdieu and Wallerstein. I am particularly skeptical toward the world society/world polity paradigm, often identified with John W. Meyer, its main, although by no means sole, proponent (Meyer et al. 1997). As Go convincingly argues, it can be seen as strongly underpinned by the assumption of the centrality of the logic of diffusion and does not seem to be a highly sophisticated tool. World society theory sees global culture as cultural templates or models to which all societies and organizations adhere and thereby reproduce. Moreover, the image projected by Mayer is of a consensual world society. In Go's perspective, which is also advocated by Larissa Buchholz and Monika Krause, world-systems theory appears overly materialistic and world polity analysis culturalist. Go and Krause note that the latter is often directly pitched as an alternative to the material focus of world-systems theory. For these authors, field theory constitutes an optimal alternative, a higher-level synthesis of all other models, especially given that it permits both the material and cultural aspects of global arenas to be taken into account. For Go (2008b: 223), world-systems theory defines the system in terms of a strict economic core-periphery hierarchy (or a "world mode of production", international division of labor, "world economy", or "hegemon/non-hegemon"), whereas the field approach alerts us to other possible global power relationship configurations. In particular, the political ecology of recognized and unrecognized areas that have traversed the economic hierarchy may be, as Go argues, better accommodated in the global field approach. As regards the existing cultural approaches, Go argues that a field approach highlights struggles for capital. Similar criticism is raised by Buchholz (2016), who details her own version of the global field approach, which is discussed in more detail below. Buchholz expands Bourdieu's field theory by arguing that it is best suited to engage with transnational or global social orders, such as legal systems (Dezalay and Garth 1996), the global social sciences system (Heilbron 1999), world literature (Casanova 2004), the global economics profession (Fourcade 2006), the colonial state (Steinmetz 2008), and empires (Go 2008a). She also refers to the work of Krause (2014), who has insightfully demonstrated how the global field of humanitarian relief emerged by defining its (relative) independence from state politics, social movements, and religious organizations. As Buchholz argues, these studies demonstrate that Bourdieu's field theory has much to offer for transnational and global sociology when extended beyond national boundaries. Buchholz holds that, in contrast to established global paradigms, field analysis offers a mode of examination of global hierarchies that can move beyond supposedly monolithic models of one overarching global system, be it the economic system represented by Wallerstein or the cultural one represented by Meyer. As Buchholz argues, it provides an approach that can map the plurality and specificity of social spheres in a global context, each revolving around semi-autonomous logics and histories (Buchholz 2016).

I fully concur that Bourdieu's field theory tries to absorb both the material and cultural dimensions of social life into a single logically integrated conceptual apparatus (Go and Krause 2016). However, I would also argue that Wallerstein's model should not be seen as a (false) alternative to the field approach, but rather as being complementary to it. Wallerstein clearly posited the world system as being predominantly a world economy, contending that the "basic linkage between the parts of the system is economic" (1974: 15). He was later to lay particular emphasis on the systemic imperative of the endless accumulation of capital (Wallerstein 2005). However, in an open reading, these are not normative statements, but rather empirical ones, as they refer to the world system as an actually existing, historical configuration of social forces and hierarchies of social logics. In other words, Wallerstein did not imply that economic logic is, or ever was, superior to every other logic and social organization hierarchy in every context. His model rather implies that empirical studies demonstrate that it has been the dominating global principle of organizing what can be considered the world system or global field since the seventeenth century. This is not to imply that the domination of economic logic is a permanent feature of the organization of modern societies. Nor does it preclude the existence of autonomous spaces and social fields that resist or even revert that logic while remaining in some way dependent on the global economic hierarchy, or at least immersed in it.

In fact, not only world-systems theory, but also the broader system theory, can be interpreted and used in two divergent ways. On the one hand, it is a rather normative or utilitarian theory of social control. This approach is de rigueur in the cybernetics applications of systems theory. Moreover, in form, they have been popular in communist countries. The organic metaphor of the system borrowed from biology may also be narrowly, normatively functionalist. In this approach, any organ is supposed to perform certain functions, and these functions usually appear as "objective" and constitutive for a given organ. But on the other hand, it is also possible to adopt a relational standpoint in which the place of an "organ" in the "body" of any system may be seen as negotiable or as the outcome of a clash with the energies, powers, or interests of the other "organs" of the system. In biophysical terms, such a relational understanding of functionality is best reflected by the construction of the brain, with its elasticity, both physical and operational. Specific parts of the brain can take over the functionality of a damaged or dysfunctional one. Other parts of the body may also play a partial role in the overall body "brain" as they have neural networks and their own "memories". Thus, they can also "think" or participate in the overall thinking of a given organism, both on a neural level, and through their physical movement, which may have an aspect of thinking (e.g. the movement of the hands and other body parts). The most extreme form of such a decentralized cerebral system known in nature is probably the octopus (Carls-Diamante 2017).

1.2.2 Relational Readings of System Theories

On a more general level of abstraction, it could be argued that what is seen as functionalist language can also be read either as the normative language of traditional functionalist theory or relationally. In particular, it can be seen on the one hand as an ideology or legitimization of the place of a given object in the wider hierarchies of a given system, and on the other as a description of actual power relations. From this later viewpoint, the function would not be given but obtained during a power struggle and would allow for specific privileges, e.g. the ability to accumulate different types of capital. On the other hand, the act of assigning a function can be seen as an act of symbolic, albeit not necessarily effective, violence. Another aspect of functional interpretations is the notion of division of labor. Buchholz's variant of the cultural world-systems perspective assumes that an abstract, formal center-periphery distinction should be seen as an appropriate tool for modeling global systems, as it does not demarcate functional roles within a system characterized by the division of labor. Buchholz understands this to designate a continuous status variable. She argues that it follows from this that centers and peripheries are no longer entities that fulfill systemic functions, but structural positions along a continuum of inequalities among countries. In my view, this outright rejection of functional thinking is not the most beneficial approach, given the possibility of a relational understanding of functionalism, as indicated above. Even if functionalist interpretations or narratives are not the most appropriate, let alone the only, interpretations of the phenomena in question, they can nevertheless be seen as useful complementary reconstructions of specific social realms. Suffice it to mention the function of "peripheral public" (e.g. the readership of a literary production), which may be seen as an important component of the global system of literature. It should also be noted that any field can be interpreted in terms of labor division. Again, it should not be taken for granted, and certainly not seen as objective, but the different sectors of any field can be seen as performing certain functions, even if they are not "objectively", but relationally defined. The simplest of such interpretations could refer to the classic opposition on any social field, i.e. between those with the highest concentration of assets, which can be seen as functionally dominating, and those with the lowest, which can be seen as functionally dominated. It is therefore fair to say that there can be different types of functionalism (e.g. relational/interpretative vs. normative/formalist, or empirical/critical vs. legitimizing/positivist).

For his part, Go argues that Bourdieusian field theory offers an alternative not only to the dualism between meaning and economic system, but also to the emphasis upon functional cohesion found in system-based theories. For Go, Bourdieu's very definition of field theory incorporates conflict, as fields are spaces where capital is contested. Go also maintains that the global field approach accounts for the fact that core states are largely shaped by, and not merely shapers of, global political culture (Go 2008b). As he claims, theorizing the global arena as a field offers a synthetic approach that simultaneously encapsulates the insights of existing approaches and expands them. He gives several examples of such mutual dependencies. One is the US domination of the Philippines. As Go explains, the preferred US solution was to use informal control mechanisms, including the secret services, rather than initiating a new round of colonization, as this would have diminished the symbolic capital of the US and possibly driven the Philippines to communism. To this end, as Go recounts, the CIA manipulated the Philippine elections to ensure the victory of the candidate preferred by the US. This can be construed as recognizing at least partial Philippine agency (Go 2008a: 223). However that might be, I see no justification for claiming that system theories replace conflict with the narratives of cohesion and harmony. As mentioned above, functions in a system can be seen as the outcomes of negotiated or imposed settlements, and these may be subject to constant review and revision as circumstances change. The mutual influence between the components of a system is also part of its philosophy. The cores of world system may be just as dependent on their peripheries and semi-peripheries as vice versa, even when the system is configured so as to be most beneficial to its core actors. It can even be argued that demarcating semi-peripheries and peripheries, and giving considerable benefits to the former, is an aspect of the fragile dependence of the core on peripheries. System theory might therefore include most of the basic relational assumptions of the critical sociology of Bourdieu.

A more general relational reading of system theory, i.e. one that is more interpretative, critical, and empirical (and consequently less normative), is another option. A system in such an approach would be understood as a self-regulating (albeit never optimally) organism, permeated with internal tensions that could always lead to a major crisis and even terminate its existence at any time. This differs considerably from cybernetic models, where a system is a perfect, rational control tool. Its locus of power is to be found in its own self-regulation mechanism. On a higher, external level it is usually controlled by a single actor or "user" of the system. The regulation mechanism may be theorized as a homeostat, which is a relational perspective best interpreted as a locus of confrontation between competing pressures and interests rather than as an ideal control machine. As mentioned above, the notion of field of power in Bourdieu's scheme may be equivalent to the homeostat in system theory. In a similar vein, the mind of any living organism can also be interpreted as an arena where the competing interests of different sub-systems of the body, or more generally, organism, are contested.

When considering the existence (or even the possibility) of a relational variant of systems theory, it can also be useful to enumerate the crucial commonalities between field and systems theories. Both can involve a relational and an interactive approach, and an empirical rather than a normative approach. What they appear to have in common is their insistence on seeing "things as a whole", i.e. as an interconnected, mutually dependent set of objects with emergent properties. This seems to constitute the essence of the contextual nature of both approaches. Each organ of a system can only be understood in the context of the system as a whole, including the relationship between that system and the overall global system, just as any position in a field can only be understood in the context of the configuration of the entire field, and the relationship of that field to the field of power. Moreover, a system may be able to recreate a lost organ, or retrieve information from its archives (or memory). The history of countries can be interpreted this way when viewed as components of the world system. Some countries not only recover, but sometimes even reappear on the world map (not only in formal but also in its economic dimension) following political and even material destruction during wars and revolutions. It could be argued that this reveals the crucial role of their structural place in the system, and that it is this that makes such spectacular "comebacks" possible. Examples include the reappearance of the FDR (West Germany) and France on the economic map of Europe after WWII. The case of West Germany is particularly striking when compared with that of the DDR (East Germany). The FDR's return to being one of the wealthiest nations of the globe, so soon after its total defeat in WWII, which wrought the near total destruction of its economy, is a fascinating study in itself. This can obviously be interpreted in several ways. One might be to stress the role of financial resources, which despite the physical devastation, remained in the hands of German capitalists, who all moved to the Western part of the country in 1945. This would underscore the importance of accumulated financial capital in the global economic system, in contrast to per capita GDP and economic growth, which are often cited to undermine the validity of the Wallerstein model, in particular in reference to the current status of Central European countries.[2] Another reading of that miraculous post-war German recovery might stress that the world system is not merely an economic system, but has other dimensions, especially a political one, and that these are of crucial importance. Therefore, while being in the core of the system might primarily be defined by being included in its financial circuits, this involves more than simply achieving a certain level of economic growth; there has to be admission into a "club", which has crucial political and cultural dimensions. Then again, it could be noted that some countries with a lower status in the world system keep returning to their "lower normal" after (usually brief) periods of spectacular growth. Poland (along with several other countries, some outside Europe, e.g. Brazil) could serve as a case in point, as it regularly suffers from deep crises that deplete its financial resources after any period (however brief or protracted) of relative prosperity.

1.2.3 Orthodox and Flexible Uses of Bourdieu and Wallerstein

On the theoretical level, Bourdieu's approach is more abstract while Wallerstein's is more empirical. Both, however, are open to diverse interpretations and creative development. It could even be argued that the system approach is equally "dynamics-friendly", as it recognizes the importance of the accumulation of resources and the concentration of assets (energies). Most importantly, however, it focuses on their flows. In any case, as mentioned above, both can be interpreted in two contrary

[2] There is an interesting debate among Polish economists, some of whom point to the considerable contrast between the country's very optimistic per capita GDP and its per capita capital stock. While the former is approaching a third of that of Germany, the average German is around 30 times wealthier than the average Pole. That latter indicator seems to reflect the deeper hierarchies of the world system resulting from long-term accumulation (Deloitte 2015).

ways. The way in which Wallerstein's world system theory is read, and the way in which Bourdieu's general theoretical model is interpreted and applied, clearly share basic commonalities. Bear in mind that Bourdieu's approach has come in for a great deal of criticism for its supposed unsuitability for the study of dynamics and alleged determinism, and that Bourdieu himself has been depicted as a reproduction theorist (e.g. Faber 2017). These charges stubbornly persist despite having been rebutted. One example was provided by Johan Heilbron and George Steinmetz, who convincingly defended Bourdieu against such accusations from Dylan Riley (Heilbron and Steinmetz 2018). This is not to say that, on several occasions, the Bourdieu approach, including his field theory, has not been used in an unambiguously normative manner. One such case was identified by myself and my colleague Tomasz Warczok in Poland. We found numerous highly normative and judgmental applications for Bourdieu in mainstream Polish sociology. In particular, as we demonstrated, several Polish sociologists had been using class and various field configurations reconstructed by Bourdieu in France as a normative point of reference for interpreting what they saw as the "backwardness" of Polish society (Warczok and Zarycki 2014). Bourdieu's models of the French social order were assumed to reflect the structures of an ideal "Western" society, with social inequalities considered "normal" or "balanced", and class relations or configurations of the field of power "harmonious" by necessary implication. It should be noted that the geometry of the fields studied by Bourdieu in France is often implicitly presented by some scholars using his work, as "natural". This is especially the case with the "middle point", or "balance point" of the field, in particular the field of power, which often defines what can be projected as a "neutral", i.e. most natural position on the field.

World system theory can be abused or perceived in ways that are very similar to the Bourdieu approach. It might not be called abuse, but it can still be used or perceived in a similar normative way, i.e. as a rigid, normative framework; one that takes no account of system dynamics. Chamsy El-Ojeili lists several of the modernist "sins" of which Wallerstein stands accused by his critics. The most damning are functionalism, teleology, and namely determinism (el-Ojeili 2015: 124). On the other hand, we can point to a relational and historicized reading of world-systems theory; one that can additionally be seen as flexible, adaptive, and focused on the spirit rather than the letter of Wallerstein writings. I would argue that most of the critiques of world-systems theory, and especially those mentioned in this chapter, are unduly normative and essentialist, and reflect a narrow and formalist reading of Wallerstein. These readings sometimes challenge the entire approach using selected empirical evidence, e.g. any unexpected, short-term system dynamics, to undermine it. But many of the apologetic readings share the same normative assumptions, which I consider to be following the letter rather than the spirit of Wallerstein. This difference may well be related to the account of dynamics in the Wallerstein model. On the one hand, the world system can be viewed as a historically established and evolving arrangement, while on the other, it is static. In the latter approach, the emphasis is on reproduction and stability, which are obviously important elements in the theories of both Wallerstein and Bourdieu. However, a normative reading of Wallenstein results in these features being construed as aspects of the model's rigid and determinist nature, and his system being understood as not only static, but also deterministic and functionalist. On this reading, countries are assigned a status of center (core), periphery, or semi-periphery, typically on the basis of some basic statistical index, e.g. per capita GDP, and this status remains fixed. That perspective deterministically and superficially assigns countries their place in the world system hierarchy. Moreover, their positions in most other spheres, including the cultural and the scientific, are supposedly inferred from these narrow readings. In short, economic relations, however narrowly measured, are supposed to define both political and cultural dependencies in such simplistic approaches. The agency of actors—not only nation-states, but also those at lower levels, e.g. regions, cities, institutions, and individuals—are assumed to be largely determined by the position of their countries in the world-system hierarchy. World-systems theory is sometimes understood and applied this way. People look for regularities and try to explain global processes in terms of per capita GDP. On the other hand, for some critics of Wallerstein, any inconsistency between per capita GDP and other national indicators, or low correlations between these and other global hierarchies, are not only problematic, but repudiate world-systems theory in its entirety. The charge of not accounting for change is also commonly made. These critics seem to understand the theory as assuming that the global system is impervious to change, as the capitalist system has consistently retained its basic core-periphery structure. At most, different countries will occupy different positions, but the slots themselves have remained fixed. Go and Krause do no more than recognize that some variants of world-systems theory see change as mostly cyclical with oscillations between hegemonic periods and competitive periods and between periods of productive and financial expansion (Go and Krause 2016).

However, as mentioned above, the theory can be interpreted in a much more flexible manner on the basis of both an attentive reading and a creative, open reinterpretation of Wallerstein's ideas. Stephan Panther, for example, has proposed his own "selective" reading of world-systems theory, which he claims is a "a very elaborate theoretical and empirical edifice which again is but a cornerstone of a paradigm of social inquiry" (Panther 2014: 8). The historical nature of Wallerstein's theory assumes special significance in his own presentation of it. First of all, it is predicated on the dynamic nature of the world system, even if the dynamics are mostly defined in *longue-durée* categories. This also means, as Wallerstein frequently emphasized, especially toward the end of his career, that as the world system in its present form has its origins in the "long sixteenth century" (1560–1640), when Europeans first circumnavigated the globe and began colonizing and exploiting other continents, it will eventually (possibly soon) dissolve and be supplanted by some other form of organizing global society. It might then return to one of the two previously existing historical types of totalities which Wallerstein singled out in addition to world-economies, viz. multiple cultures primarily united by trade and the division of labor, and world empires, which preceded world-economies and were characterized by political centralization and the forcible extraction of tribute. It should also be borne in mind that the world-systems approach can be seen as more a framework for analyzing change than a static model, as its opponents claim. Most structuralist theories attract this kind of criticism, and is it worth remembering that the Latin American dependentist school has its roots in French economic structuralism, especially in the work of François Perroux (Palma 2016). For their opponents, structuralist theories will always be deterministic, while for their proponents, they will be an indispensable frame of reference for understanding the mechanism and meaning of any social change.

1.2.4 The Multi-level Architecture of the World System

In any case, world-systems theorists generally see the system as constantly changing in several spatial and temporal scales. In particular, they recognize: short-term changes (predominantly cyclical) that potentially modify the status of some countries by moving them between peripheries and semi-peripheries; and the more permanent movement of countries across the different tiers of the world economy, Japan being a notable example (Babones 2012); and the slow but steady evolution of the configuration of the entire system. This last involves the slow movement of the core, which began in the Netherlands in the seventeenth century. This marked the epoch of Dutch hegemony. British hegemony followed in the nineteenth and US hegemony (which continues) in the twentieth. This involved movement of the center of the core first from Amsterdam to London, and then to the East Coast of the US. There have been sporadic attempts to create alternative centers, e.g. the Soviet project. The status of the communist bloc in the context of world-systems theory is open to debate, similarly to many other issues that become an object of contention among the adherents of that theoretical approach. These debates concern various issues, including the question of which economic indicators should be used to reconstruct the hierarchies and dynamics of the system. Wallerstein's wider reception and criticism might also demonstrate that his approach can be interpreted as an open-ended intellectual program rather than a rigid ideological framework. In fact, world-systems theory, similarly to Bourdieu's theory, can be seen as a field in its own right, albeit one with a large and varied group of intellectual inheritors, adherents, and users, all of whom contribute to its development and offer diverse expansion paths. There are debates in world-systems theory circles over several of its aspects. As with Bourdieu's circle, Wallerstein's group includes both orthodox followers, aspiring to defend what they see as the correct use of his theory, and those who follow him in spirit rather than in the letter.

The geographical dimension of the world system is one of the more interesting topics debated. Major debates are held between those who argue that nation-states should be the basic units of world-systems analysis and those who argue for sub-national regions, which often coincide with administrative provinces or historical areas. One of the best examples of such a sub-national region is Northern Italy, which, incidentally, may be considered the original core of the world system. Paradoxically, the history of the Northern Italian city-states was among Wallerstein's own early interests. However, he later abandoned this study, even though it could have considerably enriched his general model. Northern Italy is highly relevant in many respects. It is considered the birthplace of the modern global banking system, which is another argument for considering it to be the original core of the world system—even before the Netherlands. Moreover, the regional disparities in Italy, especially economic ones, which persist even today, seem to be one of the most spectacular examples of the importance of sub-national differentiations in Europe. It can even be argued that world-systems theory most effectively explains long-term differentiation at the sub-national level. Its interpretation of the persistent split of Italy into a poor South and a rich North is especially convincing, and certainly more substantial than those theories that seek endogenous sources (e.g. cultural, social capital) of this deep and age-old divide (Putnam 1993). The same argument can be adduced to explain persistent regional disparities in other European states. In particular, the urbanized economic zone that extends from the South of England, through the Benelux countries and Western Germany, and into northern Italy, is an enduring reminder of the early movement of the core of the world system. This Western European region, which dates back to the thirteenth century, has had a determining influence on the economic differentiation of several European states (Heidenreich 1998; Taylor and Hoyler 2000). This belt, which stretches from

Northern Italy to London, and only includes part of France (the Grand-Est) and Germany (the Rhine-Ruhr Metropolitan Region), is sometimes labeled the "blue banana". It has always enjoyed the highest per capita income and the lowest unemployment in Europe. Several other indicators of the region's economic affluence, cultural prestige, and political centrality can be cited. Among them is the presence of the head offices of major corporations and leading universities and research centers, all of which are located here. The same type of long-range spatial inequalities, which can be explained in terms of a region's distance from the blue banana core, is observable in other parts of Europe, including its peripheries. Examples include Poland, which has a very profound regional rift between its richer Western and poorer Eastern parts (Gorzelak 1996), and Spain, where Catalonia, being the region closest to the European core, has also been the wealthiest in the country for centuries. The situation in Sweden, Hungary and the former Czechoslovakia is similar. It is important to note that the relative economic position of the regions in question is not necessarily strongly correlated to their political status. This is striking in Poland, where the political and economic trajectories of the country's various regions are strongly divergent. The Western regions of what is now Poland, irrespective of their predecessor states, were usually richer than the eastern regions, which have always been distant from the Western core. Central Poland, which was under Russian rule at the turn of the twentieth century, is an exception to the rule (it was the most rapidly developing part of Poland), although this only lasted for a short time (Kochanowicz 2006). The brevity of the economic boom of the Congress Poland, as the Russian partition was called, is yet another example of the ephemerality of economic development in peripheral settings. On the other hand, some relations between these two dimensions in the history of European regions can be discerned. Regions that are privileged by virtue of their proximity to the blue banana (location rent) tend to have recurrent movements for autonomy and even national independence. Obvious examples include Catalonia and Northern Italy (e.g. the Northern League movement), while in Poland, Upper Silesia's push for autonomy may be interpreted similarly. This "new regionalism" (Keating 1998) can be seen as part of the tendency for wealthy regions to coordinate cultural and sub-national political forces with economic logic. The new regionalism ideology interprets regions as corporations and relies on decentralizing tendencies which are promoted under neo-liberal economic slogans that claim competition is the crucial aspect of a country's regional organization.

1.3 Poland as Interface Periphery: Uses and Readings of Stein Rokkan's Approach

1.3.1 Contextualization of the Role of the Nation-State and Churches

When discussing the importance of the sub-national level of center-periphery relations, a specific historical approach, which complements the Wallerstein model, is worth mentioning. What I have in mind here is another conceptual model proposed by Stein Rokkan, who was a rare example of a historical and structurally oriented political scientist. Rokkan enumerated several historical mechanisms that produced geographical center-periphery structures. These include the construction of nation-states, which attempted to unify the geographical areas they controlled. This process began in Europe in the nineteenth century and is ongoing in many parts of the world. Another mechanism was triggered by the industrial revolution, which resulted in rapid economic growth and a massive accumulation of wealth in some areas. This led to a high concentration of economic resources—mostly in certain urban areas. However, aside from the historical dimension, Rokkan's approach involved a considerable amount of structuralist thinking, some of which paralleled Bourdieu's model. Most relevant here is that Rokkan singled out three dimensions of domination that produced center-periphery relations, which he saw as mediated through various transactions between centers and peripheries. These were the military-administrative, economic, and cultural dimensions (Rokkan 1980). These three dimensions clearly correspond the Bourdieu's three types of capital, viz. political (as a type of social capital), economic, and cultural. Significantly, Rokkan's model assumes that the relations between these levels of domination are subject to change. However, Rokkan, as mentioned above, was first and foremost a political scientist. As such, his major contribution concerns the role of the state in the emergence of both regional structures and political cleavages. In particular, Rokkan studied the role of the state in centralization through nation-building (Rokkan 1969). He also theorized the processes of periphery build-up in the territories of disintegrating empires by pointing to the role of feudalization and what he called vernacularization (Rokkan 1980: 171). Most important in this context is that Rokkan's studies provide an important contextualization for the French case, which was at the forefront of Bourdieu's thinking. Moreover, it was France that through the centuries had set the pattern after which the modern nation-state was fashioned. In many respects, France pursued centralization and cultural homogenization, especially linguistic standardization, further than most. Rokkan recognizes this centrality of France, but in contrast to Bourdieu, he complements it with studies on the development of other European countries, albeit only Western ones. Another instructive thread that Rokkan followed, but which Bourdieu never took seriously, is the role of the Catholic church outside France and in a long-term historical perspective. Bourdieu studied the religious field in France (Bourdieu 1971), but he never considered the major role of the Catholic Church in the development of peripheral nationalism in Counter-Reformation Europe, including Poland. Alliances between the Church and nationalist or secessionist movements were crucially important in the conflict between the center and the peripheries. On these and several other topics, Rokkan can also be seen as a supplement to Wallerstein. Rokkan even takes Wallerstein to task occasionally, e.g. for neglecting the development of strong military centers, such as eighteenth-century Prussia, and for giving too little attention to the military factor in general. In conclusion, even if Wallerstein and Rokkan's approaches are less advanced theoretically and are less flexible than that of Bourdieu, their main advantage lies in the fact that they offer a more advanced synthesis of historical processes. This helps to better contextualize the evolution of societies other than France, and to avoid the trap of universalizing the French historical path.

The historical dimension is definitely not one of the mainstays of Bourdieu's impressive theoretical edifice. Bourdieu obviously appreciated the importance of the historical factor in sociological analysis. He offered numerous historically oriented reflections and produced several lower-scale studies on the historical development of specific social fields in France. However, these are mostly piecemeal and often unrelated. They certainly do not form a coherent whole, let alone a grand scheme. Toward the end of his career, he fully recognized the importance of historical synthesis as a context in which social processes can be understood. This is apparent in his *On the State*, posthumously published on the basis of his lectures (Bourdieu 2014). The book clearly aims to historically synthesize the history of building the French state in a comparative perspective, but falls short. Bourdieu was thus not able to offer more than a not very coherent chain of selective, even if often profound reflections on the history of the state. Similarly, he never attempted systematic studies on the territorial dimension of social processes, making only cursory remarks on regional identity and related issues, especially in the chapter of his *Language and Symbolic Power* titled "Identity and representation: Elements for a Critical Reflection on the idea of the Region" (Bourdieu 1991).

1.3.2 Integration of Wallerstein with Rokkan and Extending Bourdieu to Sub-National Level

Wallerstein and Rokkan both have significant complementary potential for extending and grounding Bourdieu's approach in a solid comparative historical framework; one that additionally accounts for the spatial dimension of social processes, especially at the sub-national and regional levels, which were so central to Rokkan's work. This was recognized but never systematically studied by Bourdieu. It is particularly relevant to Central Europe where national borders have always been in a state of constant flux. Rokkan additionally indicates how the schemes proposed by Bourdieu and Wallerstein could be moved from the state-level to lower (especially regional) levels for the purposes of studying social mechanisms. His work even demonstrates the need for such a move, primarily because, as stated above, the regional level may be considered as an important dimension of the organization of the world system. Moreover, the regional level is often of considerable importance to the organization of the field of power of specific countries, to say nothing of their fields of politics and culture, with the center-periphery cleavage playing its typically prominent role. The considerable tensions in this dimension confirm an important and deeper insight, especially salient in the Central European context, viz. that a state is essentially a temporal and spatial coalition of conflicting actors and forces, and that this coalition produces constant tension, even at the regional level. Not even the fixity of a country's borders over a period of centuries, or even millennia, necessarily implies a lack of strong centrifugal forces. Obviously, this mainly applies to the spatial dimension, as this is where empires (and other polities) are constantly threatened by fragmentation and even secession. But these risks may also concern the political, cultural, and (last but not least), economic dimensions, with crucial class tensions that often threaten to escalate into revolutionary developments. This is at least partly due to the international contexts in which individual nation-states are immersed. These trans-national contexts, which often crucially contribute to the breakups of empires and states, as well as revolutions, can be studied using Wallerstein's approach as a useful and critical global-scale framework, one focused on power relations within a complex empirical and historical setting. Its advantages are apparent in its complex and historically embedded explanation of the specificity of the social structure in Central and Eastern Europe. This interpretation, as is argued below, may account for the nature of the region's elites, the persistence of its economic position in global hierarchies, and several other structural traits. Wallerstein offers a unique insight into the historical roots of that specific position. He has inspired younger authors to interpret the location of the region in a broader European context (Gerőcs and Pinkasz 2018). Explanations based on this approach, which can be called structural, stand in contrast to the prevalent culturalist accounts of the region's "backwardness", which assign this part of Europe some pertinent "mental" traits which supposedly condemn it to be forever "catching up" with the Western core.

I would now like to discuss how the approaches discussed above might look when integrated. They are, of course, considerably asymmetrical theoretically. However, Wallerstein's approach, even if less abstract, offers a more solid historical and spatial framework for contextualizing Bourdieu's models and their applications. I propose looking at Wallerstein's system as a global-scale reconstruction of what can be framed using Bourdieu's theory as the historical and spatial configuration of the dominant mode of domination in the contemporary era, viz. the economic realm. I further believe that the works of Wallerstein, Rokkan, and probably other historically oriented sociologists, e.g. Charles Tilly and Shmuel Eisenstadt, may help to contextualize part of Bourdieu's reconstruction in broader historical and territorial contexts. There are therefore several ways in which they can be presented as compatible. However, I now wish to briefly discuss earlier attempts to link the field and world-systems theories.

Several authors working in the Bourdieusian paradigm have referred to Wallerstein, and a considerable number of those who have written about the global circulation of ideas have addressed world-systems theory. However, in most cases, they do not refer to reconstructing either the territorial or historical structures as discussed by Wallerstein. What they usually take from Wallerstein is an abstract global framework of center and peripheries. They rarely even use the notion of semi-peripheries, probably out of fear of being accused of the "sin of functionalism" discussed above. For instance, Gisèle Sapiro linked the translation market to the world system (Sapiro 2008). A version of world-system theory was the core-periphery model, which was applied to literature by Itamar Even-Zohar in his polysystem theory, which encompassed translation (Even-Zohar 1990). As Sapiro points out, this approach and field theory share a relational methodology, albeit one that is more functionalist than structuralist. The second major difference is that polysystem theory is more focused on texts than on social agents (individuals and institutions). Pascale Casanova's seminal *The World Republic of Letters* scaled field theory up to the international level by using a dominant-dominated opposition and the notion of literary capital to describe the power relations between countries. Bourdieu's inspired center-periphery model, as modified and adapted by Casanova (2004), is widely considered a pioneering effort in globally modeling the cultural realm. It embodies the centers and peripheries of the world literary field, with some "median literary spaces" in between (Casanova 2004: 277). What Casanova calls "median literary spaces" may be seen as equivalent to "semi-peripheries" in the world system approach. Cities with a high concentration of literary resources, which Casanova calls "literary capitals", stand above them in the global hierarchy (Casanova 2004: 24). Importantly, Casanova's reference to cities implies that macro forms of capital can relate to different geographic units, including sub-national regions.

Sapiro has proposed combining the core-periphery model with field theory and the economy of symbolic goods in order to describe the translation market as embedded in both the international book market and international relations. On this market, different domains (e.g. literature, and human and social sciences) enjoy relative autonomy and have their specific agents, stakes, and rules of functioning. They therefore constitute internationalized fields. At its most autonomous pole, the literary field has its specific agent writers, literary translators, literary critics, and its specific aesthetic criteria. This is also true of the human and social sciences, where scholars play a central role in the international circulation of academic materials (Sapiro 2008: 159). As Sapiro suggested, the international book market can be regarded as structured, similarly to national book markets, around the opposition between large-scale and small-scale circulation. She also noted that Bourdieu's analytical model has many advantages for studying the international publishing market. For one thing, it allows the structure of publishing in different countries to be compared. For another, it takes the specific individual and institutional agents, as well as the international circulation of publishing models, into account (Sapiro 2008: 160). Of special interest to Sapiro is the global circulation of literary translations. As she points out, translated works can be instrumentalized in the internal struggles of specific fields, such as the literary field or specific academic disciplines, in order to renew the space of possibilities. They can be used to strengthen the cultural identity of minorities, and they can reinforce the stereotypical impressions of foreign cultures. Texts can either be depoliticized or highly politicized through this process. Their political meaning might also be radically altered (Sapiro 2008: 163).

Buchholz has proposed to write about what she calls the cultural world-systems perspective. This approach would take different aspects of culture that converge in theoretical orientation into consideration. She describes this as "inserting elements of world-systems into cultural analysis" (Buchholz 2018: 20). Buchholz places this in opposition to what she has proposed to call the more traditional mode of "inserting culture into world-systems analysis". In Buchholz's preferred approach, i.e. "inserting elements of world-systems into the cultural analysis", the purpose is no longer to develop an "extended materialist theory of domination", but an account of domination within globalizing cultural realms (Buchholz 2018: 20). As she notes, the development of this thread of analysis is related to the use of the "center-periphery" model, together with some modifications of Wallerstein's original version. Buchholz insists on the relative autonomy of culture and, following Ulf Hannerz, mentions culturally specific "charismatic geography" (Hannerz 1992: 229). This position stems from the observation that hierarchies of countries in a cultural world-system can differ from those of the economic system. Buchholz gives the example of the contemporary visual arts, where Japan can be attributed to the cultural semi-periphery in view of the limited cross-border prestige of its artists, although it belongs to the global core economically and politically (Buchholz and Wuggenig 2005). However, the fact that the economic hierarchy remains the dominant one necessarily implies that it constitutes a crucial point of reference for interpreting the meanings of other hierarchies. This is not to say that they are always subordinated or similar irrespective of context. Even if the global cultural hierarchy is largely autonomous from the economic one, it should be interpreted in the context of its relationship to it, and to a lesser extent, other hierarchies as well. In a similar vein, Pascale Casanova described the literary world as being "*relatively* independent of the political and economic universe; it is by the same token relatively dependent on it" (Casanova 2005: 85). This is consonant with my view of semiosis as a function of homology (presented in the previous chapter). The homology is always partial, but all meaning is nevertheless generated on this basis. In any given context, one of the meanings, potential or implicit, of even high-status cultural production from countries low down on the economic hierarchy, is that it may be seen as having a compensatory character. This is discussed in more detail below.

The relations between the diverse modes and scales of domination mentioned above are therefore crucial to integrating the theories under discussion. On the basic level of both the Bourdieusian or Rokkanian schemes, these would be economic, political, and cultural. The relational perspective I would defend is that their interactions should not be considered as given and stable, but rather dynamic and historical, and as the direct or indirect results of competition and conflict. The argument that world-systems theory can be read with this assumption in mind is essential. World-systems theory can be seen as a reconstruction of existing power relations between diverse modes of domination, and at the same time, as being focused on attempting to empirically reconstruct the economic dimension of domination, as this is crucial in the given historical context. The complex nature of relations between the three types of domination has been discussed by several authors. Casanova, in her discussion of the global literary space, argued that "the modes of domination are thus encased within each other. Three principal forms exert themselves to differential degrees, depending on the position of the given space: linguistic, literary, and political domination—this last increasingly taking on an economic cast. The three overlap, interpenetrate, and obscure one another to such an extent that often only the most obvious form—political-economic domination—can be seen" (Casanova 2005: 86).

1.3.3 Relations Between Different Dimensions of Global Dependence

The most systematic attempt at accommodating the complexity of relations between different modes and scales of domination known to me is that proposed by Gil Eyal, Iván Szelényi, and Eleanor Townsley in their seminal *Making Capitalism Without Capitalists* (1998). The authors had only intended to theorize the evolution of social stratification in a particular part of Europe, but ended up producing a grand theoretical scheme with much broader application. They offered a framework for analyzing the relations between economic, cultural, and political (social) capital that were eminently flexible despite being historically and geographically confined. Their case studies focused on the elites of Central and Eastern European societies, in particular Hungary, Poland, Russia, and the Czech Republic, but the framework they built was extensible far beyond Central-Eastern Europe. The authors proposed to theorize the specific forms of, and mutual relations between, specific factions of the elites using the language of Bourdieu's types of capital. Thus, for example, the intelligentsia was theorized as the elite of cultural capital, while the communist nomenklatura was the elite of political capital. These types and sub-types of capital were in turn used to reconstruct specific elite configurations and broader social structures in specific countries and during specific periods. *Making Capitalism Without Capitalists* was arguably the single most sophisticated and inspiring attempt to systematically extend "Bourdieu's system" internationally. The authors succeeded in placing the relations between specific types of capital, which Bourdieu had described in France, in a more abstract framework, one in which their patterns could take on some unexpected configurations. Thus, as they argued, during the Stalinist period, it was political capital, which assumed dominance, while at other times, particularly during certain periods in the history of Poland, cultural capital has been the dominant elite resource. In this perspective, the relations between specific elites can change, and their configurations, described by Bourdieu in France, and some Western scholars in their societies, may be altered. In particular, the political and/or cultural elite may assume dominance over the economic elite. Nevertheless, this diversity of elite configurations can also be observed in other periods and other regions.

In this context of "inverted hierarchies" there lies an interesting question of their perception. Given that economic logic and economic capital are considered dominant on the global scale, other types of capital may be perceived as inferior or "compensatory". However, when we consider a highly autonomous field, or a society where economic capital is clearly devaluated, we have another frame of reference that may define economic capital as compensatory. The compensation, which may be seen as a functional interpretation, should therefore be contextually defined. This is yet another case where functionalism can be understood in different ways. What is "compensatory" depends on the results of the competition between the different types of capital (modes of domination). The dominant capital (more precisely the actors that control its critical resources) are able to assign the dominated capitals the status of "compensatory", thereby symbolically devaluing them. However, even if this status is imposed with an intention to devaluate, compensation may be still a function that some resources may effectively play in many contexts. Thus, cultural capital in particular may be often seen as a compensatory, that is secondary, but still useful resource. A dominated position in a field can similarly be perceived as "compensatory", as can the functionality of certain system components. From this perspective, peripheral actors, and even entire regions and nation-states, can be seen as betting on globally dominated capitals. This enables their entire identities and general accumulation strategies to be interpreted as compensatory (Zarycki 2007b).

The relative nature of what is perceived as "compensatory" is also related to the frame of reference used and the prevailing attitude toward what can be called structural interpretations. Structural frames can be seen as either essential constraints on activity, which deny the agency of individuals, or as a context that possibly limits activity while giving it meaning. This may be relevant to critiques of "reductionism", which can be defined as attempting to reduce complex power relation to a single logic, and which is open to being criticized as overly simplistic. As mentioned above, Wallerstein has been repeatedly criticized for his economic determinism, or what might be termed economism. In particular, Wallerstein's view that the political and cultural realms, or superstructures, are reflections of the economic base is often considered reductionist, even though he accounts for the fact that they sometimes lag behind (Steinmetz 2003). In the same way, his concept of geoculture is often perceived as unduly simplistic and largely functionalist. However, it should be noted that Wallerstein saw the normal condition of the world system to be multicentric in the purely political dimension, i.e. that it comprises a core containing several power polities that are constantly allying with, and competing against, each other. This condition was said to contrast with earlier world empires in which a single state had conquered the other core states. This thesis could be interpreted as an argument for the necessity of the world system having an autonomous political dimension, or to use Bourdieu's language, of political capital being autonomous from economic capital, even if the former has a dominated status at the system level. Wallerstein would probably have agreed with the parallel thesis that the stability of the world system is at least partially dependent on the autonomy of cultural capital. In this context, it is worth pointing out that Bourdieu analyzed the logic of legitimation chains, claiming that the longer they are, the more effective they are. This can also be construed as an argument for the importance of culture being autonomous if it is to effectively perform its legitimizing role. This may mean that world-system theory does not necessarily imply that the political and cultural dimensions have to be directly subordinated to economic logic to serve its deeper, long-term interests. Moreover, there are many situations in which it can be difficult to objectively assess the extent to which these logics support themselves. Steinmetz, following Claus Offe, has noted that government policy may be *sub-optimal* from the viewpoint of the interests of capital. Steinmetz has further noted that the "history of Nazi Germany is only the most dramatic example of an authoritarian state in a capitalist setting pursuing policies that were immensely destructive to the interests of capital as a whole in the short term, if not to many individual capitalists" (Steinmetz 2003: 335).

The relations between the three dimensions of social organization are therefore inherently ambiguous and require negotiation. Apart from economic determinism, political, and especially, cultural determinism (i.e. culturalism, economism, and statism, or politicization) can be singled out. These can be seen as biases rooted in both the theoretical frameworks used by particular authors, and the social context of the places in which they are speaking. On the one hand, most theories, including those mentioned above, have inherent biases. World-systems theory may be identified with economism, theories of globalization and IR theories with political-bias, complexity analysis with social capital-bias and post-colonial studies, or world-polity theory, with culturalism or culture-bias. On the other, we all speak from our given place in a particular society and the place of this society in the global system. Both can be seen as lenses that distort the configuration of forces between different types of capital. For example, as I reside in Poland and belong to the

Polish cultural elite, albeit a dominated and critical faction of it, I have a tendency to criticize culturalism, which I see as permeating the biases of my fellow intelligentsia members. The culturalism of the Polish elite and its mainstream academic narratives pushes me to look for a counter-balance in more materialist approaches. Some of my Polish critics would probably object on the grounds of economism, which is apparent in my reliance on world-systems theory. In any case, I am critical of what I see as the dominant culturalist narrative of the majority of the Polish elite. To my mind, it tends to overestimate the influence of intellectuals and the ability of peripheral countries to determine their own destinies, let alone impact the global system, and it largely ignores structural restrictions and dependencies. I see it as a reflection of the primary cultural nature of the dominant elite in Poland (Zarycki 2009).

1.4 Peripheral Field of Power: East European Uses of George Steinmetz's Approach

1.4.1 Contextualizing the Field of Power

Examining the variations of the configuration of the field of power in different zones of the world system may be another way of relating world-systems analysis to Bourdieu's field analysis. Eyal, Szelényi, and Townsley did not engage in field analysis, but it might nonetheless be a useful framework in this context. In fact, this seems to be the case, even though they were not studying how the changing relations between different types of capital shape specific social fields (especially the field of power), or how these fields could be seen as a mechanism through which the relations between specific capitals are in constant flux (especially their "exchange rates"). Neither did they study the autonomy of fields on a national or international scale. Still, their contribution seems useful for the development of a global field analysis. The main thing is that their model had a clearly framed historical dimension. They showed how changing economic and political relations manifested themselves in the social stratification of central European counties. They interpreted the main phases of the communist system (e.g. Stalinism, post-Stalinist

60s, and technocratic 70s) in these terms. If field theory can be problematized in the historical dimension, then this approach shows how the configurations of capitals in specific fields, especially fields of power, can change in particular countries and during particular periods.

A methodology for distinguishing between different types of fields of power would be essential in any proposed combination of the field and world-systems theories. On the one hand, there are the fields of power of the core countries of the world system with France being a classic example. As suggested above, its reconstruction by Bourdieu can be contextualized or de-universalized by accounting for the position of France in the world system. Consequently, its dominated even if the relatively strong role of cultural capital may be linked to the strength of the French state and slightly dominated position of France on the global scene. The US field of power, by contrast, lies at the very core of the global system, and economic capital plays a more prominent role, especially in relation to cultural capital. As mentioned above, the US field of power plays also partly the role of the global field of power, although it has rivals, e.g. the Davos, G 7/8, and other forums, including those organized by the United Nations. On the other hand, there are fields of power of peripheries, semi-peripheries, and sometimes regions aspiring to independence, including colonies and dependent territories. Some semi-peripheral countries can also be seen as interface peripheries in Rokkan's framework. I subscribe to the view that the peripheral location of a given society may strongly impact the nature of its field of power. In most dependent countries, the economic bourgeoisie is rather weak (assuming it is a native economic elite, which should not be confused with a managerial class employed in foreign-owned businesses) and dependent. The latter case is best illustrated by the comprador bourgeoisie, i.e. one that predominantly serves external interests, usually those of the core of the world system (as broadly defined) or one of its states. A bourgeoisie can also be weak, not so much on account of its meager financial resources, but rather its dependence on the local political elite for protection from global economic forces (by utilizing political assets, but primarily by calling on the power of the state). This is the case with e.g. Russia, where political capital clearly dominates economic capital in the field of power. Russian oligarchs are subordinated to the political elite, and they pay for the privilege of exploiting the country's precious natural resources with unquestioning and unswerving loyalty. This exemplifies well how the power of economic capital may be weakened in the non-core countries of the world system. By contrast, political and/or cultural elites can be strengthened in such a context, particularly as political, and sometimes cultural, capital are compensatory resources of the peripheral and semi-peripheral countries of the world system.

1.4.2 Field of Power in Colonial Context

I would now like to reflect on the ways in which external dependence can impact a field of power, especially in a weak country or a sub-national polity. My point of departure is Steinmetz's notion of the colonial field of power, from which I propose to move to the more general notion of "peripheral field of power", as contextualized in the perspective of world-systems theory (as broadly understood). Steinmetz's point of reference is the imperial political system. However, the classical colony is only one type of peripheral entity. His pioneering work on the modern colonial state as a field may be seen as an excellent starting point for studying the broader category of peripheral fields of power. Steinmetz argued that "if the metropolitan state is analyzed as a field or set of fields, as Bourdieu suggested, then overseas colonial states may also represent distinct field-like formations, characterized by particular forms of relative autonomy from the metropolitan state and from other fields in the colony, and by competition among colonial state agents for distinctive forms of symbolic capital. Colony and metropole are linked by additional transnational fields, such as scientific or cultural ones. Field theory can also be extended to make sense of intra-imperial relations" (Steinmetz 2014: 92). Moreover, Steinmetz claimed that the relations between the colonial state and the metropolitan state vary, and the ways in which they do so determines whether the colonial state is an extended metropolitan field, an extended sub-field thereof, or a completely distinct field of power. In arguing for viewing the colonial state as a distinct field, Steinmetz mentions that metropolitan colonial offices and overseas colonial states were organized around different forms of symbolic capital. Moreover, not all the members of the colonial state qualified as members of the metropolitan state (i.e. an official or public servant), as a member of a colonial administration was not necessarily a member of the metropolitan state. Steinmetz posited that the "colonial state field, is embedded within the colonial field of power, a space that contains both state and non-state European actors. All white residents in European colonies possessed a certain amount of 'racial' capital vis-à-vis all colonized residents due to the rule of difference and were therefore inside the field of power" (Steinmetz 2016: 108). Steinmetz further argued that the colonial "social space" encompassed both the colonized and the colonizers. The metropolitan field of power was thus transposed onto the colonies in a truncated form, but the initial triangular structure of the elite was reproduced. Steinmetz singled out three types of colonial actors, which may be seen as the cultural, political, and economic-capital elites (Steinmetz 2008). The cultural elites were *Bildungsbürgertum* in origin. They "often emphasized empathic and hermeneutic approaches to understanding the indigenes, thereby calling attention to their ability to speak exotic languages and to think their way into foreign worldviews" (Steinmetz 2008). The political elites were "colonial military noblemen", who "tended to evaluate the colonized in martial terms and to emphasize the aristocracy's hereditary specialization in the arts of physical coercion and the command of subordinates". Finally, the economic elite were "capitalist investors" and self-employed settlers, who assessed the colonized in terms of their capacity for labor. It is worth noting that the colonial cultural elite, as described by Steinmetz, can be related to the broader class of the cultural elite of peripheral countries, one incarnation of which is the intelligentsia of Central and Eastern Europe. The intelligentsia, although usually native to their countries, often aspire to become intermediaries between Western and local culture, so as to represent their countries' interests to the outside world and explain the global culture to their fellow citizens.

Steinmetz's model of the colonial state as a field moves beyond approaches that reduce state policy to extra-state determinations and suggests a way to rethink state autonomy. As he argues, state autonomy literature theories are not able to explain why different parts of the state system, including the local state and the colonial state, sometimes become partly independent of the central state but stop short of provoking state breakup or fragmentation (Steinmetz 1993: 607). Steinmetz is also skeptical of Bourdieu's speculations on the state being a "central bank of symbolic credit" (Bourdieu 1996: 376) that dominates all other fields and is constituted "as the holder of a sort of meta-capital" (Bourdieu 1999: 57). Steinmetz argues that "the state may well have the *ambition* to become a meta-field governed by a form of meta-capital, but this does not distinguish the state from the economic or religious fields, from which similarly encompassing ambitions also arise. The state contributes to the emergence of literary, scientific, and professional fields by officially consecrating their members, but these fields may themselves contribute to state formation, as shown here. Field theory explains why peripheral governments sometimes become fields in their own right, developing specific internal criteria for judging" (Steinmetz 1993: 607). These observations, based as they are on colonial empires, are perfectly adequate in the context of e.g. the fragmentation of the Russian and Austrian empires in their final years, and the breakup of the Soviet Union into its constituent republics, along with the loss of its satellite states (e.g. Poland). It was precisely in reference to late communist Poland that Jan Kubik observed that the state lost a major part of its symbolic power to the fields it had helped establish, including the cultural and academic fields (Kubik 1994).

Steinmetz also claimed that "field theory sheds light on the ways colonial states and local or regional governments resist the *centralizing* effect of the state, sometimes breaking away altogether. This theory may also illuminate the centrifugal tendencies and the breakup of non-colonial empires" (Steinmetz 1993: 608). Again, the mechanisms indicated by Steinmetz are crucially important to modeling the emergence and re-emergence of smaller Central European states during and after World War I, and the dynamics of the communist states in the region, including the Soviet Republics, which gained their independence following the collapse of the Soviet Union. Most of these processes involved strong centrifugal forces within empires that were crumbling. At the same time, however, the emerging fields of power were usually weak and fragile. Significantly, they were seldom led by economic elites, which stands in stark contrast to the classic configuration in Western Europe. In Germany, and many other Western European countries, as Go and Krause remind us, the nation-state "was ultimately a bourgeois project to restrict relevant competitors to those tied to language and territory. Some aristocrats started buying into the stakes of the emerging national fields. This did not mean that the international aristocratic field or that local fields ceased to exist, but they lost in relative importance via-a-vis national fields" (Go and Krause 2016). The aristocracy also played a more prominent role than the weak bourgeoisie in some Central European countries, e.g. Hungary and Poland, during the economic crisis that followed World War I. In any case, newly emerging states and their fields of power often differed (and still do) from the classic Bourdieu model, and not least because economic capital and the economic elite were often supplanted by compensatory capitals and elites as the dominant forces, especially in the context of the world system, assets and elites (political, social, and cultural).

1.4.3 Field of Power in a Wider Context of Dependence

The second peculiarity of the fields of power of Central and Eastern Europe is their dependence. This stems from the general fragility of the states and their relatively unfavorable economic position in the world system. Their weakness and dependence, however, should not be perceived solely in economic terms. When these countries were in the Soviet bloc, the dependence was primarily political, but this has not always been the case. They have, however, always been culturally dependent on the Western core, and the entire Soviet Block never achieved full economic autonomy from the capitalist system. Therefore there have been times, then, when the countries of Central Europe would have qualified as Rokkanian interface peripheries in that they were being pressured by two or more external centers—mainly the Western core, but also the Soviet Union and the earlier Russian and Austro-Hungarian Empires. This sometimes created complex dependence structures, which were reflected in the complexity of the structures of their social fields, including fields of power. Some of these complexities are analyzed in the next chapter. For now I would like to briefly reflect on the overall effect of external dependence.

Thus, one of the central tensions in the peripheral fields of power is tension between external and internal forces, that is, tensions between the factions of the elite oriented toward exogenous and endogenous resources and relations. In most cases, and on a very high level of generalization, exogenous resources may be seen as directly or indirectly dependent on the forces related to the global field of power. In the case of specialized fields (e.g. cultural, religious, or economic), external centers may have specific locations within their own central zones, such as the Vatican as the center of the Catholic Church or Paris as the time-honored center for several branches of the global cultural field. The main tension in what I call a peripheral field of power might not necessarily be between completely opposite types of capital (esp. economic vs. cultural capital, as in the French case), but between two variants of the same capital—specifically the local version and the external, or internationalized, version (e.g. national economic capital vs. global economic capital). As Go and Krause noted, the national fields of any state, but especially the more dependent ones, can be divided between globalizers and their opponents (Go and Krause 2016). Marion Fourcade (2006) has given several examples of how international links have played a role in the national fields of economists in peripheral countries, while Joanna Bockman and Gil Eyal have shown how international contacts have played a role in the structuring of the national fields of social sciences—not only in peripheral countries such as Hungary and Poland, but also in the US (Bockman and Eyal 2002). In the case of interface peripheries, when two external poles of the world system exert influence and compete over the same region or state, its peripheral field of power may be divided between the factions oriented toward the two external forces (Poland provides several examples of this). The structure of conflict in a peripheral field of power can be seen as reflecting the external forces acting on the given country or region. International tension is thus translated into local tension. This aspect of translating international/external pressures into local conflicts is inherent in any field of power, especially as economic capital is usually more globalized given the globalization of the world system, but the effects of such forces are particularly strong in peripheral countries.

1.4.4 Types of Peripheral Autonomy and Peripheral Duality

Now to summarize the above analysis with an outline of the two main types of reactions of subordinated fields. The first is homological adaptation, or what Steinmetz calls transposition. This involves reproducing the structures of the dominant field (the field of power, or an upper-level field, e.g. the global field of a given discipline or realm) in the lower-level or dependent field. The second is resistance to global forces, which has been discussed above. This is the component of the strategy for achieving autonomy that makes the opposition between autonomy and heteronomy most pronounced. The classical type of autonomy is built around a field-specific resource that can be perceived in such context as compensatory. However, in conditions of strong dependence, this autonomy may be built around any resource in its local form or sub-type, and this in turn will be controlled in a manner that is relatively independent of the rules imposed by any dominant, external field. Here, it is worth recalling Buchholz's distinction between two types of autonomy. The first is "functional autonomy", which "designates how social spheres are differentiated by functionally different types of interests, practices and their logics, corresponding to Bourdieu's original conceptualization" (Buchholz 2016: 41). One of Buchholz's examples is the global art field, which can be defined in relation to the global economic field. The second is vertical autonomy, which "accounts for differentiation in relation to other field-levels of social organization in the same realm of specific interest and practice" (Buchholz 2016: 41). The examples she offers under this heading include the French and Japanese art fields (Buchholz 2016). I would argue that what Buchholz calls functional autonomy can be defined in reference to the global field of power (actual or substitute, e.g. in the hegemonic state's field of power), and that what she defines as vertical autonomy can be defined in reference to a national (or lower-level, possibly regional) field of power. She is correct to avoid the notion of "national capital" and propose the more flexible notion of "macro-capital" in its stead, as this not only refers to countries, but also to cities or regions, whichever be applicable (Buchholz 2018). As she rightly states, the structure of any global or transnational field depends on the historically specific distribution of what she terms macro-capital and should therefore be studied empirically. This attempt to link Buchholz to the foregoing discussion can be taken further by pointing out that any peripheral field of power may be either functionally autonomous, i.e. dependent on the global field of power, or vertically autonomous, i.e. predominantly defined by its relations to the local (national) system of fields, whose configuration in turn may be functionally or vertically autonomous. The vertical autonomy of a peripheral field of power may be built on its homological relation to the global field of power, or a higher-level field of power, a relation which could be termed, after Steinmetz, transposition. In any case, given that any field of power is central to semiosis in any society, these crucial differences between the fields of power of core countries on the one hand, and weaker countries and smaller territorial units on the other, may result in considerable differences in basic semantic oppositions. At the same time, it should be recognized that these two forces that the global system exerts on weaker fields of power usually act synchronously and in a parallel manner. One is usually stronger than the other, so the dominant axis of a given field of power is clearly defined as an opposition between local and global (autonomy vs. heteronomy with the autonomous pole concentrated around a specific type of capital, e.g. political) or it is homological to the global field of power, with its crucial opposition between economic and cultural capital. Frequently, however, both axes are sufficiently prominent to be perceived as similarly important. This may create a dual system of symbolic coordinates, one at least partly homological to the global configuration of the field of power, and one locally specific, but generally defined by its opposition to the global core and/or possibly another strong center of international power. This duality, which is reflected in, and partly created by, local fields of power may be seen as a characteristic feature of most of the peripheral states and regions in the world system. Actors functioning in their social spaces have to navigate between these two dimensions, which have both material and symbolic aspects. Having to synchronize these two axes is often one of the key political stakes in such systems. In a typical configuration, this sort of synchronization implies attempts at aligning the cosmopolitan pole of the autonomy vs. heteronomy axis with the economic pole of the transposed classic opposition between economy and culture. Such a linkage, or alliance of "cosmopolitan" actors, including those from the cultural and economic field with the local economic elite (which, in peripheral countries of the world system, usually has a comprador elite character), does in fact occur, but is rarely complete. The natural tendency of the cultural elite to at least partly orient itself toward local, i.e. dominated constituencies, may be one of the factors preventing the two axes from being fully aligned. However that might be, the relationship between these axes, even if close, rarely attains full homology. It can therefore be argued that this relationship between the two axes of the field of power is the core of the semiosis process in many peripheral settings.

1.5 Poland as Part of the Global East: Benefits of the Poland Case Study for the Global Sociology of Knowledge[3]

1.5.1 The Global East as Context of the Polish Case Study

This chapter reflects on the broader conclusions that can be drawn from the book's Polish-centered case study. The most salient is that it illustrates alternative configurations of diverse types of social fields, especially the field of power. These configurations can be contextualized by giving due weight to Poland's semi-peripheral status in world-systems theory. However, Poland can also be understood from perspectives that are far more specific. For example, it can justifiably be seen as belonging to a specific periphery, viz. the eastern periphery of the Western European core. This potentially leads to an inspiring theoretical discussion on the place of this region in the global production of knowledge. One recent intellectual framework is Martin Müller's "Global East". This is based on analogy with the "Global South" (Müller 2020). The Global East may be defined as the Eastern and far less visible periphery of the world system.

---

[3] This section is partly based on my paper published in "Praktyka Teoretyczna" (Zarycki 2021).

It has been much less relevant globally than the Southern periphery since around 1989. The Southern periphery has become a point of reference of a wide stream of theoretical discussion in the social sciences (and beyond), especially within the framework of post-colonial theory. The study of the transformations of Polish social structures, especially the field of power, along with selected social science disciplines, over the course of the twentieth century, can be seen as part of a broader effort to ascertain the relevance of the Central and Eastern European experience to critical social theory in general and the sociology of the social sciences and empires in particular.

While I am enthusiastic about the Global East concept, I am skeptical about the prospects for its wider implementation. The main reasons for this skepticism are presented below. Briefly, they have to do with the problems associated with making peripheral or semi-peripheral regions like Central Europe clearly visible on the global academic map. Müller's proposal is founded on the observation that, after the fall of communism, what had been called the First World became the North, or Global North, and what had been called the Third World was relabeled the South or Global South. However, the former Second World, which mostly comprised "communist" or "socialist" countries, has been erased from the mental maps of both the popular imagination and global social theory. Müller makes the point that the former Second World is now too rich to qualify as South, but too poor to qualify as North. It has some, but not all, of the trappings of European modernity. It is too European to be included in the South, but not European enough to be included in the North. In other words, it is too powerful to be a periphery, but too weak to be the center. This clearly accords with the status of semi-periphery in Wallersteinian terms, and in fact a considerable portion of the region can be considered semi-peripheral. It should be noted that world-systems theory regards semi-peripheries as particularly vulnerable to global cycles. They benefit from their proximity to the core in periods of global growth, but suffer—sometimes as severely as the peripheries proper—during global downturns. These cycles can also be observed in terms of the global general and academic interest in these regions. It is especially noteworthy that during dramatic economic or political crises, including wars and periods of turmoil, Central and Eastern Europe (or at least some of its countries or regions) never fails to attract global interest. The region also frequently produces intellectuals of global standing, as well as emigrés, during these periods. These people sometimes successfully integrate into the global intellectual elite. Once these crises have passed, the region reverts to its dull, uncreative, and "gray" self. This contrasts with e.g. Brazil, which, as Müller notes, is considered "sexy", Kenya, which is "cool", and China, which is, of course, "dynamic". For its part, the East seems condemned to stasis. Interestingly, the "South" takes in Africa, America south of the Rio Grande, the Indian Subcontinent, and Iran, but not the former Central Asian Soviet Republics (e.g. Uzbekistan, Tajikistan etc.). Müller notes that Uganda is better known by many in the global centers of media and scholarship than Ukraine, Chile is more familiar than the Czech Republic, and Laos is closer to many Westerner's minds than Latvia. The names Vargas Llosa, García Márquez, and Coetzee are instantly recognized among the Western public, whereas Aleksievich, or Szymborska are not, even though, as Müller points out, all six are recent laureates of the Nobel Prize in Literature. Müller further notes that the very intermediacy of the erstwhile Communist Central and Eastern Europe contributes to the general ignorance about the region in both the core of the world system (the North) and its major periphery (the South). That ambiguous place occupied by the eastern periphery has attracted a variety of labels, including semi-alterity (Tlostanova 2015) and demi-orientalism (Wolff 1994). The region's ambiguity is partly owing to the fact that it includes both colonizers and colonies, aggressors and victims; some countries were both at the same time. The upshot, as Müller convincingly argues, is that Westerners have never regarded the East as deserving compassion or in need of global activism, and especially not since the fall of communism. Since being reintegrated into the Western-dominated global economy, it has seen no end of environmental destruction, power politics, and rampant nationalism. Alternatives to liberalism are very thin on the ground. This perception also seems to be cyclical in nature. Merje Kuus makes the observation that the peasant motif in the stereotypical images of Eastern Europe is a recurring one, as are the invariably juxtaposed images of cyclical, and putatively characteristic, eruptions of unbridled nationalism (Kuus 2007: 27). The Global East is thus seen as inferior, but not inferior enough, a subaltern in a way, but not fully. Müller considers that, ironically, never having been colonized by the British or the French has limited the East's opportunities to be heard. In his opinion, this is due to several factors, ranging from nostalgia to tangible interests, cultural heritage on both sides (knowledge of colonial cultures in the metropolises, and knowledge or awareness about colonial cultures in metropolises), through to the privileged access of at least some of the elites of former British and French colonies to the core of the world system. This involves not only a common official language (and, more broadly speaking, cultural competence), but also social capital and economic links. The elites of the East, although generally perceived as more "Western" than those of Africa and Asia, do not usually enjoy these kinds of privileged connections. If they do, it is to a much smaller degree; one which does not translate into a broader awareness of their role in the global elite.

Müller laid out a case for embracing the former Second World as the Global East, so that it can overcome this marginalization and "reclaim its place". This necessitated thinking about the East in terms of strategic essentialism, which would allow for both "re-establishing it as a pertinent preoccupation of scholarship" and "re-inscribing it with new meaning" (Müller 2020: 16). Following Spivak, he proposed to promote the political practice of mobilizing heterogeneous marginalized groups to band together. Müller has expressed a willingness to help guarantee the region's right of recognition, compile and disseminate knowledge about it, and ensure that it is free from discrimination. He has also called for decentering the West and take into consideration universal knowledge claims emerging from the perspective of the Global East, whose liminality should be reframed as the region's strength. Müller has also expressed a desire to transform the East from an object of area studies into a subject, or "indeed a method—a means of transforming knowledge production". Time and again, he sees the liminality of the region as its major asset. To refer back to the discussion of semiosis in the first chapter, this project could be thought of as an attempt to establish a new global axis that runs East–West. This can be seen as weakly homological to the old North–South axis, as both are structured around centers and peripheries. The homology has to be weak if the new East–West axis is to acquire its own significance. Establishing such an axis would also imply an attempt to replace the former opposition, i.e. the West-(Post)Communist axis, and move beyond the paradigm of area studies. As stated above, Müller's proposal to promote the Global East paradigm, in tandem with the Global South concept, is an important, and even noble, intellectual idea. At the very least, it is highly stimulating, and applicable to the context of this work. However, I primarily use it to demonstrate the structural reasons for its untenability. My intention is not so much to challenge Müller's assumptions, but rather to reinforce his basic observations.

1.5.2 Impossibility of the Global East Intellectual Project

First, I would like to give some of the reasons for rejecting Müller's proposal in the region itself. This should indicate potential problems in developing the intellectual approach adopted in this study. The liminality of the region, to which Müller draws attention, is not only an interesting feature, but a major obstacle in the given context. As many others have pointed out, whether the region belongs to Europe or the West is, politically speaking, a high-stakes issue for the actors involved. At the same time, the symbolic hierarchies in the region are crucially important in this context. Again, as several authors, e.g. József Böröcz (2006), Attila Melegh (2006), and Merje Kuus (2004) have pointed out, these are related to different degrees of supposed Eastness/Westness. Being perceived as even slightly more or less Western and/or European than other regions, particularly close neighbors, is an issue of vital concern for most actors in the East (as broadly defined), especially those located closest to the core of the European West. They often hope that they can be somehow admitted to the "Western club", and then have the doors slammed shut on their neighbors. This often makes their nearest neighbors their most ardent enemies, or at least their most fierce competitors, which in turn leads to the entire region becoming highly fragmented or "Balkanized". Lumping so many countries and nations together under the single umbrella of the Global East, as proposed by Müller, is therefore extremely unlikely to be acceptable to the region's politicians or intellectuals—even if it can be justified by analytical rationalities or the strategic interests of Western scholars. To qualify as Eastern, i.e. as less Western than our neighbors, is to be consigned to the periphery. And this is a regional stigma. "They (i.e. our neighbors—or "opponents") are eastern and peripheral, but we belong to the West", is a commonly heard refrain from "Central" Europeans. Poland is a prime example of the unacceptability of being deemed peripheral—both in public statements and academic discourse. As the labels "East" and "periphery" are generally considered pejorative, describing Poland as a periphery, even in purely analytical terms, is guaranteed to arouse controversy. Moreover, the public in countries like Poland expect social theorists to reassure them that they are recognized, even if only as junior partners of the dominant actors, as part of Western Europe. The idea of Poland as part of the Global East, while analytically defensible, will therefore not find favor with most Polish intellectuals and social scientists. It would contradict the liberal narrative that Central Europe is part of Western Europe by virtue of EU membership, as well as the conservative narrative that Central Europe is part of the West by virtue of a common Western Christian heritage. Moreover, most Poles will never accept being put into the same basket as most of their neighbors, especially, but by no means only, Russians (Zarycki 2004). An instructive recent example of this rejection of a common Eastern identity is Poland's resistance to a proposed Berlin memorial to the East European victims of Nazi Germany. Most of the Polish voices raised in opposition called for a separate monument devoted solely to Polish citizens. Placing Polish victims in the same category as other Eastern Europeans is widely seen in Poland as playing down Poland's exceptional suffering—all the more so as most Eastern European nations fought on the side of the Nazis and Polish citizens suffered terribly at the hands of Nazi collaborators among their neighbors (Haszczyński 2020).

Several vital lessons regarding potential problems with the proposal under discussion could be learned by studying earlier debates on applying post-colonial theory to Poland. First, it should be noted that post-colonial theory has never been enthusiastically accepted in Poland. It has mostly been used to attack opponents rather than to challenge Western hegemony over the region, in particular at the deep, ontological level of criticism that lies at its core. As I have argued in several places, (e.g. Zarycki 2014), post-colonial theory has been adopted by various members of diametrically opposed intellectual and political camps in Poland. These include the main adversaries in the current political conflict in Poland, viz. conservatives or Euro-skeptics and left-liberals or Euro-enthusiasts. The left-liberals consider the uses to which conservatives put post-colonial theory to be "misuses", "hostile takeovers", etc. (e.g. Snochowska-Gonzalez 2012). Both camps employ the theory to promote their political agendas and attack each other, rather than focus on critically analyzing Poland's subaltern status in deeper, structural terms. Conservatives, in particular, enlist post-colonial theory to justify their agenda by e.g. re-traditionalizing and resisting "progressive" Western ideologies. Left-liberals see conservatives as key colonial agents who have always sought to subdue the country's minorities. The risk of developing and promoting the Eastern theory within the Global East paradigm is that it will once more become an object of similar mutual "abuses" and "takeovers", i.e. that internal confrontations among Polish actors will produce at least two conflicted variants of the paradigm. It should be borne in mind, however, that this time round, these sorts of ideological wars will not be limited to a few marginal regions, e.g. Poland as seen from the perspective of post-colonial theory, but entire states that could be central to the proposed paradigm. A further result will be tensions between Western academics working on the region, and these will not always be productive. It may even result in clashes between academics in the region and turn the emerging field into a political battleground rather than an area of creative intellectual cooperation. This would stand in stark contrast to the Southern studies, or at least their idealized image, according to which they aspire and apparently succeed in synchronizing several intellectual currents and political campaigns emerging from several countries in that part of the globe (de Sousa Santos and Meneses 2020). It should also be noted that, given the relative stability and acceptance of their wide "Southern frame", Global Southern studies can place considerable emphasis on the region's diversity. By contrast, Eastern Europe has always been characterized by excessive diversity and a distinct lack of unity. A case in point is the failure of the Pan-Slavic movement at the turn of the twentieth century. Most nations, but also ethnic groups and some regions in Eastern Europe, are founded on ideologies of uniqueness and essentialized distinctions between them and their neighbors. A "Sonderweg" mindset is also prevalent in the highly nationalized historiographies (and other social sciences) of most of the nations of the region. It can therefore be said that not everyone in the region wants to be "emancipated", or at least not together with their neighbors. Nor do they necessarily wish to be emancipated with the assistance of Western scholars, such as Martin Müller, and least of all left-liberal ones.

1.5.3 Why East Europeans Don't Want to Be Emancipated?

In some respects, the current configuration of Western academia, with the dominant role assigned to "area studies" (or, in the case of Central and Eastern Europe, what used to be known as Sovietology), in which responsibilities and roles are clearly defined, with the West being in charge of the production of knowledge (especially in its more generalized and theoretical forms), on the region is more tolerable for many. Area studies are clearly separate from the national academic fields of the countries of the region. If they make any use of them at all, it is as fieldwork, with local scholars serving as informants, data-collectors, or at best junior partners. This at least avoids any potential double standards regarding the political roles assigned by Global East studies. This is because every theory devised about the East so far has been a Western theory, similarly to the Southern theory. As with Southern theory, the project is supposed to facilitate the inclusion of scholars from the region into the Western elite, or at least increase the visibility of Eastern intellectuals among the Western elite. However, who is included and who is not will still mostly be decided by the elites of Western academic institutions. What may be different in the case of the Eastern theory is the blurring of the power relations; in particular, the command the West exerts over that knowledge area will be less obvious. However, the West will still get to decide which new meanings are to be ascribed to the region—mostly by selecting those Easterners they deem worthy of speaking on behalf of the region on the global stage.

The "emancipatory power" of critical theory in this context is also debatable. I tend to have a rather dissociative view of critical theory and political action. Critical theory is predominantly a tool for describing and deconstructing different forms of inequalities. These descriptions may well become meaningful acts of socially conditioned cognition, but rarely do they automatically have a direct impact on the "real world", or, in the language of first chapter enter into a homological relationship with the major divisions of the field of power. Obviously, they may lead to some of the inequalities becoming politicized, consequently stimulating efforts to alleviate them. Nevertheless, this sort of political activity can normally be undertaken without any critical scholar having theoretically problematized any specific inequalities beforehand. Any political action would have to be conditioned by favorable configurations of political forces in the region, especially trans-national coalitions which may lead to higher-level homologies. Apart from that, the "emancipation" of the region would require the accumulation and consolidation of tangible material resources. Mere intellectual "recognition" by Western scholars will not change either the place or overall visibility of Central and Eastern Europe significantly. Nor will it compensate for the weakness of its academic institutions, as Müller points out (Müller 2020: 743). It should also be noted, as Go remarks, that Southern Theory is a relatively weak intellectual project (Go 2016). It is commonly seen as mired in multicultural identity political posturing, as almost devoid of substance, as subjectivist and epistemically relativist, and as rooting knowledge claims in racial, cultural, or geographical identity rather than in objectivity. The Southern theory has been accused of being essentialist, parochial, and epistemically relativist. No Eastern theory in the offing is likely to be any more robust than its "older sister". In considering strategies to undermine the hegemony of academics in the Global North, Go opines that works by Southern academics need to be translated, that forums where sociologists from around the world can gather on a more equal footing need to be created, and, more generally, that there has to be a more equitable distribution of resources for sociological production. In addition, however, he sees the need to alter the "intellectual substance of the sociological work" as crucial (Go 2016: 10). However, I would posit that the domination of the Northern theory over the Southern theory is not entirely due to the supposed substantial weakness of the latter. In the light of world-systems theory and the suggestions in the first section of this chapter, I would emphasize the material component of the superiority of Northern over Southern things, and ideas. This asymmetry is, of course, similar to the eastern periphery of the world system. The economic and political power of countries at the core of the global system would therefore appear to be the most crucial factor. This dominance of the core over the peripheries and semi-peripheries is homologically reflected in the global field of social sciences, in which non-core theorizing is usually marginalized with no chance of becoming universal. It is also reflected in the more specific, but nevertheless crucial, field of infrastructural investment in research and higher education. The Global South generally has much weaker and poorer academic institutions, and these are additionally devaluated in the global rankings. The weakness of the Global East in this respect is even more pronounced and may be another reason for the infeasibility of the entire Global East project. The former communist countries do not have any globally recognized academic centers, especially in the humanities and social sciences. The story of the Central European University (CEU) is very telling in this respect (Hix 2004). This Western-financed institution was among the leading Central and Eastern European universities, even ranking first in some disciplines (Guilhot 2007). However, it was never fully integrated into the academic system of the region and was relocated to the Western part of the continent (Vienna) in 2020. Its Warsaw and Prague campuses had been closed much earlier (Pospíšilová 2019). Without at least one globally visible university in the region—one which would also be embedded in its country of location in terms of financial support and intellectual integration—the Global East project seems destined to become no more than an internal dialogue in Western universities between the global academic establishment and a narrow group of scholars who hail from the region.

1.5.4 "Rule and Divide" Mechanisms in Central and Eastern Europe

As mentioned above, one of the critical mechanisms that perpetuates the region's weakness and dependency is its constant fragmentation. This effectively allows the Western core to maintain its power via a strategy of "divide and rule". Here, it is worth recalling Larry Wolf's "Inventing Eastern Europe", in which the two primary roles prescribed for Eastern Europeans by Enlightenment intellectuals are reconstructed (Wolff 1994). The first was the role that Voltaire had suggested to the Russians, in particular their leaders, which was to submit to a strong, authoritarian ruler. Voltaire saw this as the only adequate modernization model given the difficult conditions of the East. The other was proposed to the Poles by Rousseau, who supported the Polish anti-Russian rebels (in particular the Bar Confederation, which tried to regain Polish independence immediately after the country had been partitioned by Russia, Austria, and Prussia between 1768 and 1772) and exhorted them to resist Russification and remain "European at heart". The Western core has continued to support both these roles, thereby indirectly sustaining interminable regional conflict. This strategy is clearly visible in the pragmatic relations maintained by the West (especially Germany), with the regime of Vladimir Putin, the most obvious example being the Nord Stream pipeline joint venture. On the other hand, some Western countries, and some Western capitalists, simultaneously support those Ukrainian or Belarussian democratic movements that challenge Russian control over their countries. These movements, which sustain considerable loss of life and are severely persecuted (e.g. the Maidan revolutionaries in Ukraine), employ narratives that perfectly express Rousseau's idea of "being European at heart". Both these identities of, or rather strategies for, living on the eastern periphery of Europe, could obviously be forms of compensation, one based on political capital (authoritarian rule in Russia) and the other on cultural capital (being "European at heart" in Poland). Both contribute to the fragmentation of the region and help cement the dependent status of the countries there. The West deployed an even more deliberate and overt strategy to fragment Yugoslavia at the end of the 1980s. One obvious way to emancipate the region on a global scale would be to economically integrate it in such a way as to make it less economically dependent on the West. This applies *a fortiori* to Central Europe, which is primarily controlled by the West, mostly through having a large proportion of its economic assets and property owned or managed by Western interests (e.g. Myant 2018). Here, it is worth remembering that Rosa Luxemburg (Luxemburg 1898) argued that a viable Polish state, i.e. one with a healthy economic system, depends on access to the Russian market. Interestingly, this crucial observation is rarely cited today, despite there being something of a resurgence of interest in Luxemburg on the part of the Polish left.

1.5.5 Dependence, Poverty and Dullness of Central and Eastern Europe

Müller further notes that an economic perspective along the lines of world-systems analysis can help explain why the Western core takes so little interest in the East. I would argue that Eastern Europe's "dullness" should primarily be seen as structurally conditioned. The region, especially its Westernmost part, i.e. Central Europe, largely serves as a reservoir of cheap labor for the Western core, both in the form of migrant workers and workers in regional assembly plants and call centers. However, that labor force is not expected to comment on its role in the global system or the state of Western societies—either directly or through intellectual intermediaries or representatives. Müller also points out that, in contrast to the Global South, the Global East has no exotic allure. He might have added that it is not a major tourist destination either. It is worth noting that there is some structural similarity of this situation to that in which certain peripheral regions of the West find themselves. What I have in mind here are its "inner peripheries", in particular poor, working-class suburbs of metropolitan areas. These also primarily serve to provide unskilled laborers and domestics to the core. One such inner periphery is the Inland Empire (sometimes known as Inlandia). This area is adjacent to and east of the Los Angeles metropolitan area. Despite its relative proximity to Downtown Los Angeles and Hollywood, the center of the global entertainment industry, the area is almost absent from the

American, let alone global, map of "interesting" places. It is telling that none of the scenes in David Lynch's *Inland Empire* were actually shot there. Curiously, much of it was shot in Łódź, Poland, another "remote", "eastern" and "nondescript" region of little global interest. The history of Eastern Europe shows that, so long as it poses no military, economic or political threat to the West, it is dismissed as being of no global interest. The region only becomes visible during (usually short) periods of exceptional economic growth or military consolidation, as happened at the turn of the twentieth century, when the Russian empire was experiencing an economic boom. Interestingly, this was also discussed by Luxemburg (Luxemburg 1898), who argued that Poland and Russia should be seen as of relevance to the Western theoretical debates. This period of dynamic economic growth also resulted in Russian art and literature becoming more visible globally. The situation was similar in 1957–1968, when the communist bloc attracted considerable attention and interest (to the point of fascination) with what was considered to be an alternative modernization path. However, the entire region was increasingly orientalized immediately afterwards. In the same way, the West's internal peripheries remain relatively unknown globally, and, while they remain economically deprived but politically stable, are dismissed as uninteresting. They are typically seen as backwaters, where a lot of indispensable but unexciting work is carried out for the benefit of more affluent (and thus visible) regions. Central-Eastern Europe conjures up images of assembly lines in factories owned by Western European corporations, and migrant workers, especially those working in the services and social care sectors. Central European workers and care-givers can no more speak for themselves than Spivak's subalterns.

This may be put down to structural silencing, which can be inscribed in the framework of the "semiosis" model presented above. During the communist period, the region was an important pole in the symbolic geography of Europe. The main opposition was obviously East/West, or Communist/Capitalist, which gave the entire "Eastern bloc" global visibility. There was a clear homology between its economic, political, and symbolic dimensions. Poland was frequently one of the most distinctive countries within the bloc. This was mostly due to its considerable resistance toward Moscow, but also to its relative liberalism in periods like the

1960s. This special status of Poland, which made it so visible, can also be interpreted within the framework of interface periphery. Its lost visibility is, however, quite often misunderstood, i.e. its broader context is often ignored, and it is seen as a solo actor, capable (or not) of promoting itself, and asserting its self-sustained agency on the global scene. One of the best examples of this was the "Solidarity" revolution of 1980–1981. Its importance and many of its meanings were defined in reference to the country's obvious but often unstated dependence on the Soviet Union. Poland was at that time the weakest link in the Soviet Empire's chain. Without this broader imperial context, however, it would not have been as visible as it was. Poland was still globally important for some time after the fall of the Soviet Empire as it was one of the most successful countries in "transitioning" from communism to capitalism. But with time and "normalization", by which is meant successful integration into Western economic and political structures, its visibility waned. The main point to note is that location in the global semiotic field is defined by homologies with the structures of the global field of power. This strongly impacts not only the visibility of countries and regions, but the visibility of their other fields, including those of academia.

1.5.6 Specificity of the Central and Eastern Europe's Dependent Status

Despite these circumstances, which make the Global East project difficult, if not impossible, to implement, discussing it can be an intellectually stimulating exercise. One of its crucial advantages is that it can be instrumental in theorizing a qualitatively different type of dependence on the Western core from the one studied within the framework of post-colonial theory. It can also stimulate new methodologies and insights into the workings of the global economic system. The relationships between the second and third worlds (or to use the new vocabulary, between the global South and global East), might also be interesting from this new perspective, which could offer a closer look at e.g. instances of direct interaction between the two peripheries (see e.g. Ginelli 2018). Here, it is important to bear in mind that there were precursors to

Müller's project, but they all failed to attract wider interest. This is indicative of the difficulty in generating interest in these sorts of initiatives—both within and without the region. One such proposal is worthy of special attention, although it is not even mentioned by Müller, viz. Waldstein's paper, "Theorizing the Second Word" (Waldstein 2010). In some respects, it constitutes a more profound, and even more radical, intellectual framework for rethinking the region's place in the broader intellectual framework of Western social theory than Müller's proposal. It is also important to note that Waldstein has also produced other valuable works and made a greater effort to "decenter", especially when analyzing the Russian Empire (Turoma and Waldstein 2012). Of particular interest is Waldstein's suggestion to look at the second world in a more traditional way, i.e. as a model and a resource for non-essentialist and non-Eurocentric theorizing. The second world is largely coextensive with the Global East. Waldstein also argues that there is a need for a "move of the area from deep provinces of the contemporary intellectual universe to a position as one of the key 'labs' for producing nonessentialist knowledge about (not only second world) culture and society" (Waldstein 2010: 104). Moreover, he sees the region as "an obvious source of analogies, comparative cases and (…) theoretical insights that are useful for understanding not only Russia and/or Poland but other regions as well, ultimately, human society and culture per se" (Waldstein 2010: 115). Waldstein argues that there are strong grounds for believing that, if debates on the region are combined with post-colonial theory, then the field of "Global East" studies (to use Müller's notion) can be moved to the center stage of critical social science theories and polemics. That Waldstein's stimulating manifesto did not gain adherents outside the circle of Russian origin scholars is again very telling and corroborates the unattractiveness of the Global East project. More broadly, Waldstein argues that the hybridity motif that so deeply permeates the Global East contains the potential for a powerful critique of Eurocentric and essentialist bias. He proposes looking at the second word as "a model and a resource for non-essentialist and non-Eurocentric theorizing" and at "the hybridity of the second world's cultural constructs and social formations as a model and resource for nonessentialist thinking" (Waldstein 2010: 102). Waldstein calls for a wide-ranging, theoretically profound, and empirically consequential critique of Eurocentric and essentialist assumptions that drew on inspiration from the global East. He argues that the second world's motive of hybridity may acquire a completely different meaning when analyzed within the framework of such a critique (Waldstein 2010: 103). In essence, Waldstein is calling for a dialogue between post-colonial studies and second world studies. It could be argued that there already exist studies of Central and Eastern European societies which have considerable potential to be extended outside the region and serve as a theoretical framework for analyzing both Western and non-Western societies along the lines suggested by Waldstein. This framework has the potential to be extended far beyond post-colonial studies. This is especially true of *Making Capitalism Without Capitalists* (Eyal et al. 1998). By problematizing the unusual (from a Western point of view) relations between economic, cultural, and political (social) capital, the authors produced a sophisticated theoretical tool that is based on Bourdieu's ideas, but which adapts them to accommodate the complexities of diverse social systems. The book provides an excellent illustration of how the hybrid and dynamically evolving social forms observed in Central and Eastern Europe can facilitate universal theorizing. The work can therefore be construed as an argument in support of Waldstein's thesis, even if its broader potential remains unrecognized.

This dialogue has been taking place for some time, but its scale and nature are rather disappointing. Its nature is somewhat asymmetrical in that it mostly concerns the "uses" to which post-colonial theory are put in Central and Eastern Europe, and its scale is limited to the paucity of references by Western theorists. However, it could be argued that the Russian Empire, and its successor state, the Soviet Union, are exceptionally useful sources of inspiration for post-colonial studies. Neither the Russian Empire nor the wider Soviet Empire were typical imperial powers, and they are well worth studying on account of their uniqueness and liminality. Several, mostly regional, scholars have attempted to draw attention to the potential benefits of studying the Russian and Soviet empires for post-colonial theory, but again, with limited success. One such group of scholars is the editorial collective and the circle of authors of the "Ab Imperio" Journal. This is where Waldstein published the paper cited above. But there is nothing comparable to even this journal's modest success in Poland—a country whose impact on the field of post-colonial studies is even less significant. Moreover, Poland and Polish authors are mostly conspicuous by their absence in "Ab Imperio".

At the same time, it has to be recognized that Poland is a special case by virtue of having occupied a peculiar position in both empires. Poland illustrates many of the mechanisms discussed by classical post-colonial scholars, but often in unusual and intriguing configurations. Russian control of Poland was never instituted in pure colonial terms; it was a hybrid and often liberal form of colonialism under both the Russian and Soviet regimes. The imperialism of Russia and the Soviet Union, being often weak, revealed several imperial mechanisms better than others. The cyclical weakness of these empires, and especially the slow but steady collapse of the Soviet Empire, revealed the imperial interplay between crucial social processes and ideas, with modernity at the forefront. Some recent review studies by Go give an excellent overview of the key issues on which the global sociology of empires and post-colonial theory are focused. These offer insights into some crucial but naturalized (or silenced) imperial aspects, as well as the roots, of modern societies. Many of these crucial debates reviewed by Go could clearly benefit from illustrative cases from the Global East in general, and on many occasions, from Polish–Russian relations in particular.

1.5.7 Russian and Soviet Colonialism in a Comparative Context

More generally, the context of the region potentially introduces additional complexities to classic Western models of social mechanisms. This is mainly because crucial differences between categories and ideas are often more nuanced than they are in classic colonial situations. For example, the issues of race and caste are pertinent to most Global Eastern societies. However, they are not based on binary, openly racial categories, but built on subtle, relational differences, which can be quite small, and are usually expressed with euphemisms rather than formalized. Some forms of Soviet colonial control may be of relevance in this context. Go proposes to use the term "informal imperialism" with respect to the United States and the United Kingdom in his comparison of Puerto Rico and the Philippines (Go 2009). I would argue that a similar variety of Soviet colonialism can be identified. As far as the Soviet continental empire is concerned, Poland, the GDR, and the constituent Soviet Republics (e.g. Latvia, Kazakhstan) all represent different modes of Soviet rule. But to that can be added a list countries under partial Soviet control, e.g. Cuba, North Korea, and some African states (e.g. Angola, Ethiopia, Mozambique), where Soviet templates of political and economic organization were applied in an even more selective and adaptive way. Countries that were temporarily partially dependent on the Soviet Union, e.g. Austria and (especially) Finland, can also be mentioned here. Go rejects the claim that the USA's modes of colonial rule and imperial policies were determined by the country's national character, its cultural forms, and the dispositions of imperial personnel, arguing that "In both America's colonial empire and Britain's colonial empire, there were a variety of colonial regimes and governmentalities, ranging from paternalistic protection to tutelary assimilation; both imperial states deployed informal mechanisms of influence in other parts of the world, and both states later substituted some direct methods for indirect ones while undergoing various historical waves of aggression in the longer run. One is hard-pressed to find homogeneity or national essences here" (Go 2009: 240). The same could be said of Russian and Soviet colonialism, which was highly adaptive and evolved over time, with policies often being changed over very short periods. This included wild swings between oppressive, hardline Russification, especially in the Soviet Republics, and concerted, large-scale efforts to preserve or revive indigenous cultures. This latter policy may be an example of what Go refers to as "rule by natives" that sometimes morphs into "go native, adapting themselves to local conditions and to the spaces and places they seek" (Go 2009: 240). Soviet control over Central Europe also assumed a variety of forms that changed over time. For example, in the 1968 Warsaw Pact invasion of Czechoslovakia, the Soviet Union was able to coopt Poland, Hungary and Bulgaria, thereby making them complicit. Go further contended that Empires can be seen as located in wider global fields of conflict, competing with each other as they reach across and down to more localized settings of power relations (Go 2009: 241). The

Soviet Union has to be seen as a competitor as well. Its oft-repeated anti-colonial narrative, combined with specific policies in territories under its control, can be seen as part of its game plan on the global imperial field. Soviet imperialism, often disguised as anti-imperialism, had to come up with suitable descriptions of, and justifications for, its actions, and stage-manage its operations in a way that would make them acceptable internationally. Examples include "The Limited Contingent of Soviet Troops in Afghanistan" in 1979 and its invasion of Czechoslovakia in 1968, when, as described above, the Soviet Army was assisted by several other "peoples" armies. This was depicted as a spontaneous reaction on the part of "fraternal" countries to forestall an imminent Western invasion of Czechoslovakia. Another example is the imposition of martial law in Poland in 1981. Only Polish troops were involved, but the operation was strictly coordinated by Moscow. The task was entrusted to Poland to prevent economic sanctions being imposed on the Soviet Union in response to another military "intervention", as happened immediately after its military involvement in Afghanistan. The Soviets were thus emulating British and American imperialists by "modifying their imperial policies and practices depending on the character of the local fields in which they operate and the place they occupy within those fields" (Go 2009: 242). Not only that, but the global structures and agency of subject peoples can sometimes overwhelm and alter their imperial designs. The existence of a global field of imperialism also implies the possibility of fascinating instances of competing colonialisms. Go specifically discusses US-Spanish competition. The US decried Spanish rule in the Philippines (1565–1898) as an irredeemable failure with deplorable long-term effects. The Spaniards were supposedly unable to offer mass education, and Filipinos had acquired tyrannical traits as a result of Spanish rule. By contrast, US rule (1898–1946) was depicted as emancipatory. Go quotes the assertion that "the US came in in defense of liberty, justice, and humanity" to overthrow the armed authority of Spain and to give people liberty, protection" (Go 2009: 242). Poland provides several instances competing imperial actors being undermined in this way. The most salient example is World War I, when the German,

Austrian,[4] and Russian colonists were in direct conflict. The German and Austrian colonists painted Russian colonialism in Poland as uncivilized and oppressive, while the Russians tried to undermine the legitimacy of the German and Austrian presence in Poland by presenting them as exploiters and promising autonomy and self-government to all Poles in the Russian Empire. All three empires, and later Soviet Russia, presented themselves as Enlighted modernizers of the territories they controlled. These were discursive strategies very similar to those observed by Go in the Philippines (Go 2008a). As he notes, the US colonial occupation was billed as a "school of politics" for the entire nation. The colonized Filipinos were to receive "a practical political education" in self-government. A similar narrative of tutelary colonialism, motived by discipline and democratization, feature prominently in the myth of the Austrian colonization of Poland during the second half of the nineteenth century. Interestingly these myths were largely incorporated by the modern Polish historical narrative in the form of the positive images of the history former region of Galicia, or the Lesser Poland (Zarycki 2007a). Even today, many of the intellectual elite in the south-east of Poland, boast of the region's supposedly superior democratic culture, for which they credit the Austrians. Similar elements were present in the German narrative of the occupation of Poland, which began in 1915 and lasted until 1918 (Rolf 2020). These narratives are discussed in more detail in later chapters. However, the Germans were not as successful as the Austrians in getting their *mission civilisatrice* narrative incorporated into mainstream Polish historical discourse.

Go (2013b: 211) notes that the Western European colonial powers introduced concepts such as citizenship, justice, humanity, and rights, but withheld their denotata; freedom and democracy were extolled but only granted to those at the top of the imperial hierarchy. The very same mechanism was employed by the Austrians and Prussians in their partitions of Poland, but not so much by the Russians, as the Russian Empire was not all that progressive in the late nineteenth century. Still, it had at least abolished serfdom (private estates, 1861; imperial lands, 1866). It should, however, be kept in mind that it did so against the

[4] Strictly speaking, Austro-Hungarian (after 1867).

backdrop of Polish nationalists promising to emancipate the serfs in an attempt to enlist their support. In other words, emancipating the serfs was not a completely voluntary act of magnanimity. Nor was it part of any previously planned modernization program. It was a concession the colonizers felt compelled to grant the colonized. The spread of liberal values described by Go was very much in evidence in the Austrian partition, although mostly only after 1848. It should be noted, however, that Austrian liberalism was at least partly due to the serious weaknesses exposed by the Hungarian Revolution of 1948 and Austria's defeat in the Austro-Prussian War of 1866. Late nineteenth-century Austrian liberalism, which is still trumpeted in Poland, was more the outcome of external and internal pressure on the empire than any progressive agenda initiated by the Austrian government. The colonial situation in the Austrian province of Galicia from the late nineteenth century until 1914 was also similar in that most of the democratic rights granted by Vienna were originally restricted to the landed gentry and the nobility; only gradually did they trickle down to the population at large. Go further argues that even such seemingly neutral concepts as civil society bear vestiges of an imperialist mindset (Go 2009). The fall of the Soviet Union revealed that certain concepts, as they were understood during its existence, had many imperial or colonial elements and facets. This was most apparent in Central Europe. These concepts, which were subsequently shown to have contributed to imperial domination, include "socialism", "progress", "modernization", "emancipation", and even "equality" and "justice". Their meanings drastically changed after the fall of the Soviet bloc. This lay bare their political nature and their reliance on an external framework of imperial control, both of which had conditioned their imposition and colored their redefinitions after 1945. The period 1989–1991 is crucial for gaining special insight into the symbolic mechanism of imperial or colonial control. Central Europe's breakaway from formal Soviet domination and acceptance of a relatively informal dependence on the Western core revealed several aspects of social control that are transparent in most Western societies, as they had been rendered invisible by their deep naturalization and the stability of the social structures in which they were embedded, both domestically and internationally. Post-colonial scholars discuss many other mechanisms that reveal untypical, and often nuanced, aspects in the context of Central and Eastern Europe.

1.5.8 Impossibility of Cosmopolitanism on Peripheries

One specific notion, which, as post-colonial authors point out, cannot be understood outside a broad international (and in many countries, imperial) context, is cosmopolitanism. Go notes that cosmopolitanism can be seen as the new humanism (Go 2013b: 210). He also brings up a vital paradox, viz. that "empires, in fact, also facilitated the flow of ideas, not least as European colonizers touted terms like 'civilization,' 'rights of man,' 'reason' and 'liberty'" (Go 2013b: 210). Interestingly, while few Poles saw the Russian Empire playing any such role for the betterment of Poland, most saw the Austrian Empire doing so to some extent. I say "interestingly", because all three empires that governed partitioned Poland during the nineteenth and twentieth centuries (and the Soviet Union later), played a significant role in disseminating and implementing these ideas. This includes various forms of cosmopolitanism. So far as the Russian Empire was concerned, this flow of fresh ideas was largely an unintended consequence of its development; it was certainly not the result of any policy decision taken by the Empire's conservative rulers. Obviously the ideas flowing into Poland from the West were far more numerous and varied. Poland should rightly be credited as the middleman in the transfer of Western ideas into the Russian heartland. Nevertheless, there was also a flow of ideas from Russia to the Polish partitions, especially the ideas of the Russian liberal intelligentsia, the creative class, writers, philosophers, and other intellectuals. Russian Politicians also played an important part, especially those of the liberal and left persuasions. They also had Polish partners, who increased their influence on Polish society. It was not always welcome among Poles, but the fascination with Russian ideas, especially "progressive" ideas, was one of the factors that shaped Polish political life under the Russian Empire.

This heritage is, however, almost entirely forgotten. The entire political camp of Polish liberals that had been part of the wider political scene in the Russian Empire simply vanished after 1918, even though they had been in the vanguard in shaping the modern Polish brand of cosmopolitanism under imperialist conditions. Their disappearance is sometimes called "The Liberal Atlantis" of Polish politics in the Russian Empire (Jaszczuk 1999). Other forms of Polish cosmopolitanism were strongly related to Poland's imperial dependence. During the communist period, these were related to Poland's dependence on the Soviet Union. One was communist internationalism, which had the effect of increasingly involving most of the Soviet bloc in the third world. The ensuing internationalization of Poland, and many other countries of the bloc, was labelled socialist globalization (Mark et al. 2020). Much of this heritage has now fallen into oblivion. Cosmopolitanism is closely linked to modernization. The above discussion can therefore be seen as part of a broader debate on the modernization of the Global East, in particular Poland, and more specifically, on the modernization of that part of Poland which was directly governed by the Russian Empire and then controlled by the Soviet Empire. The prevailing attitude in Poland is that neither the Russian Empire nor the Soviet Union modernized Poland at all, although there is obviously a minority of scholars and intellectuals who contend that the communist system had considerable modernizing aspects. These include Jakub Majmurek (2010), Andrzej W. Nowak (2016), Adam Leszczyński (2017), Agata Zysiak and Wiktor Marzec (2020), and Piotr Koryś (2018). However, most of them evince little interest in the wider, i.e. Soviet, context of such modernization. For some of them, this thesis also has a strong political aspect. It is clearly seen as an aspect of political critique of the dominant conservative vision of Poland's recent history and often part of an attempt to develop a new narrative for the Polish left.

1.5.9 Multiple Dependencies and Dualities of Central and Eastern Europe

In his discussion of Bourdieu's early contribution to what later became post-colonial studies, Go also called attention to an essential aspect of colonialism, viz. economic exploitation, which, in the case of Algeria (studied by Bourdieu) included the confiscation of prime land (Go 2013b: 213). That exploitation, as Bourdieu noted, was often justified by bourgeois ideology. Currently, we can point to several parallel processes taking place, especially in Central and Eastern Europe. These include dispossession, which, according to David Harvey (2006), is becoming a major driver of capitalist accumulation. The privatization of state assets, often at fire-sale prices, was rampant during the "transition period", and has always been presented as a central plank of modernization and efficiency, and even civilizational progress. Ukraine is expected to follow suit and privatize its public assets, real estate, industry, and infrastructure. The selloff of these assets, unremittingly insisted on by institutions such as the IMF and the WB, is presented as a precondition for modernization, and an entry requirement to join the club of European, i.e. civilized, nations.

The Global East can contribute to the development of post-colonial theory in other ways. Go reminds us that Bourdieu proposed viewing a "colonial situation" as a field (Go 2013a). The colonial situation of Poland is more complicated than most, as it can involve competition between several actors, as discussed above. The smaller peripheral states of Central Europe, and even the Russian and Soviet empires, despite their dependence on the West, can justifiably be seen as colonizing forces, as they marginalize and/or oppress a multitude of ethnic, religious, and/or cultural minorities to a greater or lesser extent. Such minorities maintain complex, and often hierarchical, sets of relations with the state and each other. These states and empires, although perceived as dominant actors on a regional scale, are themselves "subaltern" on the global scale—so much so that Viacheslav Morozov proposed calling Russia a "subaltern empire" (Morozov 2015). These political entities exploit lower-level actors (states, communities, groups, etc.) on behalf of the globally dominant actors, especially those representing the core of the world system.

Sometimes, however, relatively stronger actors enter into alliances with these lower-level groups in order to resist some aspects of the domination of the global forces. Several instances of such alliances are visible in the activities of the Russian state on the global scene.

These intertwined structures of domination create multidimensional social systems. This does not imply that they are more complex than those of the Western core states or Western colonies in the Global South. What is crucial in this context is that such multidimensionality does not depend on any of the dimensions or social divisions being clearly dominant. This may produce a configuration of a dual social sphere, i.e. a society in which none of the two competing social logics can gain unequivocal hegemony. Bourdieu's concept of the cultural sabir is relevant here. As Go reminds us (Go 2013a), this predates the post-colonial theories of hybridity and mimicry articulated by Bhabha (1994). The sabir "constantly refers to two different and even opposing logics" locked in a "double-sidedness expressed... in all realms of existence" (Bourdieu and Sayad 1964: 164). Go argues that the nature of the sabir also bears similarities with W. E. B. Du Bois's "double consciousness" (Go 2013a: 13). Go also points out that Frantz Fanon asserted that the colonial world was bifurcated (Go 2013b: 213). This sort of duality or multidimensionality may well be even stronger in Central Europe, or for that matter, most of the peripheral countries of the Global East. This is because they are not always characterized by such a sharp and formalized hierarchization of different cultures and social logics. This notion of duality, or multidimensionality, can be related to the analysis of the structure of peripheral fields of power introduced above. There, I argue that they have a twofold (or more) logic: a local one and another produced by the external, dominating field of power (usually in the form of opposition between economic and cultural capital). Duality, or greater multidimensionality, of academic fields is also very common in the region, and several instances of it are discussed in this book. It can therefore be argued that almost every Central European scholar, especially those of the older generation, is cultural sabir to some extent. He/she is a post-communist Easterner embedded in the local networks of social relations and traditions while aspiring to be a Western or global scholar. He/she is not quite a European academic, whatever he/she thinks, but is forever torn between the local academic field and the global game. And these are often words apart.

1.5.10 The Unattainable Universalism of Central and Eastern Europe

Waldstein was hoping that the Global East would be "discovered", specifically where it intersected post-colonial theory and other anti-essentialist critiques. For him, this would have meant that not only had Western theorists discovered the region, but that they had also found that all these theories were applicable to both East and West. His hope was that such a meeting would promote the region from its backwater status in the intellectual world and make it a key "laboratory" for producing non-essentialist knowledge about the culture and society of the second world and beyond (Waldstein 2010: 104). He argues that the most pertinent factor in this context is that post-colonial theory demonstrates the relativity of the boundaries between the metropole and the colony, and then goes on to explore the "mutual constitution" from which their social and cultural identities and institutions historically emerged. He additionally claims that post-colonial theorists have relativized the distinction between nation and empire in a number of interesting ways (Waldstein 2010: 105). Significantly, Waldstein calls for the "deprovincialization of hybridity" in post-colonial analysis, explaining that this would be a step toward deprovincializing the second world, and claiming that it would make the Global East a legitimate source of "rules", by which is meant theoretical propositions and models (Waldstein 2010: 108). He follows up on this idea by proposing that Western empires be seen as "more like" Russia than the other way round. He calls this a "heuristic turn" that challenges the assumptions of conclusions of not only students of the second world, but also specialists of the "West" and "East" (Waldstein 2010: 107). As Waldstein argues, "To put it in essentialist terms, this is when less European nations (Russians) or outright Orientals (Turks) establish their imperial sovereignty over more European nations (e.g., Greeks and Poles). To put it more analytically, this is a case when centers of political power and cultural prestige do not coincide or—in even more radical cases—they are located on the opposite poles of imperial interaction axes. In a word, political authority comes from St. Petersburg while cultural influence spreads from or via Warsaw" (Waldstein 2010: 108). Waldstein compares the Romans, who were culturally dominated by the "colonized" Greeks, with the Japanese, who were culturally dominated by the Chinese for centuries without being in any way politically or economically dominated by China. He ends this line of reasoning by pointing out that the cultural prestige that French "high" culture enjoyed at various points in the twentieth century did not correspond to the high points of France's political influence. We could continue this list of configurations by adding Polish–Russian relations. Waldstein seems to concur with Go when he argues that the example of Russia might best support the contention that the "sharp distinction between colonial and contiguous empires as properly modern and not-quite-modern empires needs to be reconsidered" (Waldstein 2010: 107). This observation resonates with several of Go's reflections. As Go notes, "While empires and civic-liberal nations have been seen as opposite and even contradictory political forms, this essay argues that they are similar. Both create and depend upon hierarchical differentiation accompanied by exclusion and subjugation. Furthermore, they are logically related. The hierarchies typically attributed to empires are inscribed into the very theoretical and institutional core of civic-liberal nationhood" (Go 2017).

1.5.11 Construction of Continuities and Discontinuities

One of the most striking aspects of empires is their dynamics. George Steinmetz notes that sociologists find that focusing on empires makes it impossible to overlook world systemic and racial disparities of power (Steinmetz 2013: 373). Empires are riven by anti-imperial movements, and this prevents analysts from becoming trapped in static, functionalist models. This was, of course, also the experience of Poland as part of all the empires mentioned above, all of which eventually collapsed, and in most cases quite suddenly, which produced a fascinating problem of continuities and discontinuities, as well as that sudden rescaling's of social fields. Transitions between highly globalized imperial fields and much more narrow, often isolated, national fields were a common experience for people living in Poland in the twentieth century. This makes the fuzziness of the distinctions between diverse political bodies and scales even more apposite. One of the most spectacular transitions was the emergence of nation-states in Central Europe after World War I. This fundamentally altered the whole trans-national relations paradigm. As Juho Korhonen notes, "The trade-off was to grant nationalism the universal status formerly held by empire but based on a negation of the former neutral and mediating form of field relations. Under the nation-state, the national meta-capital regulates and defines hierarchies and divisions. Global hierarchies become inter-national, rather than intra-empire, lending legitimacy to ideas of nation-state sovereignty and citizenship" (Korhonen 2018: 18). These sorts of sweeping transformations (which also took place after World War II and after the fall of the Soviet Union), can be analyzed through the evolutions of social fields, especially those of power, and, on a lower level, social sciences. The structural evolution of these fields is both a cause and an effect of such transformations. Also interesting in this context are the discontinuities and inapparent continuities of fields of different levels and other social structures. This was a particularly pressing issue in the interwar period and it remains a fascinating theoretical problem. A scrupulous study of these transition periods in the Global East can contribute to the development of Bourdieu's inspired theory of the state, and to the fields of imperial and post-colonial studies. One of Poland's most peculiar and fateful characteristics is its location. The country long lay at the confluence of three (later two) empires and was the Westernmost, and most Western-oriented periphery of the Russian Empire, before becoming part of the Soviet Central European zone. This makes it a fascinating laboratory in which to study the emergence of a modern nation-state and its generally accepted redefinitions. This perspective enables the history of the Polish state, which disappeared and reappeared on the political map of Europe between the late eighteenth and the mid-twentieth century, to be studied in an innovative way. In particular, it lays bare the relation between the political history of statehood and the trajectories of social fields, which often predate state formation and remain autonomous from it. More tangible issues arise as a result of such discontinuity. These include the lack of any clear definitions of the Polish economy, Polish politics, and even Polish culture, although this last would seem to be the least problematic, as Polish culture can be thought of as the main dimension of continuity of what is generally considered the core of Polishness (Zarycki et al. 2017). But this still leaves a lot of ambiguities concerning the place of minorities in the Polish canon, e.g. "Polish Jews", "Polish Ukrainians", and "Polish Germans", as well as religious minorities. Although the position of these minority groups and cultures within the broader Polish cultural hierarchies has been thoroughly researched, the subject of their separate continuities has not, and it may be just as complex. The Jewish community of Poland, to take a single, albeit spectacular, example, exhibited incredible discontinuities during the twentieth century—especially its leaders active in the public sphere. Diverse definitions of continuity vie for dominance in all these dimensions—a clear sign that they are best understood and should be studied in terms of social fields.

Tensions within, and reconstructions of, these fields, in particular the central Polish field of power, are not always immediately apparent. This is because there is a firmly entrenched conviction that the Polish nation has an unshakeable and immutable core that is impervious to foreign invasions and global revolutions. One of its best known expressions of this is the first sentence of the Polish national anthem: "Poland has not yet perished, so long as we still live". Poland's discontinuity is especially well illustrated by its territorial boundaries, which have shifted so much over the centuries that it is difficult to agree on what properly constitutes the "proper" geographical extent of the country. This becomes particularly acute when trying to use any historical statistics or attempting to reconstruct dynamics of social, economic, demographic or processes considered to be Polish. A number of difficult issues are affected, e.g. how to deal with territories where ethnic Poles were economically dominant, especially in terms of land ownership (e.g. parts of central Ukraine until 1918), while most of the population were ethnic Ukrainians, as was the case from the late eighteenth century until 1918. Some argue that these regions were not Polish at the beginning of the twentieth century and are not Polish today. Others counter that they are an important part of Polish history by virtue of the wealth of the Polish elite who lived there. A separate chapter could be devoted to the problematic aspects of the (dis)continuity of Polish statehood. It could be argued that the case of Poland draws attention to the subjective and political nature of the continuity of modern society and the diverse and competing aspects that constitute it. Central and Eastern Europe certainly has no shortage of entities whose continuity is at least debatable. For instance, several of the former Soviet republics have had some of their cultural and political heritages redefined. However, what is often interpreted as the debatable continuity of these states may be primarily a sign of weakness; a weakness that calls their claims of having inherited specific traditions into question. It is intriguing that some continuities are highly desired while others are emphatically unwanted. The latter include externally imposed labels, e.g. "post-communist", which is assigned to most former Soviet bloc countries, despite their desire to be perceived as "normal" European states. By contrast, labels such as "post-Nazi" or "post-fascist" are almost never assigned to e.g. Germany or Italy. These comparatively wealthy Western European states have successfully established a universally recognized discontinuity in this respect.

1.5.12 Universalizing the Polish–Russian Relationship

Waldstein's proposal to incorporate the former second world, or the Global East, into post-colonial theory is of the greatest relevance to this work. Therefore, I would like to reflect on some more specific arguments, in particular, his focus on Polish–Russian relations as a model object for social theory. Of special interest here is Waldstein's proposal to define the Russian–Polish imperial encounter, as he put it, "… as a model—obviously not *the* model—of the imperial encounter, rather than an exception or a pathology…", suggesting that "… we may achieve at least two goals: to challenge postcolonial theorists to clarify their conceptual tools and to make a strong point for the need of dialogue between cultural and social sciences" (Waldstein 2010: 110). Poland and Polish–Russian relations are, of course, only one of the models to be found in the Global East. However, I would argue that, in addition to serving as a model, they can be interpreted as a structure that can bind any two peripheral entities and which can be configured in a number of ways. Focusing on Polish–Russian relations, provided they are seen as mutually constitutive, may complicate the image of the region, in particular the image of the history of Russia and the other Empires that have ruled Polish territory, especially Germany.

Many of the important developments in Russia's history can be interpreted as moments when the "Polish factor" had a crucial impact on the direction it subsequently took. One such moment is even celebrated as a public holiday in Russia: "Unity Day" celebrates the liberation of Moscow from Polish "interventionists" or occupiers in 1612. The nineteenth century, when Russia controlled the largest partition of what had been Poland, experienced several moments when Poland had a major impact on the course of Russian history. Important constitutional experiments in early nineteenth century Russia were first tried in Poland, the most important being the establishment of Congress Poland (formally the Kingdom of Poland) as an autonomous province of the Russian Empire, in 1815. Its autonomy, however, was curtailed following uprisings in 1830 and 1863 (Mażewski 2013). Congress Poland was a testing ground for Western constitutionalism within an otherwise unitary Empire ruled by absolute monarchy. The November Uprising of 1830 brought that period of liberal experimentation to an abrupt halt. These developments also influenced Polish and Russian academic circles. The intellectual heritage of Russia benefited enormously from having helped partition Poland. Russia gained Vilnius University, then known as the Vilnius Academy. It was renamed the Imperial University of Vilna by Tsar Alexander I in 1803. The same liberal Tsar founded the University of Warsaw in 1816. During the course of the nineteenth century, Poland benefited for a time from these liberal Russian experiments, but severe Russian reprisals, chiefly against rebellious factions of the Polish nobility, and increasingly repressive measures against the general population, followed in the wake of two military uprisings (1830 and 1863). These and related developments support the contention that the Polish factor was crucial for the trajectory of the Russian Empire and has permanently shaped many of the features of the modern Russian identity. It should also be borne in mind that as the Russian partition of Poland was the largest, Russia's population was increased, and its ethnic and social composition altered, considerably. In particular, it gained a large Jewish population. Not only that, but the numerous Polish nobility, the main agent behind the Polish uprisings, and the several strata of Polish Catholic peasants changed the social composition of the Russian Empire and its field of power. It bears repeating that the serfs were first emancipated in Congress Poland, and that this measure was primarily taken to contain Polish unrest, as the Polish insurgents had been promising to do likewise. These and other related facts are known to the specialist in Russian history. However, they are rarely used to interpreted Russian history more broadly, and least of all in Russia. This may part of the marginalization of Poland in the Russian identity. Interestingly, the Polish factor, i.e. the crucial impact of Polish actors on the trajectory of the Empire, is not only downplayed in the mainstream version of the history of the Russian Empire. The history of the German/Prussian Empire similarly downplays the role of the Polish factor, which was made up of a large Polish population and active Polish elites. While it was not comparable in scale or impact to the Polish factor in Russia, Poles nevertheless played a major role in some decisive moments in the history of Germany/Prussia. Rogers Brubaker provides a good example of this in his history of the development of the modern citizenship model in Germany (Brubaker 1992). Since the formation of Germany in the late nineteenth century, citizenship has been based on the principle of jus sanguinis that had applied in Prussia, although jus soli (under certain strict conditions) has also been recognized since 2000. Brubaker demonstrated that this was mainly driven by the fear of a large population of ethnic Poles (not only those residing in what was then Eastern Prussia, but also migrant workers from the Russian Empire), that would have been difficult to fully Germanize. This effectively denied many of them German citizenship, thereby allowing them to be deported to Russia. Poland however rates little more than a footnote in any sociological discussion on the heritage of German imperialism. For instance, Steinmetz does not include Poland when discussing German colonialism and its role in the development of German sociological theory. At the same time, it has been demonstrated that the "Polish question" has been an important factor, mostly as a threat and a negative point of reference, in the sociological thinking of Max Weber (Abraham 1991).

One of the reasons that Poland and Russia make such an interesting study from the perspective of the sociology of empires is that they are simultaneously so similar and so different, so close and so distant in so many respects. This makes the relations between the two countries highly ambiguous. In many aspects, they represent two incompatible modes of being "eastern", while being closely linked, no matter how frequently and strenuously both countries disavow this relationship and deny their closeness. Now to discuss some of these crucial differences. From the point of view of the world-systems theory, Poland is closer to a typical semi-periphery, and Russia to a periphery proper, especially in view of its stronger reliance on the export of raw materials and natural resources. Poland has recently appeared to be an internal periphery of the West, especially since joining the EU in 2004 and the Schengen Area in 2007. By contrast, Russia is an external periphery. The two countries differ much more profoundly in their social structures. As mentioned above, these can be theorized as representing two divergent modes of compensation of peripheral status, viz. cultural and political. In Poland, the main compensatory resource is cultural capital, whereas in Russia, it is political capital. This difference in strategy became clearly visible during the Bolshevik Revolution, but it began much earlier. In Poland, the broader revolutionary wave led to the establishment of an intelligentsia-led nation-state rather than a communist one, although it began with a socialist revolution. Both Poland and Russia were led by intelligentsia elites from about 1918, albeit by different factions of the intelligentsia. In Poland, the national intelligentsia founded a parliamentary democracy and respected the interests of the Catholic Church, even seeing it as an ally, while in Russia, a radical left revolutionary intelligentsia seized power and established an autocratic, highly secularized regime. When Stalin came to power (1927), even that radical left cultural elite was marginalized in the field of power, and many of its members later perished in the Great Purge (1937–1938). These differences are well reflected in the shape of the fields of power of the two countries. The Polish field of power, especially since the fall of communism, has been dominated by diverse factions of the intelligentsia and the cultural elite (Zarycki 2020). In Russia, the political elite is dominant, one to which both the native economic elite and the cultural elite are subordinate. The main division in the field of power runs across specific camps of the state elite, e.g. between those institutions that control the country's finances and those referred to as "force ministries", i.e. the security services and the military. In Poland, the field of power is divided between various factions of the intelligentsia, in particular, those with more internationalized and/or cosmopolitan cultural capital, and those with more local and traditional cultural capital (e.g. strong ties to the Catholic Church). One of the major conclusions to be drawn from this brief comparison of the Polish and Russian trajectories is that they are obviously opposing models of how the eastern peripheries of the Western core function.

1.6 Reinterpreting Poland from the Perspective of Global Historical Sociology

1.6.1 Toward Decentering of the Polish Historical Narrative

I now wish to briefly focus on Poland. I should stress, however, that, in what follows, the country is interpreted in ways that are not compatible with the dominant, stereotypical global images, or the stereotypical historiographic narratives, of Poland. The interpretations presented in this book deliberately challenge the dominant Polish-centric narratives and models, primarily by putting the Russian and Soviet empires back into the picture of the dynamics of twentieth-century Polish society.

If I had to list the authors and narratives which I aspire to challenge and oppose, it would be easiest to identify those from whom I take inspiration when aspiring to overcome the abovementioned stereotypes. This is because the vast majority of historical works on Poland, both popular and academic, are written from a decidedly Polish-centric perspective. It can be argued that, paradoxically, this is not due to the strength of

Polish science and culture. If anything, it can be seen as a result of the fact that Poland has not been able to build a larger analytical apparatus to look at its history from a broader perspective. More precisely, Poland has not been able to sustain such an apparatus. This is because it previously had one in the form of the Polish school of economic history in the 1960s (Kola 2018; Sosnowska 2018). Nor have similar efforts been made in the West over the past couple of decades. At least, there is no evidence of such attempts in the global social science fields. One of the fields that concerns us here is "area studies", in particular, "Central and Eastern European Studies". The global reconstruction of Russian and Soviet imperialism is currently being pursued by a number of prominent scholars. This has contributed to the creation of new alternative perspectives on Russian history (e.g. Morozov 2015; Miller 2008; Mogilner 2013). Similarly oriented Polish studies are few and far between. Among the few commendable exceptions to this dearth of studies on Polish history from a global perspective are the works of Maxim Waldstein, already cited here, and those of Malte Rolf and Jan Surman, which I use extensively in the second part of the book. Other authors who attempt to follow this path include Wiktor Marzec and Małgorzata Fidelis. The former has offered a global perspective on the course of the 1905 revolution in Polish lands (Marzec 2020), while the latter puts the cultural transformations in Poland in the 1960s into a global context (Fidelis 2013). However, these are the exceptions that confirm the rule.

When an anonymous reviewer of the manuscript of this book asked whether the global stereotype of Poland could be Polish-centric, I somewhat paradoxically answered in the affirmative. This is because Polish history is either marginalized or ignored altogether in global historiographical discourses, or the basic Polish-centric framework on which it is based is reproduced. Among the best known examples of such works on Polish history in the West are those by Norman Davies (e.g. Davies 1981). These seem to prove the existence of what can be called the dominant, global stereotype of Poland. This has the positive aspect of the above mentioned idealizations, but there is definitely a negative aspect in which the particularities of contemporary Polish history are found in a distinctive Polish nationalism and anti-Semitism, around which a separate current of Polish-centric Holocaust studies is currently developing (Kichelewski et al. 2019).

The work of Brian Porter-Szücs (2014), which attempts to break free of the Polish-centric paradigm, is something of an exception in this context. Porter-Szücs attempts to "disenchant" the history of Poland by pointing out that many of its typical elements appear in the histories of a number of other countries. Porter-Szücs's book, however, makes little attempt to understand the dynamics of Polish history from the vantage point of the logic of the empires and the broader global forces that control it. It can largely be read as a polemic against the more conservative interpretations of Polish history, which often pathetically emphasize the uniqueness of the Polish experience and attempt to construct an image of Poland as a land of the free and an innocent victim of empires and invaders. Recent works by Polish intellectuals belonging to the newly formed folk history school (*ludowa historia Polski*) enter into an even more open polemic with these idealized visions of Polish history concocted by political conservatives (e.g. Leszczyński 2020; Rauszer 2020 or Pobłocki 2021). These scholars attempt to write history in ways that demystify the conservative narratives of Poland as heir to the idealized pre-1791 noble republic. They do so by illustrating the plight of the popular classes and detailing the exploitation to which they have been subjected over the centuries. However, these works do very little to systematically contextualize Polish history in the comparative perspective of the logic of empires which controlled its lands. Because they are written from anti-elite positions, they also systematically depreciate the importance of elites in the making of history. This book, by contrast, foregrounds the role of elites. This is not to deny the importance of the popular classes or undermine the social importance of studying their plight. It is simply my analytical choice to focus on the non-obvious ways in which elites create the social world, and naturalize the ways in which we perceive the histories of our countries and the languages we use.

1.6.2 Polish History in Imperial Contexts

The role the two empires played in the history of Poland is often portrayed in a simplistic, entertaining, but mostly negative way. This book attempts to give a much more complex and nuanced picture. The dependence of the Global East on the global core is deemed crucial to understanding Poland's social processes. Less attention is devoted to "internal colonization", especially within Poland, although its importance is certainly not denied. Some aspects of this have been discussed in Polish debates, so they do not require a systematic discussion here. The role of the Russian and Soviet empires in shaping the structures of the modern Polish society has been much less studied within and outside Poland. The same is also true of German/Prussian influence, and to a much lesser extent, of the Austrian Empire, as its heritage is studied more often and usually interpreted in a more positive light. As a result, the impact of Austria on Polish society is better recognized. As mentioned above, however, my main focus here is on the role of Russia and the Soviet Union, as they occupy a special place in the mainstream Polish historical narrative. The Russian and Soviet empires are predominantly viewed as evil in Poland; as inveterate enemies and oppressors. Poland sees itself as a victim of their oppression, and resistance is a crucial element of the Polish national identity (Zarycki 2004). Resistance to the German/Prussian empires is also important in Polish historiography, but is less central and less naturalized than resistance to the Russian/Soviet empires. Germany/Prussia is simultaneously seen as an oppressor and a source of modernity and civilization in the Polish imagination. None of these empires, however, is considered to have had a major impact on the contemporary Polish state or on Polish society. This is because the history of Poland is written by its cultural elite, i.e. the intelligentsia, as a heroic saga of unwavering resistance to imperial powers bent on eradicating all things Polish, including its culture, economy, and traditions of statehood. In this context, many Poles would find it galling to have to acknowledge the extent to which their country is a "child of empires", to use Steinmetz's (2013) words. The role of empires in the development of the social sciences in Poland—both in terms of careers of individual scholars, and institutional structures and traditions—is particularly relevant in the context of this book. A close examination of the legacy of Poland's twentieth century academic system confirms Steinmetz's assertion that "politically objectionable regimes can be conducive to intellectual insight and creativity, especially under conditions of scientific autonomy and where resources for research are available. Empires are culturally and socially more complex than nation-states and therefore push their intellectuals to look beyond narrow ethnic and cultural categories" (Steinmetz 2013: 373). This book presents a considerable number of arguments in support of the thesis that the Russian Empire produced some of Poland's best known scholars and even entire intellectual schools. It obviously does not follow that most of them supported the Russian Empire in general or Russian rule in Poland in particular. During the 1960s, the Soviet Union similarly indirectly fostered an unprecedented outpouring of intellectual creativity in Poland. Again, however, this does not mean that most Polish scholars at the time were enthusiastically pro-Soviet.

It is crucially important from my point of view that the wider frames of the partitioning powers prior to WWI, as well as the Soviet Empire after WWII, if not earlier, are crucial contexts for interpreting of dynamics of Polish society and the Polish social sciences over the course of the nineteenth and twentieth centuries. The social sciences in Poland, however, tend to downplay that context and assign most of the agency to Polish actors during these periods. Even historically, these external frameworks were ignored by Poles. One of the most instructive examples of this is *Lalka* (The Doll), one of the major novels of Bolesław Prus (1847–1912), one of Poland's greatest literary figures. The book was written in 1887–1889 and set against the backdrop of contemporary Warsaw. However, the only indication that Warsaw was then in the Russian Empire is the currency (the ruble). In the same way, the broader context of Soviet domination is often ignored when post-war Poland is analyzed. And when it is recognized, it is often seen as no more than an imposed political and cultural framework that was actively resisted by most Poles. This dominant anti-communist paradigm prevents many of the crucial social processes in communist Poland from being interpreted as part of a broader imperial context. The "Solidarity" phenomenon is one of the most illustrative examples of this, as it can only be fully understood in the broader imperial context. The Solidarity revolution of 1980/81 was a peripheral mutiny in the Soviet Empire. And this aspect is crucial. Moreover, Solidarity's discourse cannot properly be understood without taking into account that it was expressed in the language of "socialist values" imposed by the Soviet Union. On the one hand, Solidarity's discourse was an attempt to discredit the ruling communists by relentlessly pointing to the interminable gap between socialist propaganda and the reality of life under socialism. On the other hand, however, it was an attempt to claim the rights and freedoms guaranteed by the socialist system and force the system to comply with its own officially declared ideals. Again, consider the similarity between this configuration in Polish-Soviet relations and Go's interpretation of U.S. domination over Puerto Rico (Go 2008a). Of special interest here is Go's discussion of the domestication, reproduction, transformation, and reevaluation of U.S. culture in Puerto Rico. However, as Go asserts, none of this was engineered. He argues that "after learning the Americans' meanings and models, the Puerto Rican elite turned them against their rulers… Americanization might happened in Puerto Rico, but it simply brought with it new dilemmas, contradictions and tensions that went far beyond the scopes and hopes of the imperialists' goodwill" (Go 2008a: 294). The "Solidarity" revolution can be interpreted in exactly the same way. The Polish rebels had assimilated the values and language imposed by the Soviets (although supported and "repackaged" by factions of the local elite and other classes), but they were opposed to the Soviets and the Polish ruling class, whom they considered Soviet vassals. In short, they were rebelling against the imperial power. Most other anti-communist movements, and many intellectual-critical discourses, of that period should be similarly interpreted. The prevailing interpretation of the Solidarity revolution of 1980/81, however, is that it was an expression of the idealism of Polish workers and the intellectuals who supported them. This leaves contemporary observers of Polish politics bemused as to why "the spirit of Solidarity" disappeared after 1989 and why there has been so little interest in the ideals of the Solidarity revolution since then. The paradox consists in the fact that the idealist discourse of Solidarity was only comprehensible with reference to the official ideological discourse of the Soviet empire; a discourse largely accepted by the protesting workers. However, this is mortifying to admit today, especially for those admirers of Solidarity who see it primarily as an anti-communist and anti-Soviet movement. While these were undeniably major factors driving the political dynamics of the unrest in Poland during that period, anti-communism and ant-Sovietism have to be understood in the context of Polish society as it then was. The actions of the Polish protesters only make sense when analyzed in the imperial and global context of the time. It can be argued that Solidarity's opposition to the ruling elites of the communist state widened the rift between the bulk of Polish society and that elite until the main cleavage of the field of power became homological to the global axis of opposition between Soviet communism and Western capitalism. The initial context of the revolt, however, was deeply embedded in the values and discourses of communist Poland. The seemingly inexplicable disappearance of Solidarity is a corollary of decontextualizing it and rejecting relational explanations for its institution, demands, and methods. These same images often suggest that the "essence" or "spirit" of Solidarity might be lurking somewhere beneath the surface of current political debates and is worth seeking out—partly because of the international fame that Solidarity achieved in its heyday. As mentioned above, Solidarity's erstwhile high profile was not only owing to the nature of the movement, but also to the configuration of the geopolitical forces of the time. Several examples of scholarly works predicated on these assumptions can be cited. Elżbieta Ciżewska-Martyńska's paper is a prime example. It provides a concise review of the classic works written on this topic, while analyzing the "Solidarity" phenomenon without referring to the broader Soviet context (Ciżewska-Martyńska 2015).

1.6.3 Political Contexts of Alternative History Writing

My final reflection in this chapter is that this type of analysis may also constitute a challenge from the political point of view. Most of the issues discussed in this chapter potentially have direct political implications when analyzed using the logic of the contemporary Polish political scene, as they are usually inscribed in the emotional narratives that determine the contexts, set the agendas, and dictate the terms of political conflicts. The uses to which post-colonial theory are put (as mentioned at the beginning of this chapter) eloquently illustrate just how deeply politicized these academic debates can be. It bears repeating that Russia is mostly viewed as an evil oppressor in Poland, and this perception is a central reference point for Poland's national identity. Thus, the history of Russian imperialism is often interpreted critically and emotively in Poland. As for the communist period in Poland, that is the subject of an even more emotional and explosive debate. To discuss the Russian and Soviet factors in Poland's history is to come under constant pressure to make clear moral judgments and join in the political critique. Russia's recent military assault on Ukraine has further exacerbated tensions in this regard. However, although the dominant image of Russia and the Soviet Union is a highly negative one, there are small groups of activists and intellectuals that challenge it. Some of these are mentioned above in the context of debates on the modernization of Poland. However, several of these critical scholars have agendas embedded in wider, and not necessarily academic, causes. These are generally politically motivated critiques of the conservative paradigm of Poland's history; one in which Russia plays a particularly evil role. However critical I might be of these stereotypically conservative, if not downright nationalistic, perceptions that demonize Russia and the Soviet Union, and which are indelibly inscribed in classic orientalist oppositions, such as Russia-East-evil vs. Poland-West-goodness, or communism-oppression-backwardness vs. capitalism-freedom-modernity, it would be rash to replace them with their equally simplistic opposites. Debates between these two visions may be instructive, and I have been following them with interest. However, uncritically adopting any of the positions carries the risk of preventing these wider, imperial contexts from being reconstructed in a more objective, impartial way. Judgmental interpretations of the issues under investigation are best avoided, as neither summarily condemning nor praising the Russian and Soviet colonization of Polish territory and domination of Polish society can lead to a better understanding of the complex mechanism of imperial and colonial machinery. Moreover, I see the mission of the sociology of empires as primarily an analytical deconstruction of the often highly ambiguous, entangled dependencies of imperial and colonial systems. My main aim here is therefore to promote the exchange of tools and case studies between the diverse fields of interest of sociology of empires and post-colonial studies. My deconstruction of the role of the Russian Empire and the Soviet Union in the development of the social sciences and humanities in Poland is emphatically not an attempt to further any political agenda. I do, however, reject the presentism that favors a highly politicized interpretation of the past. In short, I wish to distance myself from engaged, activist sociology, i.e. "public sociology", in favor of "critical sociology", to borrow Michael Burawoy's terms (Burawoy 2005). This work is addressed to an academic readership and is focused on deconstructing the dominant intellectual framework and reconstructing the long-neglected complex historical context, that has produced the social sciences in Poland as we know them today.

References

Abraham, Gary A. 1991. Max Weber: Modernist Anti-Pluralism and the Polish Question. *New German Critique* 53: 33–66.

Alford, Robert R. 1962. A Suggested Index of the Association of Social Class and Voting. *Public Opinion Quarterly* 26 (3): 417–425.

Althusser, Louis. 1994. Ideology and Ideological State Apparatuses. In *Mapping Ideology*, ed. S. Žižek, 100–141. London: Verso.

Anderson, Perry. 2016. *In the Tracks of Historical Materialism.* London: Verso Books.

Babones, Salvatore. 2012. Position and Mobility in the Contemporary World-Economy: A Structuralist Perspective. In *Routledge Handbook of World-Systems Analysis*, ed. S. Babones and C. Chase-Dunn, 327–335. Oxon: Routledge.

Bertalanffy, Ludwig v. 1973. *General Systems Theory: Foundations, Development, Applications.* Harmondsworth: Penguin.

Bhabha, Homi K. 1994. *The Location of Culture.* Milton Park: Routledge.

Bockman, Johanna K., and Gil Eyal. 2002. Eastern Europe as a Laboratory for Economic Knowledge: The Transnational Roots of Neoliberalism. *American Journal of Sociology* 108 (2): 310–352.
Böröcz, József. 2006. Goodness Is Elsewhere: The Rule of European Difference. *Comparative Studies in Society and History* 48 (1): 110–138.
Boucher, Geoff. 2009. *The Charmed Circle of Ideology: A Critique of Laclau and Mouffe, Butler and Zizek*. Melbourne: re. press.
Bourdieu, Pierre. 1971. Genèse et structure du champ religieux. *Revue française de sociologie* 12: 295–334.
Bourdieu, Pierre. 1977. *Outline of a Theory of Practice*. Cambridge and New York: Cambridge University Press.
Bourdieu, Pierre. 1983. The Field of Cultural Production, or the Economic World Reversed. *Poetics* 12: 311–356.
Bourdieu, Pierre. 1986. The Forms of Capital. In *Handbook of Theory and Research for Sociology of Education*, ed. J.G. Richardson, 241–258. New York, Westport, Connecticut, and London: Greenwood Press.
Bourdieu, Pierre. 1988. *Homo Academicus*. Stanford, CA: Stanford University Press.
Bourdieu, Pierre. 1989. *La noblesse d'Etat. Grandes ecoles et esprit de corps*. Paris: Minuit.
Bourdieu, Pierre. 1991. *Language and Symbolic Power*. Cambridge, MA: Harvard University Press.
Bourdieu, Pierre. 1996. *The State Nobility. Elite Schools in the Field of Power.* Stanford: Stanford University Press.
Bourdieu, Pierre. 1999. Rethinking the State: Genesis and Structure of the Bureaucratic Field. In *State/Culture: State-Formation after the Cultural Turn*, ed. George Steinmetz, 53–75. Ihaka: Cornell University Press.
Bourdieu, Pierre. 2005. *The Social Structures of the Economy*. Cambridge: Polity.
Bourdieu, Pierre. 2014. *On the State: Lectures at the Collège de France, 1989–1992*. Cambridge: Polity.
Bourdieu, Pierre, and Abdelmalek Sayad. 1964. *Le déracinement: la crise de l'agriculture traditionnelle en Algérie*. Paris: Éditions de Minuit.
Bourdieu, Pierre, and Loïc J. D. Wacquant. 1992. *An Invitation to Reflexive Sociology*. Chicago: University of Chicago Press.
Bourdieu, Pierre, and Loïc J. D. Wacquant. 1993. From Ruling Class to Field of Power: An Interview with Pierre Bourdieu on La Noblesse d'État. *Theory, Culture and Society* 10 (3): 19–44.
Brubaker, Rogers. 1992. *Citizenship and Nationhood in France and Germany*. Cambridge: Harvard University Press.

Buchholz, Larrisa. 2016. What Is a Global Field? Theorizing Fields Beyond the Nation-State. *The Sociological Review* 64 (2): 31–60.

Buchholz, Larissa. 2018. Rethinking the Center-Periphery Model: Dimensions and Temporalities of Macro-Structure in a Global Field of Cultural Production. *Poetics* 71: 18–32.

Buchholz, Larrisa, and Ulf Wuggenig. 2005. Cultural Globalization Between Myth and Reality: The Case of the Contemporary Visual Arts. *ART-e-FACT: Strategies of Resistance.* http://artefact.mi2.hr/_a04/lang_en/theory_buchholz_en.htm(4).

Burawoy, Michael. 2005. 2004 American Sociological Association Presidential Address: For Public Sociology. *The British Journal of Sociology* 56: 259–194.

Campbell, Angus, Philip E. Converse, Warren E. Miller, and Donald E. Stokes. 1960. *The American Voter*. New York: Wiley.

Carls-Diamante, Sidney. 2017. The Octopus and the Unity of Consciousness. *Biology and Philosophy* 32: 1269–1287.

Casanova, Pascale. 2004. *The World Republic of Letters.* Cambridge: Harvard University Press.

Casanova, Pascale. 2005. Literature as a World. *New Left Review* 31: 71–90.

Ciżewska-Martyńska, Elżbieta. 2015. The Meaning of the 1980s' Anti-Politics' Legacy Within the Contemporary East-Central European Civil Societies. *Intersections. East European Journal of Society and Politics* 1: 37–58.

Coulangeon, Philippe, and Julien Duval, eds. 2013. *Trente ans après La Distinction, de Pierre Bourdieu.* Paris: La Découverte.

Danilova, Elena. 2014. Neoliberal Hegemony and Narratives of "Losers" and "Winners" in Post-Socialist Transformations. *Journal of Narrative Theory* 44 (3): 442–466.

Davies, Norman. 1981. *God's Playground. A History of Poland*, 2 vols. Oxford and New York: Clarendon Press, Columbia University Press.

Deloitte. 2015. *Czy oszczędności krajowe będą w stanie finansować długoterminowy wzrost gospodarczy w Polsce?* Warszawa: Deloitte Polska.

Denord, François, Paul Lagneau-Ymonet, and Sylvain Thine. 2018. Primus Inter Pares? The French Field of Power and Its Power Elite. *Socio-Economic Review* 16 (2): 277–306.

de Sousa Santos, Boaventura and Maria Paula Meneses. 2020. Introduction: Epistemologies of the South—Giving Voice to the Diversity of the South. In *Knowledges Born in the Struggle: Constructing the Epistemologies of the Global South*, ed. B. de Sousa Santos and M.P. Meneses, XVII–XLIII. London: Routledge.

Dezalay, Yves, and B.G. Garth. 1996. *Dealing in Virtue: International Commercial Arbitration and the Construction of a Transnational Legal Order*. Chicago: University of Chicago Press.
Dobry, Michele. 1992. *Sociologie des crises politiques: La dynamique des mobilisations multisectorielles*. Paris: Presses de Sciences Po.
Éber, Mark Á. 2011. Which Duality? Why Dual? On the Gains and Losses of Hungarian Sociology's Dual-Society Theories. *Review of Sociology* 21 (4): 44–20.
el-Ojeili, Chamsy. 2015. *Beyond Post-Socialism: Dialogues with the Far-Left*. Houndsmills: Palgrave Macmillan.
Enyedi, Zsolt. 2005. The Role of Agency in Cleavage Formation. *European Journal of Political Research* 44 (5): 697–720.
Even-Zohar, Itamar. 1990. *Polysystem Studies*. Durham: Duke University Press.
Eyal, Gil, Iván Szelényi, and Eleanor R. Townsley. 1998. *Making Capitalism Without Capitalists. Class Formation and Elite Struggles in Post-Communist Central Europe*. London and New York: Verso.
Faber, Agoston. 2017. From False Premises to False Conclusions: On Pierre Bourdieu's Alleged Sociological Determinism. *The American Sociologist* 48 (3–4): 436–452.
Fidelis, Ma.łgorzata. 2013. The Other Marxists: Making Sense of International Student Revolts in Poland in the Global Sixties. *Zeitschrift für Ostmitteleuropa-Forschung* 62 (3): 425–449.
Foucault, Michel. 1980. *Power/Knowledge. Selected Interviews and Other Writings, 1972–1977*. New York: Pantheon Books.
Fourcade, Marion. 2006. The Construction of a Global Profession: The Transnationalization of Economics. *American Journal of Sociology* 112 (1): 145–194.
Gandelman, Claude. 1988. The Dialectic Functioning of Mukařovský's Semiotic Model. In *The Prague School and Its Legacy in Linguistics, Literature, Semiotics, Folklore, and the Arts*, ed. Y. Tobin, 265–273. Amsterdam: John Benjamins.
Gerőcs, Tamás, and András Pinkasz. 2018. Debt-ridden Development on Europe's Eastern Periphery. In *Global Inequalities in World-Systems Perspective: Theoretical Debates and Methodological Innovations*, ed. Manuela Boatcă, Andrea Komlosy, and Hans-Heinrich. Nolte, 131–153. New York: Routlegde.
Ginelli, Zoltán. 2018. Hungarian Experts in Nkrumah's Ghana: Decolonization and Semiperipheral Postcoloniality in Socialist Hungary. *Mezosfera.org*. http://mezosfera.org/hungarian-experts-in-nkrumahs-ghana/.

Go, Julian. 2008a. *American Empire and the Politics of Meaning: Elite Political Cultures in the Philippines and Puerto Rico During U.S. Colonialism.* Durham: Duke University Press.

Go, Julian. 2008b. Global Fields and Imperial Forms: Field Theory and the British and American Empires. *Sociological Theory* 26 (3): 201–229.

Go, Julian. 2009. The 'New' Sociology of Empire and Colonialism. *Sociology Compass* 3: 775–788.

Go, Julian. 2013a. Decolonizing Bourdieu: Colonial and Postcolonial Theory in Pierre Bourdieu's Early Work. *Sociological Theory* 31: 49–74.

Go, Julian. 2013b. Fanon's Postcolonial Cosmopolitanism. *European Journal of Social Theory* 16: 208–225.

Go, Julian. 2016. *Postcolonial Thought and Social Theory.* New York, NY: Oxford University Press.

Go, Julian. 2017. Myths of Nation and Empire: The Logic of America's Liberal Empire-State. *Thesis Eleven* 139: 69–83.

Go, Julian, and Monika Krause. 2016. Fielding Transnationalism: An Introduction. *The Sociological Review* 64 (2 Suppl): 6–30.

Gorzelak, Grzegorz. 1996. *The Regional Dimension of Transformation in Central Europe.* London and Bristol: J. Kingsley, Regional Studies Association.

Grabowska, Mirosława. 2003. The Post-communist Cleavage on the National and the Local Level in Poland. In *Regional Issues in Polish Politics*, ed. T. Zarycki and G. Kolankiewicz, 157–178. London: School of Slavonic and East European Studies, University College London.

Guilhot, Nicolas. 2007. Reforming the World: George Soros, Global Capitalism and the Philanthropic Management of the Social Sciences. *Critical Sociology* 33: 447–477.

Hannerz, Ulf. 1992. *Cultural Complexity: Studies in the Social Organization of Meaning*. New York: Columbia University Press.

Harvey, David. 2006. *The Limits to Capital*. London and New York: Verso.

Haszczyński, Jerzy. 2020. Polski pomnik, niemieckie wątpliwości. *Rzeczpospolita*, June 11.

Heidenreich, Martin. 1998. The Changing System of European Cities and Regions. *European Planning Studies* 6 (3): 315–332.

Heilbron, Johan. 1999. Towards a Sociology of Translation: Book Translations as a Cultural World-System. *European Journal of Social Theory* 2 (4): 429–444.

Heilbron, Johan, and George Steinmetz. 2018. A Defense of Bourdieu. *Catalyst* 2 (1): 35–49.

Hix, Simon. 2004. European Universities in a Global Ranking of Political Science Departments. *European Political Science* 3: 5–23.
Hjellbrekke, Johs, and Olav Korsnes. 2018. A Place at What Table? An Analysis of Symbolic Capital Hierarchies at the Annual Dinner of Norges Bank (the Norwegian Central Bank). In *New Directions in Elite Studies*, ed. O. Korsnes, J. Heilbron, J. Hjellbrekke, F. Bühlmann, and M. Savage, 93–112. Oxford and New York: Routledge.
Hjellbrekke, Johs, Brigitte Le Roux, Olav Korsnes, Frédéric. Lebaron, Lennart Rosenlund, and Henry Rouanet. 2007. The Norwegian Field of Power Anno 2000. *European Societies* 9: 245–273.
Jaszczuk, Andrzej. 1999. *Liberalna Atlantyda: Główne nurty liberalizmu polskiego 1870–1939 r.* Warszawa: KB.
Jessop, Bob. 1985. *Nicos Poulantzas: Marxist Theory and Political Strategy*. New York: St. Martin's Press.
Jessop, Bob. 2012. Social Imaginaries, Structuration, Learning, and 'Collibration': Their Role and Limitations in Governing Complexity. *Zarządzanie Publiczne* 1: 71–84.
Jones, Gareth Stedman. 1996. The Determinist Fix: Some Obstacles to the Further Development of the Linguistic Approach to History in the 1990s. *History Workshop Journal* 42: 19–35.
Keating, Michael. 1998. *The New Regionalism in Western Europe: Territorial Restructuring and Political Change*. Cheltenham: E. Elgar.
Kichelewski, Audrey, Judith Lyon-Caen, Jean-Charles Szurek, and Anette Wieviorka. 2019. *Les Polonais et la Shoah: Une nouvelle école historique*. Paris: CNRS Éditions.
Klimczuk, Andrzej, and Magdalena Klimczuk-Kochańska. 2019. Core-Periphery Model. In *The Palgrave Encyclopedia of Global Security Studies*, ed. S. Romaniuk, M. Thapa, and P. Marton. Cham: Springer International Publishing.
Kochanowicz, Jacek. 2006. *Polish Kingdom: Periphery as a Leader*. Paper Prepared for the XIV International Economic History Congress in Helsinki, Finland, August 21–25.
Kola, Adam F. 2018. Marian Małowist's World History and Its Application to World Literature. In *The Routledge Companion to World Literature and World History*, ed. M. Hawas, 57–68. London: Routledge.
Korhonen, Juho. 2018. The Relation Between Imperialism and Nationalism in the German Empire and Its Aftermath: A Bourdieusian Field Theoretical Perspective. *Studies in Ethnicity and Nationalism* 18: 167–187.

Koryś, Piotr. 2018. *Poland From Partitions to EU Accession: A Modern Economic History, 1772–2004*. Cham: Palgrave Macmillan.

Krause, Monika. 2014. *The Good Project: Humanitarian Relief NGOs and the Fragmentation of Reason*. Chicago: University of Chicago Press.

Kubik, Jan. 1994. *The Power of Symbols Against the Symbols of Power: The Rise of Solidarity and the Fall of State Socialism in Poland*. University Park: Pennsylvania State University Press.

Kuus, Merje. 2004. "Those Goody-Goody Estonians": Toward Rethinking Security in the European Union Candidate States. *Environment and Planning D-Society and Space* 22: 191–207.

Kuus, Merje. 2007. *Geopolitics Reframed: Security and Identity in Europe's Eastern Enlargement*. Basingstoke: Palgrave Macmillan.

Lahire, Bernard, ed. 2001. *Le travail sociologique de Pierre Bourdieu: Dettes et critiques*. Paris: La Découverte.

Lakoff, George, and Mark Johnson. 1980. *Metaphors We Live By*. Chicago: University of Chicago Press.

Lebaron, Frederic, and Brigitte Le Roux. 2018. Bourdieu and Geometric Data Analysis. In *The Oxford Handbook of Pierre Bourdieu*, ed. T. Medvetz and J. J. Sallaz. Oxford: Oxford University Press. https://www.oxfordhandbooks.com/view/10.1093/oxfordhb/9780199357192.9780199357001.9780199350001/oxfordhb-9780199357192-e-9780199357122.

Leezenberg, Michiel. 2001. *Contexts of Metaphor*. Oxford: Elsevier Science.

Leezenberg, Michiel. 2002. Power in Communication: Implications for the Semantics-Pragmatics Interface. *Journal of Pragmatics* 34 (7): 893–908.

Leszczyński, Adam. 2017. *Leap into Modernity—Political Economy of Growth on the Periphery, 1943–1980*. Frankfurt am Main: Peter Lang.

Leszczyński, Adam. 2020. *Ludowa historia Polski*. Warszawa: Wydawnictwo WAB.

Lipset, Seymour M., and Stein Rokkan, eds. 1967. *Party Systems and Voter Alignments: Cross-National Perspectives*. New York: Free Press.

Luhmann, Niklas. 2012. *Introduction to Systems Theory*. Cambridge: Polity Press.

Lunding, Jacob Aagaard, Christoph Houman Ellersgaard, and Anton Grau Larsen. 2021. The Established and the Delegated: The Division of Labour of Domination Among Effective Agents on the Field of Power in Denmark. *Sociology* 55 (1): 110–128.

Luxemburg, Rosa. 1898. *Die Industrielle Entwicklung Polens*. Leipzig: Duncker & Humblot.

Majmurek, Jakub. 2010. PRL jako projekt modernizacji peryferyjnej. In *PRL bez uprzedzeń*, ed. J. Majmurek i P. Szumlewicz, 65–88. Warszawa: Instytut Wydawniczy "Książka i Prasa".
Mark, James, Artemy M. Kalinovsky, and Steffi Marung. 2020. *Alternative Globalizations: Eastern Europe and the Postcolonial World*. Bloomington: Indiana University Press.
Marković, Mihailo. 1985. Transfer of Concepts from Natural Sciences into Social Sciences. *Scientia Yugoslavica* II: 99–113.
Martin, John Levi. 2003. What Is Field Theory? *American Journal of Sociology* 109 (1): 1–49.
Marzec, Wiktor. 2020. *Rising Subjects: The 1905 Revolution and the Origins of Modern Polish Politics*. Pittsburgh: University of Pittsburgh Press.
Mażewski, Lech. 2013. *System polityczny, prawo i konstytucja Królestwa Polskiego 1815–1830: w przededniu dwusetnej rocznicy powstania unii rosyjsko-polskiej*. Radzymin: Wydawnictwo von borowiecky.
Mazur, Marian. 1966. *Cybernetyczna teoria układów samodzielnych*. Warszawa: Państwowe Wydawnictwo Naukowe.
Mazur, Marian. 1976. *Cybernetyka i charakter*. Warszawa: Państwowy Instytut Wydawniczy.
Medvetz, Thomas. 2012. *Think Tanks in America*. Chicago: The University of Chicago Press.
Melegh, Attila. 2006. *On the East-West slope: Globalization, Narration, Racism and Discourses on Central and Eastern Europe*. Budapest and New York: CEU Press.
Meyer, John W., John Boli, George M. Thomas, and Francisco O. Ramirez. 1997. World Society and the Nation-State. *American Journal of Sociology* 103 (1): 144–181.
Miller, Alexei. 2008. *The Romanov Empire and Nationalism. Essays in the Methodology of Historical Research*. Budapest and New York: Central European University Press.
Mogilner, Marina. 2013. *Homo Imperii: A History of Physical Anthropology in Russia*. Lincoln: University of Nebraska Press.
Morozov, Viatcheslav. 2015. *Russia's Postcolonial Identity: A Subaltern Empire in a Eurocentric World*. Houndmills, Basingstoke, and Hampshire: Palgrave Macmillan.
Müller, Martin. 2020. In Search of the Global East: Thinking Between North and South. *Geopolitics* 25: 734–755.

Myant, Martin. 2018. Dependent Capitalism and the Middle-Income Trap in Europe and East Central Europe. *International Journal of Management and Economics* 54: 291–303.

Nowak, Andrzej W. 2016. Tajemnicze zniknięcie Drugiego Świata.... O trudnym losie półperyferi. In *Polska jako peryferie*, ed. T. Zarycki, 86–104. Warszwa: Wydawnictwo Naukowe Scholar.

Osiatyński, Wiktor. 1996. The Roundtable Talks in Poland. In *The Roundtable Talks and the Breakdown of Communism*, ed. Jon Elster, 21–68. Chicago: The University of Chicago Press.

Palma, José Gabriel. 2016. The Dependency School and Its Aftermath: Why Latin America's Critical Thinking Switched from One Type of Absolute Certainties to Another. In *Handbook of Alternative Theories of Economic Development*, ed. E.S. Reinert, J. Ghosh, and R. Kattel, 386–415. Edward Elgar Publishing.

Panther, Stephan. 2014. *Institutions in a World System: Contours of a Research Program*. Berlin: desiguALdades.net International Research Network on Interdependent Inequalities in Latin America.

Pêcheux, Michel. 1982. *Language, Semantics and Ideology*. New York: St. Martin's Press.

Pieterse, Jan Nederveen. 1988. A Critique of World System Theory. *International Sociology* 3: 251–266.

Pobłocki, Kacper. 2021. *Chamstwo*. Wołowiec: Wydawnictwo Czarne.

Porter-Szücs, Brian. 2014. *Poland in the Modern World: Beyond Martyrdom*. Chichester: Wiley.

Pospíšilová, Tereza. 2019. Giving and Refusing: The Contested Reception of Transnational Philanthropy in the Case of the Prague Central European University Foundation. *Nonprofit and Voluntary Sector Quarterly* 48: 266–282.

Pouliot, Vincent. 2010. *International Security in Practice: The Politics of NATO-Russia Diplomacy*. Cambridge: Cambridge University Press.

Putnam, Robert D. 1993. *Making Democracy Work: Civic Traditions in Modern Italy*. Princeton: Princeton University Press.

Rauszer, Michał. 2020. *Bękarty pańszczyzny: Historia buntów chłopskich*. Warszawa: Wydawnictwo RM.

Rokkan, Stein. 1969. Centre Formation, Nation-Building and Cultural Diversity: Report on a Symposium Organized by UNESCO. *Social Science Information* 8 (1): 85–99.

Rokkan, Stein. 1980. Territories, Centres, and Peripheries: Toward a Geoethnic-Geoeconomic-Geopolitical Model of Differentiation Within

Western Europe. In *Centre and Periphery: Spatial Variations in Politics*, ed. J. Gottmann, 163–204. Beverly Hills, London: Sage.
Rolf, Malte. 2020. *Russian Rule in the Kingdom of Poland (1864–1915)*. Pittsburgh: University of Pittsburgh Press.
Roueff, Oliver. 2013. Les homologies structurales: une magie sociale sans magiciens? La place des intermédiaires dans la fabrique des valeurs. In *Trente ans après La Distinction, de Pierre Bourdieu*, ed. P. Coulangeon and J. Duval. Paris: La Découverte.
Sapiro, Gisèle. 2008. Translation and the Field of Publishing. *Translation Studies* 1 (2): 154–166.
Saussure, Ferdinand de. 1983. *Cours de Linguistique Générale*. London: Duckworth.
Schmitz, Andreas, Daniel Witte, and Vincent Gengnagel. 2017. Pluralizing Field Analysis: Toward a Relational Understanding of the Field of Power. *Social Science Information* 56 (1): 49–73.
Silva, Elizabeth, and Alan Warde. 2012. *Cultural Analysis and Bourdieu's Legacy: Settling Accounts and Developing Alternatives*. London: Routledge.
Skocpol, Theda. 1977. Wallerstein's World Capitalist System: A Theoretical and Historical Critique. *American Journal of Sociology* 82: 1075–1090.
Snochowska-Gonzalez, Claudia. 2012. Post-colonial Poland—On an Unavoidable Misuse. *East European Politics and Societies* 26: 708–723.
Sosnowska, Anna. 2018. *Explaining Economic Backwardness: Post-1945 Polish Historians on Eastern Europe*. Budapest: CEU Press.
Steinmetz, George. 1993. *Regulating the Social: The Welfare State and Local Politics in Imperial Germany*. Princeton: Princeton University Press.
Steinmetz, George. 2003. The State of Emergency and the Revival of American Imperialism: Toward an Authoritarian Post-Fordism. *Public Culture* 15 (2): 323–345.
Steinmetz, George. 2008. The Colonial State as a Social Field: Ethnographic Capital and Native Policy in the German Overseas Empire Before 1914. *American Sociological Review* 73 (4): 589–612.
Steinmetz, George. 2011. Bourdieu, Historicity, and Historical Sociology. *Cultural Sociology* 5 (1): 45–66.
Steinmetz, George. 2013. A Child of the Empire: British Sociology and Colonialism, 1940s–1960s. *Journal of the History of the Behavioral Sciences* 49: 353–378.
Steinmetz, George. 2014. The Sociology of Empires, Colonialism, and Postcolonialism. *Annual Review of Sociology* 40 (1): 77–103.

Steinmetz, George. 2016. Social Fields, Subfields and Social Spaces at the Scale of Empires: Explaining the Colonial State and Colonial Sociology. *The Sociological Review Monographs* 64 (2): 98–123.

Suckert, Lisa. 2018. Unravelling Ambivalence: A Field-Theoretical Approach to Moralised Markets. *Current Sociology* 66 (5): 682–703.

Sum, Ngai-Ling, and Bob Jessop. 2013. *Towards A Cultural Political Economy: Putting Culture in Its Place in Political Economy*. Cheltenham: Edward Elgar Publishing.

Taylor, Peter, and Michael Hoyler. 2000. The Spatial Order of European Cities Under Conditions of Contemporary Globalization. *Tijdschrift voor economische en sociale geografie* 91 (2): 176–189.

Tlostanova, Madina. 2015. Between the Russian/Soviet Dependencies, Neoliberal Delusions, Dewesternizing Options, and Decolonial Drives. *Cultural Dynamics* 27: 267–283.

Tuan, Yi-Fu. 1977. *Space and Place: The Perspective of Experience*. Minneapolis: University of Minnesota Press.

Turoma, Sanna, and Maxim K. Waldstein. 2012. Empire De/Centered; New Spatial Histories of Russia and the Soviet Union. In *Empires and the Making of the Modern World, 1650–2000*. Farnham: Ashgate.

Wacquant, Loïc, Tom Slater, and Virgílio Borges Pereira. 2014. Territorial Stigmatization in Action. *Environment and Planning A* 46 (6): 1270–1280.

Waldstein, Maxim. 2010. Theorizing the Second World. *Ab Imperio* 1: 98–117.

Wallerstein, Immanuel. 1974–1989. *The Modern World System*, 3 vols. New York: Academic Press.

Wallerstein, Immanuel. 1989. The French Revolution as a World-Historical Event. *Social Research* 56: 33–52.

Wallerstein, Immanuel. 2005. *Africa. The Politics of Independence and Unity*. Lincoln: University of Nebraska Press.

Wang, Yingyao. 2016. Homology and Isomorphism: Bourdieu in Conversation with New Institutionalism. *The British Journal of Sociology* 67 (2): 348–370.

Warczok, Tomasz, and Tomasz Zarycki. 2014. Bourdieu Recontextualized: Redefinitions of Western Critical Thought in the Periphery. *Current Sociology* 62 (3): 334–351.

Wierzbicka, Anna. 1992. Semantic Primitives and Semantic Fields. In *Frames, Fields, and Contrasts: New Essays in Semantic and Lexical Organization*, ed. A. Lehrer and E. Feder Kittay, 209–227. London: Routledge.

Wolff, Larry. 1994. *Inventing Eastern Europe: The Map of Civilization on the Mind of the Enlightenment.* Stanford, CA: Stanford University Press.

Wróbel, Szymon. 2012. Efekt interpelacji. *Teksty Drugie* 3: 22–33.

Zarycki, Tomasz. 1998. Towards the Theory of Structures of Maps in Their Deep Concept Dimension. *Cartographic Journal* 35: 71–78.
Zarycki, Tomasz. 2002. Four Dimensions of the Centre-Periphery Conflict in the Polish Electoral Geography. In *Social Change: Adaptation and Resistance*, ed. T. Klonowicz and G. Wieczorkowska, 19–38. Warszawa: Wydawnictwa Instytutu Studiów Społecznch UW.
Zarycki, Tomasz. 2004. Uses of Russia: The Role of Russia in the Modern Polish National Identity. *East European Politics and Societies* 18: 595–627.
Zarycki, Tomasz. 2007a. History and Regional Development. A Controversy over the 'Right' Interpretation of the Role of History in the Development of the Polish Regions. *Geoforum* 38: 485–493.
Zarycki, Tomasz. 2007b. An Interdisciplinary Model of Centre-Periphery Relations: A Theoretical Proposition. *Regional and Local Studies Special Issue* 2007: 110–130.
Zarycki, Tomasz. 2009. The Power of the Intelligentsia. The Rywin Affair and the Challenge of Applying the Concept of Cultural Capital to Analyze Poland's Elites. *Theory and Society* 38 (6): 613–648.
Zarycki, Tomasz. 2011. From Soviet to a Western-Dominated Political Scene: The Geopolitical Context of Politics in Poland. In *Cultural Transformations After Communism: Central and Eastern Europe in Focus*, ed. B. Törnquist-Plewa and K. Stala, 82–108. Lund: Nordic Academic Press.
Zarycki, Tomasz. 2014. *Ideologies of Eastness in Central and Eastern Europe*. London: Routledge.
Zarycki, Tomasz. 2015. Class Analysis in Conditions of a Dual-Stratification Order. *East European Politics and Societies* 29 (3): 711–718.
Zarycki, Tomasz. 2020. Crises of the Communist and Neoliberal Orders 30 Years Later: Structural Comparison of Poland 1975 and 2019. *Social Science Information* 59 (3): 484–504.
Zarycki, Tomasz. 2021. Comments on Martin Müller's "In Search of the Global East: Thinking Between North and South." *Praktyka Teoretyczna* 38: 191–201.
Zarycki, Tomasz, and Andrzej Nowak. 2000. Hidden Dimensions: The Stability and Structure of Regional Political Cleavages in Poland. *Communist and Post-Communist Studies* 33 (3): 331–354.
Zarycki, Tomasz, Rafał Smoczyński, and Tomasz Warczok. 2017. The Roots of Polish Culture-Centered Politics: Toward a Non-purely Cultural Model of Cultural Domination in Central and Eastern Europe. *East European Politics and Societies* 31 (2): 360–381.
Zysiak, Agata, and Wiktor Marzec. 2020. Historicizing the Asynchronous Modernity in the Global East. *Eurasian Geography and Economics* 61: 663–685.

2

Structural Reading of the Poland's Nineteenth- and Twentieth-Century History

2.1 Introduction

2.1.1 Basic Assumptions

This part of the book offers an interpretation of the contemporary history of Poland, especially of its elites; one based on a relational reinterpretation of the dynamics of its field of power. This analysis begins in the early nineteenth century and ends with recent developments in Polish history. It serves as a reference point for analyzing the dynamics of the Polish field of linguistic and literary studies, which is presented in the following chapter. One of the critical features of the present interpretation is that it is framed in a broader international context. It includes a discussion on how the Polish national proto-field of power was formed within the empires that governed Poland in the nineteenth century and how its dynamics were related to the dynamics of the fields of power of those empires. This chapter further demonstrates that the history of nineteenth-century Poland is a fascinating but still largely unexplored subject from the point of view of "imperial history" (e.g. Gerasimov et al. 2005; Steinmetz 2014). It has to be borne in mind that Poland did

T. Zarycki, *The Polish Elite and Language Sciences*,
https://doi.org/10.1007/978-3-031-07345-8_2

not formally exist as a state from 1795 until 1918. During this period, Polish political, cultural, and economic life continued within three separate state organisms. As such, these constituted three parallel social spaces that were relatively independent of each other, but at the same time strongly interconnected, thereby comprising a relatively homogeneous whole. The emphasis on the word "relatively" seems appropriate here since it is difficult to speak of unity or cohesion of the Polish nation and its elites, especially in the institutional dimension. On the other hand, "Polishness", with all its internal differentiation, remained a common game—initially for the Polish elites within the three empires, and later for the lower classes of the emerging modern nation as well. It may be recalled that at the time of the "partitions" of Poland, the lower classes had very little sense of national identity in this part of Europe, and there was a great deal of religious, linguistic, and ethnic diversity. Still, the Polish-Lithuanian Commonwealth, when it was progressively partitioned by Austria, Russia, and Prussia in, 1772, 1793, and finally 1795, had a distinct identity and fairly well-developed political and cultural traditions. Its political and economic elites, because of their roots in that state organism, did not easily blend into the three empires. To be sure, some of these elites were in favor of partitioning the country, or at least did not oppose it, entering into political deals with the partitioning powers, especially Russia and its imperial elite. They were hoping to boost their carrier prospects and further their political and economic interests. They also assumed that Poland, the lion's share of which was to be incorporated into Russia, would maintain a significant level of political and cultural autonomy. For many rich aristocratic families, this hope was more or less fulfilled. Their economic interests were satisfied, and many of them built political careers in the imperial administrations.

2.1.2 The Polish Proto-Field of Power

The history of the Polish polity of the nineteenth century can be interpreted in terms of what can be called the Polish proto-field of power. I am here using the notion of a "proto-field" as it refers to a weakly institutionalized, transnational field, which cannot be considered as fully equivalent to the fields of power of modern nation-states. Still, it can be argued that it not only existed as a social platform for the interaction of the Polish elite, but also had a crucial impact on the social processes in what can be called Polish lands. Even if it was technically composed of three sectors (Russian, Austrian, and Prussian), on a higher level, it formed a single whole within which members of the Polish elite communicated and circulated. It is worth noting that the dynamics of the relationship between such a proto-field of power and the fields of power of the individual Empires was highly variable and at many times considerably desynchronized. The main factor to be taken into account here is the degree of Polish autonomy within the confines of the individual empires. This, in turn, was to a large extent a result of the policy and state capacity of the partitioning states and the dynamics of the relevant sectors of the Polish field of power.

One important aspect of the functioning of the Polish proto-field of power was the de-territorialization of Polish social networks in the nineteenth century. This was conducted on a large scale in the Russian Empire, especially in the major cities, where Poles settled in search of carrier prospects, including Siberia and other remote regions, where they were exiled. At the same time, several regions of what was then Western Russia had sizable Polish populations. The identity of these theoretically Polish people was complex and varied over time and from place to place. An analysis of nineteenth-century Polish political history shows that the Polish social space was similarly indeterminate and relative. Polishness had confessional, family, language, and social status aspects, all of which were subject to constant change and renegotiation. There were also dynamic flows between the cultural, political, and state spheres at every level of social organization. People frequently changed their cultural, national, political, and religious identities, to say nothing of citizenship. These dynamics obviously affected the Austrian and Prussian partitions as well. It was particularly pronounced among the elite, where the transitions from Polishness to Russianness intensified in the late nineteenth century, although there are counterexamples. At the same time, the assimilation of Jews into various national cultures, especially Polish, Russian, and German, continued apace. National movements, e.g. in Belarus, Lithuania, and Ukraine consolidated, and similarly drew part of the elite, especially the intelligentsia, into their spheres. As a result, many people formed changeable and multidimensional identities, which really only began to crystallize after World War I. Contemporary national historiographies, however, tend to downplay the significance, or even the visibility, of these ambivalences. Instead, they clearly distinguish the "national" (here Polish) from the "foreign" (here, the dominant imperial cultures) and the cultures of smaller minorities. One of the best illustrations of this phenomenon is that historical monographs, particularly national histories, rarely emphasize the fact of multiple or changing identities and state affiliations of many members of the elites of countries in the region. Their affiliations other than primary are treated as transitory or marginal. Permanent changes in national affiliations are sometimes presented as acts of treason. The scale of the phenomenon is also ignored, e.g., the fact that the transition of intelligentsia members between Polish, Russian, German, and other cultures was very frequent at the turn of the century.

However, the most salient feature of the period was not the significant flows between cultures, as important as this was to the various elites and several other social strata. In addition to this was the simultaneous participation in different cultural worlds. Multilingualism was common, and among the elites, particularly academics, there was considerable international mobility. The region, at present, could not be more different. It is therefore often impossible to assign people active in this period to a single national group or state identity. They were active in several national and international fields (e.g. the scientific field or the communist movement) simultaneously. An interesting vestige of these ambiguities are the biographical entries of historical figures in the multilingual variants of Wikipedia. The various language versions of several entries describe the same person as belonging to different nations. We can find people who are simultaneously assigned Polishness (for instance by virtue of family origin), Belarusianness (e.g. by virtue of being born in what is now Belarus), Lithuanianness (for example by virtue of spending most of their careers in, e.g. Vilnius), and finally, Russianness (by virtue of citizenship to give just one example). Therefore, an analysis based on social fields, especially the field of power, seems a particularly promising tool for studying Central European social elites at the turn of the twentieth century. It allows for a redefinition of the history of countries such as Poland, which did not appear on the political map of Europe until 1918. These countries can be analyzed through the prism of their proto-fields of power, which seem to frame "national games" much better than formal institutional frameworks or purely cultural or ethnic criteria.

The notion of a proto-field of power is likewise well suited to dealing with several other ambivalences. Both these concerning the level of individual biographies, which could combine the presence of specific persons in the orbit of two, and in times of turbulent changes, such as World War I, even more than one field of power. On the other hand, it allows the history of countries that lacked full statehood to be reconstructed in a coherent way. It can also be helpful to avoid a "secondary nationalization" of that history, i.e. reading it in terms of the frameworks of contemporary nation-states. The nature of the "essence" of Poland, and other contemporary countries in the region at the turn of the twentieth century can be debated. Some will point to the role of institutions which had a national or proto-national character; others to the importance of national cultures. However, the core of the national game seems to be best captured by the notion of proto-fields of power, which can also be analyzed in terms of networks. By this is meant transnational, and often geographically dispersed, but nevertheless relatively coherent, networks of social elites acting in a relatively coordinated way, the essence of which is defined in relation to "national cause". This cause could be understood by different members of the network as anything from an uncompromising fight for independence to cautious efforts to achieve a degree of cultural autonomy. However, autonomy was always understood within the context of political spaces of empires. Representing specific ethno-religious groups in the international sphere was an aspect of, or stake in, the activity of these proto-fields of power. This involved certain para-diplomatic activities when circumstances permitted, and diverse cultural activities. Developing canons of national culture and patterns of national languages was an important aspect of this. As in all fields of power, direct contact between its members was crucial. Most of them, even if they did not know each other personally, were aware of each other's existence. It seems fascinating that they were able to act in a coordinated way and remain in active contact despite the political barriers and distances that separated them. In some respects, the partitions of Poland had a beneficial effect, even though the three empires coordinated and harmonized their Polish policies, and cooperated in implementing them, especially when it came to cross-border activity on the part of Polish insurgents. These imperial borders, however, could not prevent every Polish insurgent, or activist who had been repressed or restricted in his/her endeavors, from escaping imprisonment and other forms of punishment. Some of them were even able to save part of their property or continue academic careers that had been cut short. This sort of tactical shifting between the three empires was common for the Polish political and economic elite, who could be seen as a hidden dimension of the interface periphery effect.

Cities where the Polish diaspora congregated were important nodes of this proto-power field. For the Polish elites, the key node was Paris, but the capitals and large cities of the partitioning powers (e.g. Berlin, St. Petersburg, Moscow, and Vienna) were also important.[1] In this context, it should be noted that the most important function of any power field is to enable different elites to interact and negotiate the terms of converting their capitals. This also appears to be the key function of proto-fields of power. It should further be borne in mind that most empires managed the territories under their control in a multidimensional and partially independent manner. Policies pertaining to religion, education, agriculture, property, aristocratic titles, press freedom, and taxation were all pursued separately. In line with, or in competition to these processes, any proto-field of power attempts to coordinate these dimensions within a selected area of the empire, or even across its borders, in a ways that is independent and bypasses imperial capitals. It should also be noted that the competing imperial and national proto-fields often functioned in completely different dimensions and manners. For example, the Polish proto-field of power was able to wield a lot of influence on the peasantry through the Catholic clergy. This allowed that field to link the various

[1] Examples of studies devoted to the importance of Polish diasporas in the great metropolises of Russia and Germany at the beginning of the twentieth century include such works as (Bazylow 1984; Miśkowiec 2013; Nałęcz-Dobrowski 2015).

dimensions of social life, making it more difficult for the policies of the partitioning states, especially those aimed at assimilating the peasantry, to win them over with gestures such as the enfranchisement reforms.

2.1.3 The Intelligentsia in a Peripheral Proto-Field of Power

I now wish to look at the broader geopolitical context of nineteenth-century Poland and consider whether the notion of interface periphery is applicable to it. The three empires that divided Central-Eastern Europe and set up new political systems in the region after 1815 were at the peak of their powers. It was not in the interest of these powerful and mutually wary empires to stir up Polish aspirations on neighboring territories, as it carried the risk of revived Polish nationalism or aspirations for political autonomy spilling over onto their own territory. The situation would have been different had an overwhelming majority of ethnic Poles found themselves in one of these three empires. Such was the case with Finland, which had been completely incorporated into Sweden, and after 1809, Russia. As stated above, Russia had by far the largest proportion of Polish territory. However, it established the Kingdom of Poland (Congress Poland), and initially endowed it with considerable autonomy. Russia offered a compromise to the Polish elites in the early nineteenth century, which was, in principle, accepted. The coronation of Tsar Nicholas I as King of Poland in Warsaw in 1809, which was attended by key members of the Polish elite, was a symbolic moment. The most spectacular example of a Pole becoming a member of the establishment of the Russian Empire was that of Prince Adam Czartoryski (1770–1861), who was appointed Minister of Foreign Affairs under Alexander I. Czartoryski's stint only lasted two years (1804–1806) and he later emigrated after supporting the failed insurrection of 1831. As brief and rare as his ministerial stint was, it was by means the only example of a Pole occupying a senior administrative position in the Russian Empire. Throughout the nineteenth century, and until 1917, many other Poles, especially aristocrats and nobles, enjoyed successful careers at all levels of the Russian administrative hierarchy. What made Czartoryski's career special was that it was part of Alexander's policy of co-opting a generous portion of the Polish elite into the Russian imperial system, or the core of its field of power. That project proved to be short-lived; Poles in St. Petersburg were later perceived as individuals rather than the political representatives of any wider Polish circles or interests. A telling insight into the complexity of the interaction between the Russian rulers and the Polish elite can be found in a recent study of Russian rule in nineteenth-century Warsaw by Malte Rolf (2020). The author maintains that there had always been a close relationship between the Russian high aristocracy and selected Polish aristocratic families. Some were even close to the Tsar's court. This was apparent at the beginning of the nineteenth century, and especially during the reign of Alexander I (1801–1825). One of the Tsar's most surprising appointments was that of General Józef Zajączek (1752–1826) to the position of *Namiestnik* (Viceroy) of Poland in 1815. Zajączek had fought against Russia in the Polish–Russian War of 1792, and the Kościuszko Uprising (1794), before serving as a general under Napoleon during the French invasion of 1812. He held the post of Viceroy until 1826.

The Kingdom of Poland was an autonomous administrative unit of the Russian Empire. It was created at the Congress of Vienna in 1815 (hence the name "Congress Poland"). Its predecessor was the French client state, the Duchy of Warsaw, which had existed from 1807. Until the November Uprising (1830–1831), the Kingdom had its own constitution (which Russia itself lacked until 1832), parliament, army, and currency. Polish was its official language, and the Polish language Royal University of Warsaw was founded in 1816 by Alexander I. What connected Poland with Russia was the person of the monarch (the Russian Tsar was simultaneously the King of Poland) and foreign policy, which was a royal prerogative. Nicholas I abolished the Constitution of the Kingdom in February 1832 in retaliation for the November Uprising. He replaced it with the Organic Statute, which abolished the Polish parliament and army, and integrated the Kingdom more firmly into the Empire while preserving its administrative autonomy. Poles found it difficult to obtain senior administrative positions after the November uprising, and even more so after the January Uprising (1863–1864). Despite that, however, they often made significant careers in the Empire.

Some of them consequently dropped out of Polish social and political life. Their careers are therefore not often visible to Polish historiography, even though they maintained a cultural identity.

With some simplification, it could be said that the anti-Russian uprisings, which are primarily construed as a struggle for national emancipation in Poland, were connected with the rise of a new social class specific to this part of the continent: the intelligentsia (Gella 1976). The development of the intelligentsia can be explained primarily as a result of the weakness of economic development combined with the processes of modernization, in particular the evolution of empires, which increasingly aspired to cultural homogenization following the French model, but whose capacities constrained them to remain multinational. The main source of recruitment to the intelligentsia was the impoverished, primarily petty, nobility (Jedlicki 2008). This class was in reduced straits as a result of the slow but steady collapse of the manor economy and by the anti-Polish policies of the Tsarist government, which was geared toward expropriating them and stripping them of their titles. Another element of this policy was to enlist the support of the peasantry in the hope of gaining its political loyalty, and in the longer term, to culturally assimilate it. This policy was largely effected in the Russian and Austrian partitions by means of the enfranchisement reforms. It is worth noting that analogous reforms had been introduced in Prussia earlier and were less political in nature (Wajda 1990). They significantly reduced the political influence that went with owning a large estate (i.e. belonging to the Junker class), by abolishing serfdom and enabling peasants to purchase land, but these estates remained largely intact in Pomerania and East Prussia until 1945 (Schoer 1976). Austria and Russia did not deprive landowners of their property or political influence—least of all those with aristocratic titles, as they were seen as potential political allies. The wealthy Polish landed gentry became a key government ally in the Austrian province of Galicia (today SE Poland and SW Ukraine) after the political reforms of 1867. As for the Russian partition, although the alliance of the Polish aristocracy with St. Petersburg never assumed an institutional or openly political framework, it was maintained on the individual family level until the monarchy was abolished in 1917 (Szwarc 1991). The Polish uprisings of the nineteenth century can therefore be seen as revolts of the emerging "young" intelligentsia elite against the "old" aristocratic and landowning elite, which collaborated, or at least peacefully coexisted, with the imperial regime, and against the emerging bourgeois elite. At the same time, it is important to distinguish between the intelligentsia and intellectuals. While intellectuals played a social role, the intelligentsia was an element of the class structure; one that can be seen both in terms of class and strata, as its status relied on both social and cultural capital (Zarycki 2015).

The Central European intelligentsia can then be seen as having been created, albeit unwittingly, by the empires that governed the region. It is within the networks of primary and secondary schools established by these states that subsequent generations of the intelligentsia were educated and thus reproduced. Their elite also found their way to the imperial public universities, although it is worth noting that student bodies were, in any case, very cosmopolitan before World War I. However, the large number of students who attended universities outside Poland was also due to the limited number of universities in the partitions. University education was usually financed by the wealthy families of the students. These families typically owned landed estates that provided them with substantial incomes. In other words, Austria, and even more so, Russia, provided the structural framework for the reproduction of both factions of the Polish elite. On the one hand, the intelligentsia were educated in public schools and many of them were employed in public institutions. On the other, the Polish aristocracy mingled with the landed gentry, politically supported by the monarchical elites of both countries. Within this second group, significant transformations were taking place. Some old families were financially ruined while "parvenus" became immensely wealthy. As ancient aristocratic families were bankrupted, they were replaced by rich bourgeois, who purchased property, and sometimes titles, especially at the Austrian court. This faction of the elite was supported by the authorities in St. Petersburg and Vienna primarily as an ally in controlling the populace. Educating the intelligentsia in public schools, and promoting them in the public service, was explicitly intended to assimilate them into the dominant or state culture (Russian and German, respectively). This, however, was only partly successful. Assimilation was greatest in the Prussian partition, but even there, it was not complete. The confrontation between the cultural (the intelligentsia) and economic (the landed gentry and the bourgeoisie) factions of the Polish elite was never openly institutionalized but was increasingly visible. The balance of forces was gradually tilted in favor of the cultural faction. The two factions corresponded to the "White" and "Red" parties during the January Uprising. The Reds were influenced by the revolutionary movements in Italy and the Balkans and heartened by Russia's loss in the Crimean War. They gradually leaned toward armed rebellion against Russian rule. The Whites represented the aristocracy, the bourgeoisie and wealthy landowners, such as those in the Agricultural Society (*Towarzystwo Rolnicze*). They were centered around conservative magnates such as Agricultural Society founder and chairman, Count Andrzej Zamoyski (1800–1874), and wanted the pre-1831 Constitution reinstated. For his part, the Tsar solicited the loyalty of the Polish peasantry by moving to abolish serfdom. It is worth noting that the Whites did not support armed rebellion, but promoted "organic work". Their relations with the authorities of the Empire were usually much better than those of the Reds. As stated above, the Whites had also suffered after the failed November uprising through the collapse of the manor economy, Tsarist expropriations, and Polish nobles being stripped of their titles. Another failed uprising threatened to be ruinous for the landowning and bourgeois elites, and they were understandably reluctant to commit to it. Lech Mażewski appositely describes this as the "insurrectionary blackmail" of the authorities (Mażewski 2004). The "revolutionary" factions, who pushed for insurrectionary uprisings, endured extreme hardships, and many paid the supreme price. However, they suffered much smaller structural losses. First, they always had a large reserve to draw on, as there was no shortage of petty noblemen and people who had recently left this stratum. There were also more educated people from other classes in the Russian and Austrian partitions than there were available public service positions. Second, the material and political losses of the emerging intelligentsia were not all that significant, as these people had much less to lose in this respect. The bloody clashes of the uprising, and the brutal retaliation and repression that followed, gave rise to martyrological legends that became the core of the symbolic capital of the growing intelligentsia myths. Thus, the intelligentsia elite grew in strength, and the establishment of the Second Republic in 1918 can be regarded as its final victory over the economic elite.

The key mechanism transforming the Polish proto-field of power in the Russian Empire were the uprisings discussed above (1830–1831 and 1863–1864), and the participation of part of the Polish elite in the French Invasion of Russia (1812) and the revolution of 1905. In this last case, the contribution of the lower strata of Polish society was also significant (Marzec 2020). It is worth noting that while Polish autonomy in Russia was severely restricted after the failed uprisings of 1831 and 1864, it was increased after 1815 and 1905. The increased imperial pressure on the Polish polity after the failed uprisings drove the radicals (who supported armed struggle for independence) further from the center of the field of power. Their relative marginalization cleared the way for cleavages homologous with the imperial fields of power. These initially included the opposition between cultural capital (the intelligentsia or educated elites), and later, political and economic capital. Those periods during which imperial policies were liberalized made room for peripheral cleavages based on the opposition between "pragmatists", i.e. those willing to cooperate with the imperial powers, and "patriots", i.e. those calling for more direct action towards regaining national independence.

Thus, the Russian sector of the Polish field of power in the nineteenth century had a dominant axis based on the dispute between the faction of the elite that supported compromise with the Russian authorities and that which favored confrontation. Importantly, there was a clear homology between the internal and external dimensions discussed in the theoretical chapter. By "internal dimension" is meant the classic opposition between the pole of economic capital, represented by the landowners and the emerging bourgeoisie, and the pole of cultural capital represented by the contemporaneously emerging intelligentsia. The external dimension of the proto-field of power was the division between those who favored cooperation with Russia and those who favored armed rebellion against it. There was a fair degree of homology between the two dimensions; it was mainly the intelligentsia calling for armed rebellion. Paradoxically, the intelligentsia was also the structural beneficiary of those failed anti-Russian uprisings. Russia punished those members of the elite who took part in the uprisings by confiscating their property and abolishing their political privileges (e.g. stripping them of their titles, although the Polish proto-field of power continued to recognize the Polish peerage, albeit informally). The minor, propertyless nobility and the intelligentsia also suffered as a result of the failed uprisings. Many of them were killed in the fighting, or executed, imprisoned, exiled (especially to Siberia), or forced to emigrate in the aftermath. As already mentioned, however, they gained a relative advantage as a class. This is because the uprisings strongly strengthened the moral (symbolic) capital of the intelligentsia, and reinforced the symbolic importance of their martyrdom and the variant of Polish culture based on it. The loss of property often irreversibly deprived aristocrats and landowners of their social status and forced them to move over to the intelligentsia, and build careers and obtain status primarily by means of cultural capital.

2.2 Austrian and Prussian Poland

The capacity of the state to achieve political and cultural integration was a major factor in determining the scope of Polish autonomy in the Austrian and Prussian partitions. It is worth noting that Prussia offered relative political and cultural autonomy in its partition of Poland in the first half of the nineteenth century. Austria, by contrast, attempted to build a unitary empire based on the French model. This involved an ambitious policy of developing a common school curriculum with German as the sole language of instruction, and restricting the influence of the Catholic Church and native non-German elites. In the Austrian partition, known as Galicia, universities were Germanized, a network of German-language public schools was established, and the political role of the Polish aristocracy and nobility, who had been subordinated to imperial officials, was significantly reduced. This ambitious Austrian policy, however, was not entirely effective due to the Empire's limited resources and inefficient administration. Austria initially felt so strongly about this that it refused to offer the Polish elites any form of self-government or compromise. Its unitary monarchy strove to integrate Poland into its increasingly national state model, which was defined by German culture.

In short, Vienna at first aimed to completely Germanize its Polish territory according to the standards of modernization set by France. This policy was completely reversed after the Revolutions of 1848 (sometimes known as the Springtime of Nations) and the Austro-Prussian War (1866), which weakened the Empire and ultimately led to the establishment of the dual Austro-Hungarian Monarchy in 1867. For Galicia, this meant significant and gradually increasing (as the imperial center in Vienna weakened) autonomy. Additionally, the Polish elite, led by their feudal faction, gained in influence. In effect, the Polish landed gentry and clergy became Vienna's main political partners. The previous strategy of fostering peasant revolts against the Polish landowners and attempting to replace them as the elite with German-speaking officials who had settled there was suspended. The main reason for this about-turn was insufficient resources to sustain the previous policy. The peasantry was once more placed "under the protection" of the Polish nobility, especially its wealthier elites, and was systematically "nationalized" within newly Polonized public institutions, and especially the church (Łuczewski 2012). The weak bourgeoisie remained strongly dependent on the landed gentry and mixed with it socially and economically. This was fostered by the very weak dynamics of economic development in Galicia.

In the Prussian partition, on the other hand, the dynamics were somewhat reversed. This was largely due to the shift in the balance of power between Austria and Prussia in favor of the latter during the course of the nineteenth century. In 1815–1830, the as yet unconsolidated Prussia retained some degree of autonomy in the former Polish province of Greater Poland, which had come under its jurisdiction. This was mainly effected through a policy of Polish–German bilingualism, which was formally declared in 1815. This policy, however, explicitly provided for Polish being gradually supplanted by German. One indication of the relative autonomy of the Grand Duchy of Posen was the appointment of Prince Antoni Henryk Radziwiłł (1775–1833) as governor. Moreover, the Radziwiłł family enjoyed a close relationship with the Royal House of Prussia. For example, Wilhelm Radziwiłł (1797–1870) and Bogusław Radziwiłł (1809–1873) were both Prussian generals and politicians. However, this model of using the Polish aristocracy as intermediaries or partners in the management of Polish estates only lasted from the Congress of Vienna to the November Uprising. Radziwiłł was then dismissed and the office of governor abolished. At the same time, Polish lost its theoretically equal status with German, and Germanization was accelerated, reaching its apogee in 1871. Some autonomy in the cultural sphere was allowed, but only private associations were permitted to use the Polish language in Prussia. However, even they began to experience increasing pressure after the creation of the German Empire in 1871 (Surman 2018b). The autonomy granted to Poles was therefore clearly dependent on the shifting power relations between Austria and Prussia as the two powers vied for hegemony over a united Germany. Austria's defeat might have benefitted Poland, as it indirectly resulted in Polish cultural autonomy in Galicia. This region was much larger and more populous than Greater Poland. It can be speculated that Poles would have remained relatively autonomous in Prussia had that empire lost to Austria and been unable to politically and culturally consolidate its territories.

The most determined Germanization efforts were made by Prussia and later the German Reich. They made considerable progress in linguistic Germanization of the Prussian administration on the former Polish territories and almost completely Germanized the education system as far as the language of instruction was concerned, with the exception of religious education. It is worth noting that Prussia did not establish any higher education institutions, not even German-language ones, in the former Polish territories. This suggests that the Prussian partition had a more peripheral status in relation to the national center than in the other two partitions. Janusz Hryniewicz has posited that Wielkopolska, as an internal German periphery, despite a fairly high level of general infrastructure development compared to the other partitioned territories, developed more slowly than it might have done had it remained in the Polish state. This obviously assumes that the Polish state would have continued to exist throughout the nineteenth century (Hryniewicz 2003). What is particularly interesting, however, is that despite the effectiveness of the Germanization campaign, neither the strength of the Prussian and German states, nor the relative ease of culturally assimilating the smallest Polish-speaking population of the three partitions, sufficed to achieve this crucial German goal. The inability to completely

Germanize Wielkopolska significantly influenced the formulation of the contemporary definition of German citizenship (Brubaker 1992). In particular, the French model, viz. *ius soli* (right of the land), under which anyone born on the territory of the state is granted citizenship, was not followed in Germany. This was because it assumed the ability to culturally assimilate anyone living within its borders. Germany opted for *ius sanguinis* (right of blood), which remains in force (with certain exemptions) and was originally intended to allow persons deemed not German to be deported. Such deportations were infrequent. According to Rogers Brubaker, they did not apply to unassimilated German Poles, but mainly to economic migrants to Germany from the Russian Empire. However, the mere existence of legislation which provided for ethnic cleansing is evidence of the country's limited integration capacity, compared with the much more efficient France, which was considered a paragon of modernity. As Brubaker points out, another problem with the "Polish element" in that part of Poland in Prussia/Germany was the large outflow of surplus labor, mainly to Western Germany and the United States. Poles also emigrated, but not to the same extent as ethnic Germans. This made it more difficult to Germanize the eastern periphery of Germany. Incidentally, it is worth noting that the structures of migration streams, especially circular migrations, from Polish lands to Germany did not change for over 100 years, which goes some way to explaining the stability of the structures of economic geography in this part of Europe (Kępińska and Stark 2013). However, the most important conclusion to be drawn from the above observations is that the economic peripherality of what was an inner eastern periphery of the German state made full cultural and political integration impossible. One of the key reasons was the outflow of German-educated cadres, especially those with a high level of cultural capital and social status. These people played a crucial role in transferring the dominance of German culture at the regional level into its practical hegemony, which meant displacing Polish culture and language from everyday use. To some extent, a westward outflow of German-speaking cadres, especially to the imperial capital, also affected Galicia, which was perceived by most Austrian officials as an unattractive place of permanent residence. The region appeared to most German-speaking elites coming from Austria proper as a civilizationally backward, "wild" eastern periphery of the country and the continent. As a result, Galicia was often treated as a place of forced "exile" from which it was difficult to return (Röskau-Rydel 2011).

2.3 Russian Poland

2.3.1 The Intelligentsia in the Russian Poland

The Russification of all the territories subordinated to the Empire was also a political desideratum for the Russian authorities. As an ideal, it became more prominent in the second half of the nineteenth century, and as a matter of policy, its implementation became more urgent in Poland after the January Uprising. Russification, however, proved to be an unattainable goal. The main reason was Russia's failure to build a universal system of elementary education, possibly because it feared the "revolutionary" effects of such an institution—literacy would have exposed the masses to rebellious sentiments. On the other hand, St. Petersburg fought the petty Polish gentry as a potentially rebellious force (e.g. by stripping them of their titles), thereby forcing them to move to the cities to escape the devastation of the small farm economy (Sikorska-Kulesza 1995). However, as mentioned above, the Russian imperial authorities simultaneously cooperated, although not within an institutionalized framework, with the elite of the Polish landed gentry. In any case, the Tsars supported wealthy Polish landowners in their struggles against peasant rebellions. Russia was therefore too weak and had more ethnic Poles than it could efficiently culturally and linguistically integrate (i.e. Russify), to say nothing of religiously integrating them into the dominant Russian Orthodox Church, which was completely unattainable. A major stumbling block in this process was that Poles in Russia could use the symbolic capital of their putative "Westness". This could be seen in their higher overall educational aspirations, their Roman Catholic religion, and their language's use of Latin script and copious borrowings from Germanic and Romance languages, as well as in their social capital, which included connections in Western countries, although these were primarily important to the elite.

The relation of the Polish field of power to the broader field of power of the Russian Empire may appear paradoxical in several respects. On the one hand, the Russian Empire, especially in the second half of the nineteenth century, was intended to limit Polish autonomy and gradually Russify its Polish territory. On the other hand, it often contributed to strengthening the autonomous fields of Polish activity, which was mostly an unintended consequence of its actions or the development of several partly autonomous fields of the Empire. Most importantly, the Empire educated and employed a significant portion of the Polish intellectual elite. As Jerzy Róziewicz points out, Poles held more professorial positions in Russia at the end of the nineteenth century than they did in the Prussian and Austrian partitions combined. As mentioned above, it was much easier for Poles to build such careers in Russia proper than it was in the Kingdom of Poland (Róziewicz 1984: 174). On the other hand, the Empire often tried to limit the educational opportunities available to Poles. The Polish language universities in Warsaw and Vilnius had already been closed following the November Uprising, as had the Lyceum in Krzemieniec Podolski (an elite Polish semi-high school). In the second half of the nineteenth century, attempts were made to limit the access of Poles to universities in Russia proper, where many of them had been educated. For example, Poles comprised up to 30% of the student body of the University of Moscow in 1861. This was a matter of concern among the Russian establishment (Róziewicz 1984: 215). At Kyiv University, which was founded as a replacement for Vilnius University, and was intended to be a school for the Russian elite, the authorities were also concerned about the large proportion of Polish students. After the January Uprising, all Polish students were expelled and new enrollments could include no more than 20% of Poles. However, this did not diminish the importance of the Polish intelligentsia at Kyiv University or in Kyiv itself. This is something of a paradox, as Kyiv had ceased to be part of the Polish-Lithuanian Commonwealth in 1686.

To briefly recapitulate, the development of the Polish intelligentsia and the Polish national movement in its various forms, as well as the maintenance of Polish identity on a popular level, during the nineteenth century was largely made possible by the weakness of the partitioning empires. This was patently true of Russia and (especially) Austria, which had been considerably weakened by Prussia and had to abandon its ambition to build a homogenous empire. These visions had to be relinquished in favor of more realistic projects to build de facto or even de jure multinational and multi-confessional empires kept in relative equilibrium by the balance of competing interests between various national, class, and other groups. As a result, Poles played a crucial, and (for most of the time) even dominant, political role in Austrian Galicia for its entire existence, except at the very end, when Vienna chose the Ukrainian option. Polish culture was also, in effect, regionally dominant, and at least from 1870, even hegemonic. This is amply illustrated by the fact that a significant portion of the German-speaking settlers became socialized into Polish culture. It also remained the key model and point of reference for the Ukrainian national movement (Hrycak 2009). From the January Uprising until the 1905 Revolution, the Russian authorities severely downgraded the political position of the Polish elites, while the theoretical autonomy of the Kingdom of Poland was revoked. In practice, however, the Kingdom was a completely separate administrative unit of the Empire right up until Russian rule ended in 1915, and its public service, with the exception of the top echelons, was dominated by Poles and had very few Russified management levels. As Andrzej Chwalba (1999) convincingly demonstrates, contrary to the dominant stereotypes still present in Polish historiography, the administration of the Polish Kingdom was dominated by Poles throughout its existence, although few of the most senior positions were available to them. However, there was no policy of purging Poles from the rank and file (especially not those earning less than 1000 rubles p.a.), where ethnic Poles dominated. Their proportion in the administration was obviously less than the proportion of Poles in the overall population of "Vistula Country", but within the Empire as a whole, the proportion of Poles in the public service was much higher than the proportion of Poles in the population and much higher than the corresponding ratios in Austria-Hungary and (especially) Prussia. These many public service clerks, teachers, and doctors—all earning average salaries—comprised the elite of a Polish intelligentsia that was gradually but systematically gaining in strength. This tension between the economic and cultural poles of the nascent Polish field of power, i.e. between that part of the elite which owed its status to its economic capital and that which was employed in the public service and the liberal professions, continued in a way that was reminiscent of similar processes in Western countries right up until the October Revolution. The divisions on the Polish and Russian political scenes of the time can also be interpreted in terms of this opposition. On the one hand, there were liberal and partly conservative parties respectively oriented toward the elites of the old and new economic capital. These corresponded to the rich landed gentry and the emerging bourgeoisie. On the other hand, intelligentsia parties, both on the right and (especially) the left, steadily grew in strength. They represented milieus oriented toward status building in terms of cultural and social capital, but at the same time, they could function in the field of power through the material income they earned in public jobs, especially clerical jobs. Because of the close ties between the Russian Empire's economic system and its governmental and administrative structures, the economic pole of the Polish proto-power field was much more pro-Russian than the intellectual-cultural pole, which was heavily involved in nationalist activities. At the same time, it is worth recalling that Austria, while granting broad self-government to Poles, including the incorporation of Polish conservatives into the ruling elite of the Empire, also consciously supported the development of the Ukrainian national movement. This was a strategy calculated to create a counterbalance to Polish influence and to limit Russian influence on the Ruthenian population.

2.3.2 Restriction of Autonomy After the January Uprising

Now to discuss the situation in Russian Poland at the turn of the twentieth century in more detail. First, as mentioned above, the Kingdom of Poland enjoyed progressively less autonomy between the January Uprising and the Revolution of 1905. Tsar Alexander II responded to the January Uprising by revoking the Polish Constitution and incorporating the territory directly into Russia. At the same time, martial law, which had been introduced in 1861, was never lifted. The Russian Partition of Poland therefore remained in a state of constant military or "special" rule. This can also be interpreted as a special status; one which may have some similarity to colonial status, which usually implies that the legal system of the metropole is not fully in force. In fact, the extent to which extent the Russian Partition could be considered a colony is an open question, as it involves several ambiguities. First, the very nature of "colonial" status that this question presupposes may be doubtful. As Steinmetz notes, "A careful examination of modern colonial history reveals that the vaunted distinction between indirect and direct rule mapped only loosely onto actual governing practice" (Steinmetz 2014: 91). Second, as will be argued here, Russia, similarly to Austria and Germany (although only until the outbreak of WWI), made every effort to avoid giving the impression that its Polish territories were colonies.

Let me now return to the gradual restriction of the autonomy of the Russian-controlled Kingdom of Poland after 1864. This is especially interesting on account of its paradoxical nature; one which can also be framed in terms of its unintended consequences. The formal, legal restriction of the political autonomy of the Kingdom resulted in its becoming more autonomous in other dimensions. Some of them can be seen as compensatory, but again, as mentioned in the introductory chapter, what is seen as "compensatory" may be very relative. Thus, a special committee for abolishing institutions in the Kingdom was convened in St. Petersburg. The processes which the committee oversaw included abolishing the office of the Secretary of State of the Kingdom of Poland (1866), and replacing it with "His Imperial and Royal Majesty's Own Chancellery for the Affairs of the Kingdom of Poland" (Rolf 2020). In the same year, the Council of State of the Kingdom of Poland was abolished, and the budget of the Kingdom was subsumed into the general budget of the Empire. In 1885, the Bank of Poland, located in Warsaw, was subordinated to the Russian Ministry of the Treasury, although it was made a branch of the Russian State Bank. In 1867, the Committee on Religion and Public Enlightenment and the Justice Committee were abolished. In 1874, the office of Governor-General was redefined to include authority over the Russian civil administration and the military. A new administrative division was introduced in 1866, and in 1867, the Kingdom of Poland was renamed the Vistula Country (*Privislinsky krai*), although the use of the former name was not formally forbidden). The use of Polish had been forbidden in all administrative proceedings in 1863. The Main School (*Szkoła Główna*) in Warsaw was closed in 1869. This was a small tertiary institution with a Polish curriculum that had been established in 1857 as a Polish language medical-surgical academy and restructured in 1862 as a general educational institution. In its place, the Imperial University of Warsaw was founded in 1870. This is discussed below.

In this context, it should be noted that the autonomy of universities in Russia was fiercely debated for the entire second half of the nineteenth century. Liberals were in favor of expanding the network of Russian universities and modeling them after the Western European pattern. They believed that this was a prerequisite for modernizing Russia and for maintaining its international competitiveness. Conservatives feared that expanding the Russian university network was politically dangerous, as it would educate an increasingly radical youth and create and spread discourses that undermined the symbolic monopoly of the traditional elite of the Empire (Schiller 2008). In practice, although universities continued to operate in line with Western models, their autonomy was limited and they were subjected to political control. The process of institutional expansion was slow and piecemeal, not only for political reasons but also due to financial constraints. The Provisional Government finally granted fully autonomy in 1917, although they obviously did not enjoy it for very long.

Significantly, landowners' self-government (the *Zemstva*) was not introduced in Russian Poland, although it was elsewhere in Russia, in 1864. The same was true of urban self-government, which pointedly excluded Poland when it was introduced in 1870. Both exceptions, even if related to the martial law mentioned above, did not only apply to the Kingdom of Poland, but also to territory that had belonged to Poland prior to 1791 (i.e. the first partition of Poland). This fear of Poles, who not only resided in some parts of these areas, but owned or managed a considerable number of large estates (although they could only inherit, not purchase, land in Russia's Polish territories), can be construed as a manifestation of the weakness of the Russian Empire. The exceptional nature of Russian rule in the former Polish territories, which required special legislation, was ambiguous and quasi-colonial. As suggested by

Malte Rolf (2020), the Russians themselves were embarrassed by the inconsistency of their policies. There was a clear will in St. Petersburg, as Rolf argues, not to assign colonial status to the Russian Partition of Poland and not to make Warsaw a colonial city. Russian imperial policy was to unify the Empire. To this end, it mandated uniform laws. Assigning a "special", i.e. colonial, status to the Polish provinces, and Warsaw in particular, would (and to some extent did) demonstrate the weakness of the Russian state machinery and its ability to integrate its "Western provinces". Russian policies stood in stark contrast to Austrian, and (especially) German policies, which not only eschewed legal exceptionalism for their non-German territories, but also special measures that would have undermined their visions of a uniformity of laws throughout the state. Germany's approach to Germanizing the Polish minority in the former Polish region of Wielkopolska (Posen Province) was to adapt German law to meet the challenge rather than making exceptions for the region. An example of this, already discussed, was basing citizenship on ius sanguinis. This made it possible to selectively grant it to Poles, so as to be able to deport those not wanted.

2.3.3 Ambiguities of the Russian Rule in Poland

The Rolf (2020) study is truly inspiring for its reconstruction of the complex Polish–Russian relations in Russian Poland. Rolf assumes a position contrary to national historiographies, which he defines as being based on the assumption that historical processes on the territories of future nation-states inevitably lead to the aspiration for independence. This position is certainly prominent in Polish historiography. For Rolf, imperial rule is best interpreted as defining the background against which the struggle for nationhood is waged. However, he also indicates other perspectives, reconstructing the complex geometry of the field of power of the former Russian Empire. One example of the multi-layered perspective approach adopted by Rolf is the perspectives of Polish and Jewish circles for which the Empire opened carrier opportunities. These people did not view Russia as a repressive occupying power. However, such voices are much less prominent in the accounts of contemporary

Polish historians. Rolf also offers a fascinating anthropological study of the Warsaw elite during the second half of the nineteenth century. He points out that Russian city officials and imperial administrators often deferred to the interests of Polish landowners. This was particularly apparent in urban planning, where the Russian officials sided with the Polish landowners to the point where some modernization projects had to be abandoned. At the same time, senior Russian officials saw themselves as part of the social fabric of the city—certainly not as occupiers. The impressive economic and social development of Warsaw as the third metropolis of the Russian Empire at the turn of the twentieth century can therefore be seen as the product of fruitful Polish–Russian cooperation. This perspective is conspicuously absent in Polish textbooks.

On the other hand, Rolf may be decentering the Russian imperial perspective in the context of writing a general history of the Russian Empire, and looking at Russian policies primarily from the standpoint of the Polish question. Thus, he points to the role of Russian officials, having returned to St. Petersburg after their stints in Warsaw, in provincializing Imperial discussions. He is aware that his work may be inscribed into the new imperial history, even though he does not fully exploit this avenue. Moreover, his study offers insights into the semi-colonial forms of rule that are of interest to post-colonial sociology. Among the fascinating topics he examines are the Russian discussions on the various ways of integrating Polish territory. Rolf suggests that "Vistula Country" could be seen as a laboratory of Russian (semi-)colonial rule, similarly to colonial countries, where colonial techniques and knowledge, as well as the notions produced in colonies, have sometimes been tested in selected territories with special status. This later impacted the metropolis, not only in how the other provinces were governed, but also in how the metropolis defined itself. By virtue of its strategic location, Warsaw had a strong impact on the Russian center. It was also an interface periphery, and for the Russians, a window to the West. Several trends, especially in urban politics, came to Russian metropolitan centers via the mediation of Warsaw. Jan Surman makes a similar observation regarding the Habsburg empire, stating that his "work suggests the necessity of greater inclusion of peripheral histories in the general narrative of the Habsburg Empire, which also means rethinking it from a spatial perspective" (Surman 2018a: 279). Moreover, he suggested that "one cannot understand processes in Vienna without looking at Cracow or Chernivtsi, just as one cannot understand processes taking place in Lviv without knowing about Graz; similarly, one cannot understand Vienna without taking Berlin into consideration, nor Lviv without Kyiv" (Surman 2018a: 280).

In any case, as Rolf argues, there were two conflicting visions for governing Polish territory within the Russian establishment. On the one hand, a unitarian imperial structure along French lines was considered to be the most modern. However, in the 1860s and 1870s, Tsar Alexander II enacted reforms that provided for limited self-organization of civil society and invited it to help govern the Empire. However, the Tsar regarded Poland as an obstacle to these liberalizing policies. Thus, as mentioned earlier, because of the state of emergency, no zemstva (local government institutions) were established in Poland. In effect, as Rolf argues, the intentions of Russian policies after the January uprising were unclear. On the one hand, anti-Polish discourse steadily gained traction. Common themes included the inferiority of Polish noble culture, and the backwardness of the romantic nature of Poles. The term "Pole" became a common pejorative epithet in official Russian discourse. Rolf argues that Russian policies were initially not so much aimed at Russifying as de-Polonizing the Kingdom of Poland. One means of achieving this was to reserve senior posts for non-Catholics. However, many Poles were nevertheless appointed to such positions due to there being insufficient funds to attract non-Polish candidates, as well as a shortage of suitably qualified candidates. These ineffectual policies obviously reflected the weakness of the Russian Empire.

Interestingly, as Rolf explains, the "Polish question" sparked a debate in Russia as to what "Russianness" meant. The difficulty probably lay in the fact that Poles constituted a distinct ethno-religious group and, until recently, had had a large and powerful nation of their own. This made their integration a thorny issue. Another aspect of the influence of Poland on Russian affairs was the role played by many former University of Warsaw professors, who, after returning to central Russia, often became experts on nationality issues. They comprised something of an informal think tank in which new imperial policies were conceived. Through these debates, the voices and interests of the Polish field of power were transmitted to the imperial center. Rolf argues that this Polish impact on central Russian debates was not unique, although it was probably the strongest of the imperial peripheries. When the State Duma (parliament) was established, Russian nationalist forces came into conflict with diverse national movements from the Russian peripheries. The Western peripheries of the Empire were overrepresented in the nationalist parties as well as the Union of October 17, a moderately right-wing, anti-revolutionary party. And this was no accident. It could therefore be said that the peripheries partially took over the center—at least as far as the agenda of the political debate was concerned. Rolf suggests that, in this context, Saint Petersburg, unlike Vienna, was incapable of acting as an ostensibly neutral mediator between the conflicting interests of the peripheral and central actors. This inability to effectively "divide and rule" can be interpreted in terms of Russia's fecklessness in legitimizing control of its peripheries. The weakness of the Austrian Empire could probably be interpreted as being somewhat more self-reflexive and kept under tighter control. This was in particular manifested in the institutionalization of the autonomy of Polish culture and the attempts to counterbalance it with other peripheral forces, especially the Ukrainian nationalist movement. The situation in the Russian Empire could also be interpreted in terms of a specific type of coupling of the imperial field of power with the peripheral proto-field of power. In Austria, including Polish conservatives in the imperial field of power produced a homology based on the diffusion of the core cleavage, one based on the opposition between economic and cultural capital, to the peripheries, whereas in Russia, government officials returning from Poland produced a homology based more on the diffusion of the peripheral cleavage to the core of the Empire, one based on the opposition between proponents of centralized, or even autocratic, rule (this reinforced political capital) and proponents of the autonomous development of the provinces of the Empire and its diverse social fields (based on the accumulation of cultural or economic capital).

2.3.4 The Main Cleavages of the Polish Proto-Field of Power in Russia

I now wish to focus on the configuration of the Polish proto-field of power within the Russian Empire. Of special interest to me is its liberal pole, one organized around the intelligentsia and the bourgeoisie, and which favored cooperation with their counterparts in the Russian imperial field of power. Many of them were educated at the Main School, which, as Henryk Wereszycki (1990) noted, was by no means a center for the anti-Russian insurgency movement, although it was nonetheless closed on the wave of post-uprising policies aimed at restricting Polish autonomy. The Main School in Warsaw was dominated by positivists—with such notable students as Henryk Sienkiewicz (1846–1916) (the Noble prize laureate who, however, later rejected positivism), Bolesław Prus (1847–1912), another famous Polish writer, and Aleksander Świętochowski (1849–1938), a social thinker and one of the leaders of the liberal political movement. Last but not least, there was Jan Baudouin de Courtenay (1845–1929), the famous linguist whose role in the development of Polish linguistics is discussed at length later in this book. Moreover, as Wereszycki reports, a sizeable portion of the student body of the Main School did not participate in the January uprising. In the academic year 1862/63, only 13 out of 728 students of the school joined the fight. Of special interest is one of the leaders of the positivist movement Aleksander Świętochowski, who proposed an apolitical program of organic work, was anti-clerical and critical of the role of nobility. They were characterized by positivism, scientism, Western civilization, and self-enrichment. Surman sees Polish positivism as "a threshold epoch between Romanticism and modernism, thus denoting a period that in other literary cultures is referred to as 'realism' or 'naturalism'". As he notes, Polish positivism is also commonly linked to the failure of the January Uprising. Its defeat spawned a crisis in Romantic messianic ideology, which had previously dominated the Polish language media (Surman 2018b: 241). An important component of positivism was "organic work", understood as the opposite of the irresponsible romantic politicking that led to the failed January Uprising.

Positivists produced several streams of loyalist political and intellectual life in Poland. One of the more conservative was the *Stańczycy* (Court Jesters) circle, which represented the interests of the upper classes (especially landowners) and strove to reconcile the elites of the three empires. Surman, however, argues that the Polish positivism that developed under Russian rule had little in common with the positivism that developed under Austrian rule. Galician positivism was especially weak, while Varsovian positivism was a strong and lasting current that had an enormous influence on Polish philosophy and literature from 1870. According to Surman, neither the Stańczyk movement, which was a typical conservative circle, nor the Kraków historical school, can justifiably be associated with positivism, although they both condemned the "irresponsible romanticism" of the uprisings. Varsovian positivism, however, was much more progressive and anti-clerical. Its exponents usually endorsed the Darwinist approach whereas the Stańczyk movement typically rejected it. That is why, according to Surman, Jan Baudouin de Courtenay, and Julian Ochorowicz (1850–1917), graduates of the Main School, were not well received in Kraków. Both left the university under a cloud. As Surman points out, in 1895, Kazimierz Twardowski (1866–1938), a student of Franz Brentano (1838–1917), was appointed professor of philosophy in Lviv, and was later instrumental in establishing the famous Warsaw-Lviv school of analytical philosophy. The Warsaw positivists closely followed Western intellectual trends. Their "West" was, however, limited to France and Great Britain, and pointedly excluded Germany, which they associated with Hegelianism, spiritualism, and metaphysics. The extent to which a specific political and intellectual current, such as positivism, can be seen as a single phenomenon in the wider Polish proto-field of power, or as a set of separate, albeit related, streams typical of imperial sectors of that field, is obviously a political question. This was a stake in political games and in attempts to coordinate the Polish proto-field of power, and it remains topical in current historical debates on how to interpret the heritage of nineteenth-century politics. It can obviously be perceived from the Polish, as well as the three imperial, perspectives.

Now to focus on the Russian sector of the Polish proto-field of power. One of the leaders of the broadly defined positivist and loyalist camp was Włodzimierz Spasowicz (1829–1906), an attorney and literary critic (actually, an independent literary scholar) from St Petersburg. Spasowicz advanced a tri-loyalism program, i.e. loyalism of the Polish elite to the three empires, coupled with a program to promote the ideal of Swiss federalism in them. Spasowicz edited the journal *Kraj* (Country) together with Erazm Piltz (1851–1929). Kraj was published in St. Petersburg and offered a relatively free forum for political debate, as it was not subjected to the same stringent censorship that was in force in the Polish provinces of the Empire. At the same time, Spasowicz, as a member of the Russian elite, and who could even be considered a member of the broader field of power of the Empire, enjoyed additional political capabilities and resources. A similar political position was enjoyed by another loyalist party established in 1905, viz. the Party of Real Politics (UPR). On the other hand, the loyalist movement also had a more liberal wing. It gained strength in tandem with the growth of the Russian liberals. In 1905, the Progressive Democratic Party was founded under the leadership of Aleksander Świętochowski, who had close ties with an analogous Russian party, the Constitutional Democrats. The party demanded the political autonomy of the Kingdom. The National Democratic Party, known as Endecja, was initially fairly loyalist as well, even if predominantly for tactical reasons and only toward the Russian Empire. In any case, the National Democratic Party became the major player on the right wing of the political scene in Russian Poland, as it was able to attract wider popular support. Given its increasingly nationalist orientation, it was able to exploit an upsurge of antisemitism that had emerged in Poland during the latter half of the nineteenth century. This wave of antisemitism was partly triggered by a large influx of Jews from Russia proper, mostly as a result of the establishment of the Pale of Settlement in 1791. Much of this was inside Polish territory. A considerable portion of the Polish elite resented the influx of Jews ("Litvaks") from the depths of Russia. Jews were also attracted to the Polish part of the Russian Empire, as they had equal rights in the Kingdom of Poland by virtue of the Aleksander Wielopolski's reform of 1862.[2] At the same time, on the left of

[2] Aleksander Wielopolski (1807–1877) was the head of Poland's Civil Administration within the Russian Empire.

the political spectrum, there was a major division between the communist and social-democratic parties. In institutional terms, this translated into opposition between the internationalist Social Democracy of the Kingdom of Poland and Lithuania (SDKPiL) and the Polish Socialist Party (PPS), which called for Polish independence.

Now to examine the situation in Russia proper. The country had a large Polish community, especially in Moscow and St. Petersburg. Polish political organizations in Russia were predominantly of three persuasions during the first few years of the twentieth century (Ponarski 2003). First, there was the old Polish emigration (both liberal-left and conservative). A considerable number tended to cooperate with the liberal, bourgeois Russian opposition, in particular with the Cadets (the Constitutional Democratic Party, which had a plurality in the First Duma), the Essers (the Party of Socialist-Revolutionaries, of which Alexander Kerensky was a member), and sometimes even with the Mensheviks. Second, there were the "realists", who regarded the ruling elite of the Russian Empire as the only reliable political partner in Russia. They were mainly represented by National Democracy (Endecja). Their ranks were bolstered after the outbreak of World War I, when numerous Polish war refugees, mainly from the Kingdom of Poland, arrived in Moscow and other Russian cities. The third orientation was represented by the PPS, which was initially pro-Austrian. The communists should also be mentioned here. They were obviously in favor of working with the Bolsheviks.

2.4 Higher Education and Research in Russian Poland

2.4.1 Alternative Educational Paths for Polish Students in Russia

As the only university in the region was the Imperial University of Warsaw, this section primarily discusses this institution, and largely relies on Rolf's study (Rolf 2020). The university was established in 1870 as a purely Russian institution with a clear mission to Russify the Kingdom of Poland. Initially, Poles mainly studied law and medicine, only comprising 3% of the student body at the historical-philosophical department. The restrictions imposed after the January Uprising, along with certain religious criteria, made the university unattractive for Poles. The Historical-Philological department mostly educated future Russian officials. Most of the professors were ethnic Russians, but there were also Germans from Germany, Lithuania, Latvia, and Estonia, as well as Czechs and Poles. Subjects related to Russia were privileged in the curriculum. Disseminating knowledge about Russia was one of the crucial tasks of the university. The Russian language was obligatory in informal contacts between students in public places. On the other hand, some of the professors, mostly recent arrivals to Warsaw, were interested in interacting with Poles. Nikolay Karayev, in particular, was looking for contacts with Poles, but was isolated, and not many Poles were interested in maintaining contact with him. One exception in this regard was Aleksander Świętochowski, but Karayev left Warsaw in 1885. No chair of Polish philology was ever established, even though it had the support of the General-Governor of Warsaw, as St. Petersburg was implacably opposed. Thus, Russia abandoned any attempt to regulate Polish language and culture through the main university on Polish territory. One of the reasons was that the ruling elite of Russia attributed the January Uprising to their lenient policies of 1861–1862.

Polish students had, however, several alternative educational paths, which included universities in the Austrian province of Galicia (Lviv, Kraków), and in the West, in particular Germany, France, and Switzerland. Another popular choice was universities in Russia proper, as they generally had more liberal professors and no anti-Polish bias. However, the resulting influx of Polish students to universities in central Russia induced the authorities to impose a *numerus clausus* that limited Poles to 20% of the student body. This seems to have helped push some of the younger Polish intelligentsia toward diverse associations and illegal revolutionary parties. In any case, restricting the access of Poles to institutions of higher education added to the antagonism between society and the state, which was especially fraught in the Polish part of the Empire. In particular, University of Warsaw professors were constantly at loggerheads with Polish society. This tension only led them to double down on their ideological convictions. Several of them came up with highly conservative conceptions of the state organization of the Russian Empire. These stood in stark contrast to universities in Russia proper and even to the views of the General-Governor of the Kingdom of Poland and other members of the Russian imperial elite stationed in Warsaw. Interestingly, Rolf reports that Warsaw General-Governors tended to view University of Warsaw professors as parvenus. This was most evident in the case of Alexander Bagration-Imeretinsky (1837–1900), who was a Georgian Prince, an Imperial Army General, a war hero, and one of the empire's many non-ethnically Russian aristocrats. For him, the attempts by these professors to reduce the multinational and multi-confessional Empire into a homogeneous Russian nation-state were emphatically not acceptable. This internal conflict among the Russian elite of Warsaw is a fascinating example of how the all-imperial field of power and the Polish proto-field of power were intertwined.

2.4.2 University of Warsaw as an Outpost of Russification of Poland

This attitude of University of Warsaw professors resulted in the university becoming a center of *slavianovieddienie* or general Slavic studies, a center of pan-Slavism and Slavophilsm. Warsaw historians Petr Lavrovski (1827–1886) and Iosif Perwolf (1841–1891) were interested in the early Slavs, while others were preoccupied with the unattested proto-Slavic language. Following Johann Gottfried Herder (1744–1803), they shared a belief that the essence of national character can be inferred from lifestyle and folklore. Thus they had a tendency to essentialize cultural values, and they assumed that small nations, such as Lithuania, would naturally integrate into Russian culture. Politically, several professors at the University of Warsaw were active in the Russian Borderlands Society (*Russkoie Okrainnoie obshestvo*). This was established to defend Russian culture and interests on the Western frontier of the Empire. Russian language professors, e.g. Platon Kulakovsky (1848–1913), were especially active members. Several professors were also active in the 3rd and 4th Dumas. While there were a couple of liberals among them, most were associated with the Octobrist Party (The Union of October 17)—an anti-revolutionary, right-wing constitutionalist party.

Warsaw university professors did sometimes resort to outright colonial discourses, e.g. there was a proposal for Poles who resisted Russification to be denied education and confined to special zones under military rule. More importantly, most of the Russian professors were staunchly opposed to the use of Polish as a language of instruction, arguing that Russian bestowed universal status on the university. In this context, Jan Baudouin de Courtenay, a proponent of equality of languages in the empire, was deemed a threat to the Warsaw professors. The Revolution of 1905 brought a student boycott and resulted in the university being closed for three years. When it reopened, it was even more Russian and less international. Most of the few pro-Polish professors, e.g. Alexander Pogodin (1872–1947), left for central Russia. Only Russian, or more precisely, Orthodox, professors remained. The isolation of the university from the city increased. The Polish chair remained vacant until it closed in 1915.

In 1915, the University of Warsaw, along with most government and public institutions, was evacuated deep into Russia ahead of the advancing German army. However, the library and some of the professors remained in Warsaw. The decision to leave or stay was largely a religious one. As a rule, the Orthodox professors, staff, and students left, while the Catholics remained. Most of the evacuees ended up in Rostov-on-Don, where the University formally settled permanently and operated under the name of the University of Warsaw in Rostov-on-Don until 1917, when the Provisional Government changed its name to Rostov University. Thus for some time, there were de facto two parallel universities of Warsaw: the Russian university in Rostov-on-Don and the Polish language university in Warsaw under German tutelage. This connection between Rostov University (now the Southern Federal University) and the University of Warsaw is largely forgotten, especially in Warsaw. However, a large delegation from Warsaw traveled to Russia in 2015 to celebrate the 100th anniversary of the evacuation that marked the beginning of the University of Rostov-on-Don.

The entire history of the University during the Russian period, i.e. when it was the Imperial University of Warsaw, was until recently marginalized in its public image. This sort of "cleansing of history" is typical of how the Russian contribution to the heritage of nineteenth-century Poland is treated in contemporary Poland (Zarycki 2012). A case in point is the University of Warsaw's website, which only listed its rectors from the Russian period (1869–1915) a few years ago, i.e. the Russian period of the university was not even recognized as part of its history. Moreover, it is sometimes not considered to be a part of the history of Polish science either. For example, the well-known researcher of Polish higher education, Dominik Antonowicz, observes: "There was a Russian university in Warsaw, but it was boycotted by Polish society in 1905–1908 and cannot be considered part of Polish higher education" (Antonowicz 2016: 136–137). This changing way of presenting the Russian element in Poland's history may be seen as an aspect of the constantly contested boundary between "Polishness" and "Russianness".

2.4.3 Independent Polish Sector of Higher Education and Research

However, the Imperial University of Warsaw was not the only academic institution in the region. A parallel system of not only academic research institutions but also other organizations that were independent of the Russian state emerged in the Kingdom of Poland in the second part of the nineteenth century. They were mainly funded by donations from the Polish bourgeoisie. Scientific research was supported by supposedly non-academic institutions, such as *Bank Handlowy* or *Bank Dyskontowy Warszawski* (unofficially called "asylum of Polish science"). Bankers and industrialists (such as Stanisław Rotwand [1839–1916] and Jan Bloch [1836–1902]) supported the Warsaw Institute of Technology (*Politechnika Warszawska*). Separate institutions were supported by specific political forces. The National Democrats supported the *Macierz Szkolna* while socialists maintained the "University for All". Apolitical institutions, however, were far more important, especially the Józef Mianowski Foundation (*Fundacja Mianowskiego*), which was founded in 1881 by a number of wealthy bankers (e.g. Stanisław Kronenberg [1846–1894] and Jakub Natanson [1832–1884]), industrialists and intellectuals. This supported a wide array of scientific activities and academics with grants and fellowships. The Private Museum of Industry and Agriculture was another such institution (Marie Skłodowska-Curie conducted her first research here before leaving for France). Private donors financed encyclopedias, dictionaries, and rural mythologies. The *Zachęta* Museum and Exhibition Hall, as well as the Warsaw Philharmonic, was financed by Leopold Kronenberg (1849–1937) and Ignacy Paderewski (1860–1941). Several Jewish capitalists, in an effort to prove their loyalty to the Polish nation, financed Polish academic institutions. A network of Polish schools, along with the Warsaw Scientific Society (*Towarzystwo Naukowe Warszawskie,* TNW) and the Public Library Society, were established in the more liberal climate engendered by the Revolution of 1905. A wider Polish cooperative movement also developed in the countryside as well as the city. Rolf argues that the establishment of the Warsaw Scientific Society in 1907 was the turning point in the duality of the Warsaw academic scene, likening it to the division between the German and Czech academic spheres that had existed earlier in Prague, when the Charles University was divided into Czech and German parts. In Warsaw, however, the division was along state (Russian) vs. privately financed (Polish) lines. The latter was part of a system of self-organization that included associations, cooperatives, hospitals, farmers' associations, etc., and was consonant with the idea that the intelligentsia had a responsibility toward the lower classes. The Russian government exercised less control of the private sector than the public sector. Rolf gives the example of numerous private schools being set up, subjected to less state intervention than public ones, and permitted to teach in Polish. Several other non-governmental institutions came to be dominated by Poles over the course of the nineteenth century. These included the General Welfare Council (*Rada Główna Opiekuńcza*) and the Medical Council (*Rada Lekarska*), which was dissolved in 1870. As Rolf argues, the chronic weakness of the Russian state structure was one of the key reasons why the Empire had to rely on Polish parallel institutions and tolerate Polish co-participation. This situation was summarized as follows by Władysław Grabski (1874–1938, Prime Minister of Poland 1920, 1923–1925): "In the former Kingdom of Poland, the main social institutions had created a kind of national civil government for economic and social affairs, as was evident in the extraordinary activity of the citizens' committees during the war" (Grabski 2016: 89). This duality which emerged in Poland seems to call for the recontextualization of Bourdieu's reflections "On the state". The Russian partition of Poland at the turn of the twentieth century was a highly complex national field of power with a network of several related social fields that were able to develop largely independently of the state (Empire), although the crucial role that the state (Empire) played in its functioning cannot be denied. The main sources of this configuration were a weak state, the interface periphery effect, and an autonomous elite and cultural system.

2.4.4 Main Polish Political Camps in Russia

Homologous to the divisions on the political scene in Russia, distinct ideological currents were forming in the social sciences. At one extreme, there was the imperial science clearly associated with the ruling conservatives. This was premised on a normative approach to Russian culture as the dominant culture of the Empire, and the corresponding depreciation of the Empire's other cultures. As far as social sciences were concerned, this was often linked to traditional historicism. This was the kind of science supported directly by the state administration, and it was mostly developed at the Imperial University of Warsaw. The scholars who founded this school of thought usually supported Tsarist autocracy, Russification, and the dominant role of the Orthodox Church. Two main currents emerged in opposition to it. First, the liberal current, politically associated primarily with the Cadet Party. This was based on a desire to democratize the Empire, and transform it into a multinational, and even federal, state based on the rule of law, civil rights, and civic equality of rights and duties. This current of Russian science remained under strong Western influence and was quite well integrated with the world science of the time. Internally it remained quite diverse, both politically and ideologically—politically in that a variety of sympathies from conservative-liberal to social-democratic could be distinguished; ideologically in that it ran the full gamut from strong religious inspirations to openly avowed atheism. What united these approaches, however, was respect for positivism in the broad sense of the word and an equally broadly understood republicanism (which included supporters of constitutional monarchy). The third strand was a diverse leftist current, associated with a wide spectrum of parties from socialists to radical Bolshevik revolutionaries. In the multicolored plethora of circles and personalities that made up this current, constructivist approaches were clearly visible, i.e. approaches that abstracted from history and tended toward class analysis and formalism. The whole broad and creative milieu related to that current was also inspired by the Western avant-garde of various trends. It drew its support from the institutionally marginalized intelligentsia, who had revolutionary sympathies, and were politically, academically, and artistically radical.

The wide array field of national movements within the Empire is a peculiar fourth option. These movements can be considered a separate social world, oriented toward their own fields (or national proto-fields) of power. It is probably useful here to maintain the dual optics of our analysis, as it corresponds to the competing imperial and national political reference systems. A whole spectrum of orientations can be distinguished in national movements. This is broadly similar to the intra-imperial political spectrum discussed above. On the one hand, there will be national-conservative circles, which, in the academic dimension, promote essentialist and primordial approaches to the historical legacy of their nations—usually with a strong normative accent and political overtones, and in conjunction with religious forces. At the other extreme, there are leftist forces of various orientations—national liberals among them. This spectrum is homologous to both the national proto-fields of power and the imperial field of power. Each of these sectors of the cultural and academic field had complex relations between their imperial component and the emerging national sectors. Additionally, the various national minority groups entered into complex relationships with one another. These spanned the full spectrum of interactions from fierce hostility to close cooperation and political coalitions. One example, in the Austrian partition, was a coalition formed by a component of the Polish field of power with a section of the Ukrainian national movement (in the framework of the "new era" agreement). At the same time, a major part of the Polish field of power in the Russian sector opposed the given alliance. It is also possible to speak of these relations at different levels. In particular, one view holds that, at the macro-imperial level, there was a structural coalition of national and autonomist movements, and in particular their intelligentsia elites, and the radical intelligentsia of the Empire itself, especially in Moscow and St. Petersburg. It was this broad coalition that led to the two revolutions in 1917, as Andrzej Nowak (2009) argues. The second time that such a coalition emerged (in Nowak's view) was during the crisis that led to the collapse of the Soviet Union, and the Soviet Empire more broadly, in the late 1980s. Lower-level and short-lived coalitions were also formed between a number of other forces, especially between ideologically close sectors of the field of power, e.g. Russian and Polish liberals, as well as Jewish liberals, who ran under blocked lists in some Duma elections. The communists provide an extreme example of transnational cooperation, as they did not build alliances so much as international organizational networks, and many of them were antithetical to the concept of the nation-state, which set them apart from the socialist movement. Even within the communist movement, however, there were significant fractures due to the influence of individual power fields. For example, the German field of power exerted a powerful influence on the activities and organization of the communist movement, and also provided resources. This is amply illustrated by the career of Rosa Luxemburg, who was born into a Jewish family in the Russian partition of Poland, but was mostly involved in German political life. The competition between the Russian and German fields of power also influenced politics in other ways in Poland. While many Poles were engaged in Russian political and intellectual debates, German or, more broadly, Western European, debates were much more attractive. There were obviously counterexamples, e.g. the National Democrats were more interested in participating in Russian politics and intellectual debates, albeit only for a brief period and mainly for pragmatic reasons. It should be kept in mind that these included a geopolitical conviction that Russia was a "safer" partner for the Polish national movement in view of its relative weakness, which would never allow the Russian Empire to fully integrate Polish lands and elites. Given the political weight of the Endecja, its active involvement in the Russian imperial field of power moved the Polish field of power closer to Russia than Germany.

2.5 Polish Academic and Intellectual Field Under the Austrian Rule

2.5.1 The Early Stages of History of Galicia

The organization of the Polish proto-field of power in the Austrian partition was no less interesting or complex. The Austrian partition had been known as Galicia (*Galicja*) since the first partition in 1772. The name, which is still used occasionally, was derived from the official name of the Kingdom of Galicia and Lodomeria, which was established at the same time. After the Third Partition of Poland in 1795, it formally became a kingdom under Habsburg rule. In 1804, it became a crownland of the Austrian Empire, and in 1867, after the creation of the Dual Monarchy, it became a crownland under the Cisleithanian part of Austria-Hungary, with some degree of Polish autonomy, until its dissolution in 1918. The configuration of this sector of the Polish field of power was markedly different from that in the Russian Empire. The pro-Vienna, loyalist sector was the strongest, and sometimes even the dominant, part of the Polish field of power in Galicia. It strongly relied on its participation in the ruling coalitions in the Austrian parliament and on having been relatively well integrated into the Empire's central administrative system. Polish conservatives provided several ministers, and even prime ministers, to the imperial government in Vienna. For example, the Habsburg ministers of finance were usually Poles. Thus, the formally conservative faction of the Polish elite effectively jointly governed the entire Empire. Despite this, however, they had relatively little political clout in the empire. Polish politicians in Vienna were always minor players and loyal subjects of the emperor. Nevertheless, the fact that they occupied such prestigious posts helped them win the support of their Polish constituents (mostly those belonging to the upper echelons of landowners and aristocrats) and, to some extent, gain wider support among the Polish population. This co-optation of Polish conservatives into the Austrian ruling class brought benefits to the Polish field of power, not least in ensuring the autonomy of the cultural and educational fields of Galicia, where Poles were clearly dominant. In contrast to what was happening in the Russian partition, universities in Galicia (Kraków and Lviv/Lwów/Lemberg) were considerably Polonized in the latter half of the nineteenth century. The development of a network of schools that taught in Polish, especially high schools (gymnasia), proved essential to this.

2.5.2 Main Polish Political Camps in Austria

There were some interesting developments on the left of the Austrian sector of the Polish field of power. The Polish Social Democratic Party was created as part of a wider social-democratic movement in Austria that was formed under the banner of universal suffrage following a rally in Vienna in 1890. This marked the creation of a kind of federation: the Polish Social-Democratic Party of Galicia and Cieszyn Silesia, and a separate Ruthenian then Ukrainian, Social-Democratic Party in 1897. The Polish socialist movement led by Ignacy Daszyński (1866–1936) was frowned upon by both the Austrian authorities and the dominant faction of the influential Catholic church. Galicia was also famous for its peasant movement and parties. The Christian People's Party, led by Fr. Stanisław Stojałowski (1845–1911), was opposed to the government. The People's Party (PSL) was founded in 1895, but after it declared its loyalty to the Empire, a considerable part of its more nationalist electorate defected to the National Democrats. Wincenty Witos (1874–1945) emerged as the leader of PSL after the split in 1913.

The National Democrats (Endecja) were initially marginalized and ineffectual in the Austrian partition because the progressive intelligentsia and bourgeoisie mostly supported Polish Democracy, a party that had existed since the struggle for autonomy in Galicia began. Endecja entered the region through its involvement in grassroots organizations. Among them were the Association of People's Schools (*Towarzystwo Szkoły Ludowej*) and the Falcon (*Sokół*) Gymnastic Association. Endecja also attracted such famous Polish intellectuals and writers as Jan Kasprowicz (1860–1926) and Mieczysław Gwalbert Pawlikowski (1834–1903). Another strategic step was the purchase of the *Słowo Polskie* newspaper in Lviv. This was used to win over several additional influential intellectuals, including some professors of the University of Lviv who were unsympathetic to Ukrainian culture and aspirations. Anti-Ukrainian riots at the university increased this tension and led to polarization, which was beneficial to Endecja. Thus 1905 saw the final formation of the National Democratic Party in Galicia. It is, therefore, possible to talk about a successful import of the party from the Russian sector of the Polish field of power into the Austrian sector. At the same time, there was an interesting divergence between the National Democrats and the Socialists in Galicia. While Endecja supported separation (full autonomy) from Austria, the Socialists opposed this for economic and socio-political reasons, as they saw important benefits from Galicia, the poorest province of the monarchy, remaining strongly tied to Austria. This invites speculation as to whether centrifugal anti-imperial dynamics, along which the Russian field of power was organized to a much stronger degree, was actually what was imported.

There was also a clear rejuvenation of the conservative movement in Galicia at the end of the nineteenth century. This was made possible by a consolidation of agricultural interests, and a coalition of the landed gentry and the wealthier peasantry. They sought support from the conservative forces of the Austrian-Hungarian monarchy. The Podolaks, as they were known, were the most reactionary group of landowners in the Empire. They were a faction of the old elite that had not even begun to transform itself into the intelligentsia, i.e. to convert their economic capital into cultural capital. This was very much in contrast to Stańczyk's more intelligentsia-based conservative faction. This coalition of conservative forces, based primarily on the economic interests of the elite of this backward, mostly rural, region was the dominant political force until the end of its existence.

Another layer of complexity of the political life of the province was added by its ethnic composition. For example, the settlement worked out in the late 1890s by the Prime Minister of Austria, Kazimierz Badeni (1846–1909), an ethnic Pole, broke the prevalence of Moscow-oriented activists among the Ukrainian intelligentsia and facilitated, through language reform, the development of Ukrainian cultural institutions, to a far greater extent than had been done in Russia. For a time, the dominant forces of the Austrian sector of the Polish field of power appeared to be allies, if not co-creators, of modern Ukrainian nationalism. Badeni also introduced language reform in Bohemia by enacting Czech–German bilingualism in the province. This resulted in a wave of protests from German-speaking Austrians and led to the fall of Badeni's government. This can be interpreted as the result of the tension between the core and the peripheries of the Habsburg Empire. At the same time, Polish conservative loyalists were trying to weaken liberal influences, postpone democratization, strengthen the autonomy of Galicia, and turn Austria against Germany and Russia, as they believed that Poland could only be re-recreated as part of Austria. Their ideal was a Triple Monarchy: the Austro-Hungarian-Polish Empire. Thus, the Treaty of Brest-Litovsk, signed in March 1918, pursuant to which, inter alia, Austria recognized the independence of Ukraine, but not Poland, was a strong blow to Stanczyk's political position.

2.5.3 Higher Education System in Galicia

I would now discuss the university system in Galicia in more detail. This section primarily relies on an impressive study by Jan Surman titled *Universities in Imperial Austria 1848–1918* (Surman 2018a). Galicia had two institutions that were crucial from the standpoint of the Polish cultural and academic field, viz. the universities of Kraków and Lviv/Lwów/Lemberg. Jan Surman points out that these institutions did not seem all that impressive at the beginning of the nineteenth century. The Cracow Academy (*Akademia Krakowska*, later renamed Jagiellonian University) was no more than a provincial university (*Landesuniversität*) in 1805–1817, while the University of Lviv was closed during this period, leaving only a lyceum operating in the city. In 1817, Cracow was made a free city and Lviv's lyceum was granted the status of a university under the name Francis I University. German was made the language of instruction at the University of Lviv, and the curriculum, administration, etc., organization did not differ from that of other universities in the Empire. The Kraków Academy had a semi-autonomous status being under the joint supervision of Austria, Prussia, and Russia. This allowed it to admit students from those three countries until the November

Uprising. With only 200 students, it was much smaller than the University of Lviv, which had about 1400 students. After 1846, Kraków was incorporated into Austria, and the Academy was reformed according to the model of other Austrian universities. After 1849, this process was completed, and teaching was conducted in Latin and German (in the Faculty of Philosophy) (Surman 2018a: 34). Shortly afterward, universities began to have their autonomy restored. This was presented by its supporters as a precondition for modernization. At the same time, teaching in Polish began to be slowly allowed, although initially only in Lviv. University management passed largely into the hands of full professors, although a key role was still played by the ministry, which could determine the actual scope of a university's autonomy. The German university system was a model for these reforms, although in Austria, an attempt was made to keep the system separate by, e.g. privileging Catholicism. The reforms of Governor Franz Stadion in the late 1840s contributed to Polish-Ukrainian tensions, as Polish nationalists strongly opposed the introduction of Ukrainian (known as Ruthenian in Galicia) into high schools and universities. Polish activists believed that the Ruthenians needed the Polish language as a carrier of culture and civilization and should be taught in it. This argument was, incidentally, identical to that used against the Poles when Austrian officials defended the use of German in Galician universities.

The admission of Ruthenians and the introduction of the Ruthenian language into higher education was, however, supported by some Poles. For example, Józef Dietl (1804–1878), later the rector of the Jagiellonian University, supported both the development of education and learning in Polish and Ruthenian. He saw them as a way to strengthen Slavic cultures in Galicia and believed that they could act as a counterbalance to Germanization when combined. However, the Polish voices opposed to the development of education in the Ruthenian language were much stronger. In any case, this was a period of neo-absolutism and it brought renewed attempts at Germanization. In 1853, German was restored as the official language in the universities of Lviv and Kraków for the duration of martial law, which was, however, suspended in 1854. According to the authorities' vision, "Polish science" was to be practiced, not in public universities, but in separate institutions, of which the

Kraków Scientific Society was a typical example. This might have been an attempt to revive the dual system that had existed in Galicia in the early nineteenth century. The first Polish regional scientific society was the Ossoliński Scientific Institute (*Ossolineum*), founded in 1827 by Count Józef Maksymilian Ossoliński (1748–1826). However, after it printed materials on the November Uprising, it was placed under police control until 1848. As Surman (2018a: 25) points out, the Ossolineum operated in conjunction with similar institutions in the other partitions, which bore testimony to the existence of a cross-border network of scientific institutions associated with the Polish proto-pole of power. These institutions were active before 1863 in the Grand Duchy of Posen and the Kingdom of Poland. Prior to the liberalization of Austria, it was easier to conduct Polish language academic and cultural activities in Prussia and Russia than in Austria, which was controlled by Klemens von Metternich's government and subjected to tight censorship. Meanwhile, chairs of Slavic languages were established in Berlin and Breslau (Wrocław). Another more powerful institution in Galicia was the Academy of Arts and Sciences (*Akademia Umiejętności*), which was founded in 1872. The Academy grew out of the Kraków Learned Society (founded in 1815 and renamed the Academy of Learning in 1871). From the outset, the Academy aspired to serve as a learned and cultural society for the entire Polish nation. However, its ability to accept members deemed "foreign" was restricted by Austrian law. Hence, it was sometimes criticized for its dearth of Polish scholars, which curtailed its ability to represent the entire field of Polish science (Surman 2018a: 12). It was renamed the Polish Academy of Arts and Sciences [*Polska Akademia Umiejętności*] in 1919.

In the mid-nineteenth century, Polish language chairs at Galician universities were a thorny issue for the Austrian authorities and remained vacant until 1856. The authorities tried to bring in Polish luminaries from outside Galicia for these positions, but for a variety of political reasons, this was not easy. This was mainly due to a potential lack of loyalty or proficiency in German. Józef Ignacy Kraszewski (1812–1887), a well-known Polish writer, and journalist, although given preliminary approval by the authorities, was not allowed to come by the Russian authorities, who refused him an exit visa (Surman 2018a: 78).

Conversely, scholars who met the requirements of the Austrian authorities were difficult for the Polish community to accept. For example, Karol Mecherzyński (1800–1881) was nominated to the Jagiellonian University, and Antoni Małecki (1821–1913) to a chair in Lviv. Both were trained as grammarians, had studied Old Church Slavonic, and employed a dated formalist approach to textual analysis. Mecherzyński was interested in the influence of Latin and German on the Polish language. This kind of historical interest ensured that he and his colleague would not conduct "nationalist propaganda". The period of neo-absolutism in the Habsburg Empire ended in 1860, when the powerful Austrian Minister of Education and Religion, Count Leopold von Thun und Hohenstein (1811–1888) had his office abolished. Shortly thereafter, multilingualism was permitted at Austrian public universities. Initially, there was a bilingual Polish–German system of teaching in Kraków from 1861, while in Lviv it was Polish-Ruthenian. At the same time, some of the power in Galicia was returned to the local aristocracy. In 1869, formal trilingualism was introduced at the University of Lviv, but this did not completely satisfy the ambitions of either Polish or Ukrainian hardliners.

The closure of the Warsaw Main School in 1869 meant that there was no "Polish" university in Europe. Fearing political unrest from the Polish elite, the authorities in Vienna announced that, starting in 1870, Polish would be the only language of instruction at the Jagiellonian University. A year later, the status of Polish and Ruthenian was equalized at the University of Lviv, which meant that Polish became the primary de facto language of instruction. From 1879, Polish became the administrative language of the university in Lviv, which made the situation more difficult for anyone not fluent in Polish. As a result, it was easier to get a job in Lviv for a socialist and even an Austrian who knew Polish than it was for a Rusyn or a Jew. Formally, the Polish language gained this status in 1882, when it was only possible to teach in Ruthenian additionally and with the permission of the ministry (Surman 2018a: 97). Curiously, the Charles-Ferdinand University in Prague was formally divided into Czech and German parts in the same year. In Lviv, Ruthenian circles mobilized to defend their interests. They mostly attempted this through a network of independent institutions, especially educational associations, such as

Prosvita and the Schevchenko Scientific Society. Pro-Russian organizations, such as the Russophile Kachkovskyi Society, were also active. In general, political mobilization was growing, and the influence of the previously non-confrontational and conservative Greek Catholic Ruthenians, who did not identify with the Ukrainian movement, was waning. They were displaced by a younger generation of Ukrainian activists who were antipathetic toward Poles.

Surman points out an interesting aspect of the Polish-Ruthenian disputes over the language of education in Eastern Galicia, particularly in Lviv (Surman 2018a: 98). On the Ukrainian side, the argument was that a "national" Ukrainian university would be a tool for achieving the cultural development of the nation, not a result or aspect of it, as the Poles had argued in the case they presented for maintaining of Polish character of the University of Lviv. Polish activists argued moreover that the relative weakness of the Ukrainian intelligentsia, and a shortage of suitably qualified academics, militated against a Ukrainian university in Lviv. Similar arguments to those adduced by the Ukrainians had previously been put forward by the Czechs in Prague. Thus, from the Ukrainian point of view, the cultural development and education of the Ruthenian community were being blocked by Poles. This made Poles appear as imperialists and hypocrites. After all, Poles were bitterly complaining about the limitations on the development of Polish institutions and Polish education in the Province of Posen in the Prussian partition (the Second Reich after 1871).

However, the political rapprochement between factions of the Polish and Ukrainian elites (known as the "New Era") during the 1890s under the Badeni government is worth mentioning once more. The Ruthenian language was allowed to be used in administration and was codified, becoming Ukrainian. This was a crucial victory for the Ukrainian faction of the Ruthenian movement over its pro-Russian opponents. Part of the compromise was the creation of a department of Ruthenian history at Lviv University. In the official description, this chair was to be a bridge between Eastern and Western cultures. However, Mykhailo Hrushevsky (1866–1934), who was nominated for the chair, was quite radical. He refused to speak Polish, which brought him into conflict with the Dean, Kazimierz Twardowski (1866–1938). His uncompromising stance, however, won him many admirers in Ukrainian social circles. Their support later resulted in Hrushevsky's appointment as head of the Central Rada (the parliament of the short-lived Ukrainian People's Republic) in 1918. In the meantime, the New Era came to an end in 1894.

2.5.4 The Kraków-Lviv Opposition

Surman (2018a: 209) points out an interesting contrast between the universities of Lviv and Kraków in the second half of the 1890s. Polish-Ukrainian relations at the university of Lviv were very tense, but they were much more harmonious at Jagiellonian University. In 1894, a department of Ruthenian literature was created, and the importance of the Ruthenian language to Poles was emphasized. However, the research conducted in that department turned out to be very Polish-centric. Another interesting difference between Lviv University and Jagiellonian University was that the former was much more loyalist to the Austrian authorities and politically more conservative. It only slowly began to open up to liberal and socialist currents in the 1890s. Stanisław Tarnowski (1837–1917), a well-known literary historian and later rector was a characteristic figure for this school. His promotion at the university was connected with emphasizing the nobility of his aristocratic family, which had worked and sacrificed itself for Poland "for generations". The personnel policy of both the Jagiellonian University and the Academy of Arts and Sciences in Kraków was criticized as being guided by political considerations and the social contexts of the conservative circles of Kraków's elite rather than by substantive criteria. A visible manifestation of the dominant role of this elite in the city was the ceremonial funerals held for its members, including university professors.

The staffing policy at the University of Lviv, politically speaking, was much more liberal. Progressives and socialist academics were hired as early as the 1870s. This difference was also evident in the interpretations that prevailed in the various disciplines. Among the most characteristic were the historical sciences in which Kraków professors presented a critical view of the Polish-Lithuanian Commonwealth and were oriented toward descriptive political history. In Lviv, historians were more likely to pay attention to the unfavorable broader context of the geopolitical configuration that contributed to the downfall of the First Republic. However, as Surman contends, the memoirs of Kazimierz Nitsch do not describe a simple opposition between the two universities. This is because Nitsch was convinced that he was denied a professorship in Lviv in 1908 because of his socialist, anti-clerical, and pro-Ruthenian stance. However, he was made an assistant professor at the Jagiellonian University in 1910 and a full professor in Lviv in 1914. Jagiellonian University also witnessed attempts by politicians to break the dominant tendencies. For example, in 1906, the Governor of Galicia, Andrzej Potocki (1861–1908), supported Szymon Askenazy (1865–1935), a Polish Jew and liberal, as a candidate for a history professorship at Jagiellonian University, in contradistinction to the dominant conservative professorship.

2.5.5 Internationalization of Galician Universities

Interestingly, Surman reports that lecturers were transferred from other Hapsburg universities to Galician universities more frequently than they were to Czech universities. Some professors even came from Russian and German universities. Quite a few came from the Warsaw Main School, although there were also scholars from Kazan and St. Petersburg (exemplified by Jan Baudouin de Courtenay). In this respect, the Habsburg universities in Kraków, Lviv, and Prague were far much more international than those in Vienna or Graz, where only scholars from German-speaking institutions were employed. However, in the Second Polish Republic, the universities in the former Galicia became much more local, if not downright parochial, compared to the Austrian period, as Surman argued. This was due both to a reduction in international recruitment and an outflow of faculty to other regions of the newly formed Poland. One of the paradoxes was that the relatively high internationalization of Galician universities prior to 1918 was related to the fact that, at the end of the nineteenth century, they mostly recruited ethnic Poles. These candidates, however, came from different parts of Europe, from different universities, and from different theoretical and linguistic traditions. Surman sees this as one of the reasons for the success of the Warsaw-Lviv School of Analytical Philosophy. Much of its strength lay in the cooperation between scholars educated not only in the Habsburg Empire, but also in the German Empire and France (Surman 2018a: 269). Thus, an elite dispersed in several countries produced an intellectually diverse faculty, despite a nationally defined personnel policy. The University of Vienna, by contrast, although more restricted in this respect, as it only recruited from the German linguistic sphere, was a more politically tolerant place, open to scholars with widely differing views, although this often led to heated disputes. However that may be, after the Galician universities were taken over by the newly founded Polish state in 1918, they became the main providers of academic staff for most of the other universities in Poland. Sunman reports that in the new Polish state, faculty at most universities primarily came from "Galician institutions (44 percent), while 17 percent (186 people) came from the Vistula Land and a further 11 percent from the Russian Empire. For example, 34 scholars in the humanities and natural sciences in interwar Poland had previously taught at universities in the Russian Empire, with eleven professors from St. Petersburg and Kyiv. (…) In Warsaw, 30 percent of the instructors up to 1927 had previously taught in Cracow or Lviv; The dominance of Galician scholars at the universities in Poznań and Vilnius, as well as the private Catholic University in Lublin (*Katolicki Uniwersytet Lubelski*, KUL), was even greater, ranging from around 30 percent to 100 percent (…). This dominance is even more evident if one considers that these statistics include scholars in all posts and that Galician scholars were mostly full professors" (Surman 2018a: 259).

It can therefore be said that while the Russian Empire educated more of the Polish intelligentsia, particularly in its grammar schools, the Austrian Empire educated more Polish scholars for the Second Republic. This was the fruit of the strategy of fostering the development of minority cultures in the final decades of the Habsburg Empire. The opposite strategy was deployed in Russia, but it did not hinder national movements. This was because antagonizing successive generations of the intelligentsia only spurred them on to fight for national independence. The composition of the elite of the independent Second Republic can therefore be seen as an unintended consequence of, and a side effect of the interplay between, these two policies. German policy, particularly during World War I, also played an important role, but it was mostly Austria and Russia that shaped the elite of the fledgling state and gave it its political identity.

2.5.6 Image of Galicia as a Creative Periphery

Surman attempts to explain the role of Galician universities in the development of Polish culture, as well as their contribution to European science, by pointing to the paradoxical role of their peripheral location. He cites Lotman and argues: "Yuri (Juri) M. Lotman, for whom the periphery is a space of increased intellectual productivity because it lacks the homogenizing power of the center, thus enabling cross-boundary relations impossible in the center, provided a metatheory for such conceptions of circulation. Below I privilege Lotman's view over that put forward by Michel Foucault, for whom space was controlled by the center, while peripheries had only limited possibilities for innovation" (Surman 2018a: 16). However, this simple interpretation of the significance of the peripheral location of these universities (in relation to Vienna) seems inadequate as a universal theory. Any periphery, in order to be seen as creative, must also have adequate resources, and enjoy some degree of autonomy, as well as an element of relative diversity. In the Russian Empire, this condition was more or less fulfilled in many universities. This includes those that, despite being located in the center of the country, remained partly peripheral in terms of a more abstract dimension of location in relation to the dominant forces in the Russian field of power. However, the University of Warsaw turned out to be the one in which these conditions were fulfilled to the smallest extent. This made it impossible for the university to take advantage of its otherwise unique geopolitical and cultural location. It is obviously worth remembering that the University of Warsaw was not as conservative and uncreative in other disciplines as it was in the social sciences. The science departments often conducted advanced research, e.g. Prof. Mikhail Tsvet (1872–1919) almost won the Nobel Prize in chemistry for his discovery of chromatography. As for Galicia, there was curiously more creative tension at the University of Lviv than at the University of Kraków. The latter, although on the periphery of the Austrian Empire, adopted a loyalist strategy and also became a center for fairly conservative Polish thought. It can thus be said that it occupied a position in the field of higher education homologous to the dominant conservative positions in both the field of power of the Habsburg Empire and the Polish proto-field of power. The University of Lviv adopted a more autonomous position. It was possibly an indirect beneficiary of the cultural and ethnic diversity, and even the political tensions, of the region as well, as they provided creative impulses which stimulated the academic field.

2.6 Poland Under German Rule

Having discussed the dynamics of the Polish proto-field of power up until 1914, with a special focus on its dominant sector, viz. the Russian-dominated polity, I would like to devote some attention to World War I, and in particular, to the dynamics of the Polish polity under German rule. Jesse Kauffman's (2013) fascinating study provides insight into the dynamics of the short but consequential period of the German occupation of central Poland between 1915 and 1918. His study focuses on the person of the German Governor of Poland, Hans Hartwig von Beseler (1850–1921). Kauffman meticulously reconstructs not only the workings of the German administration under his command, but also his perception of Poland and Poles. What is particularly interesting is that von Beseler appears not to antipathetic toward Poles, but rather sees them as "immature". This might have reflected the general German view of the recently conquered Polish territories and the Polish elite. At one point, Kauffman quotes von Beseler's exclamation directed at a member of the Polish elite: "You're a bunch of poets!". In the mouth of the German military governor, this can be seen as articulating the perfect mixture of fascination and disdain. It could even be said that the concept of Poland as a "nation of poets" was a natural consequence of its being viewed as a satellite of Germany, governed largely autonomously by Polish intellectuals. The Germans might even have seen these intellectuals—the elite of the Polish intelligentsia—as creative, but by no means would they have seriously considered them a serious political, let alone economic, partner. Thus, the use of the term "poet" seems to express admiration for the erudition, imagination, and even courage of the Polish cultural elite, but at the same time, it connotes irresponsibility, a detachment from economic fundamentals, and a refusal to face up to the cold, hard facts of life. Von Beseler arguably implemented a program to transform Central and Eastern Europe according to Friedrich Naumann's concept of *Mitteleuropa*. It is telling that von Beseler considered Naumann's 1915 work on *Mitteleuropa* the best book on Poland ever. According to Kauffman, Max Weber spoke positively about Naumann's views. This is one of the many aspects of the curious contribution of classical sociological thought in ideologically supporting German activities on Polish territory aimed at completely subordinating the population to the consolidating German state. The Polish majority saw von Beseler's policy of qualified support for Polish autonomy as a poor camouflage for German conquest and exploitation. As is often pointed out, its original intention was to strengthen the German army with Polish recruits. For their part, conservatives and nationalists in Germany viewed von Beseler as being too conciliatory toward the Poles. Von Beseler, in turn, appealed to Berlin to ease the economic exploitation of Poland, as this threatened to increase anti-German sentiment.

A very interesting, although now mostly historical, aspect of German policy on Polish territory was Germany's rivalry with Austria. According to Kauffman, German policy was partly driven by an undisguised contempt for the Austrians and a strong reluctance to transfer any Polish territory to the Habsburgs. According to Kauffman, von Besseler felt a strong animosity toward Catholic Austria. This was characteristic of the Prussophilic German Protestant nationalism that attracted German elites in the era of unification. Therefore, he did everything possible to prevent Austrian influence on Polish territory. This also applied to the University of Warsaw, which officially opened on November 15, 1915. Although the initial plan to hire professors from Prussia was dropped, von Besler refused to allow professors to be recruited from Austria. He approved a curriculum that was ostensibly apolitical, i.e. which avoided any emphasis on Polish independence or German domination. One of its aims was to prepare the Polish elite for a partly independent, i.e. autonomous, Poland. The university was also tasked with educating the cadres of a new self-governing Poland, an elite that would shape Poland's legal and civic culture.

Another layer of the interface periphery of Poland is visible in the German–Austrian tension reconstructed by Kauffman. As mentioned earlier, the main aspect of this double periphery from 1915 was the confrontation between Germany and Russia. Although Prussia/Germany and Russia had been uncompromising and unrelenting in their suppression of the Polish independence movement for the previous half-century, in 1914, they began to outbid each other to buy Polish support with promises of supporting Polish independence. The Russians, as they appeared to be in a weaker position since losing their partition of Poland, began to make more generous promises (e.g. the Manifesto to the Polish Nation of Grand Duke Nicholas of August 14, 1914). German policy, particularly as coordinated by von Beseler as Governor-general of the *Generalgouvernement Warschau*, was also largely formulated under the pressure of having to compete with Russia for Poland with Russia, and as Kauffman shows, with Austria as well. As a result, when it formally declared its independence in November 1918, Poland relied in part on institutions created by the Germans, including institutions of higher learning such as the renewed Polish language University of Warsaw and the Warsaw Polytechnic. However, similar mechanisms in other parts of German-occupied Central Europe no longer necessarily worked in favor of the "Polish cause". General Erich Ludendorff (1865–1937), second-in-command to *Ober Ost* supreme commander General Paul von Hindenburg (1847–1934), supported pro-Lithuanian forces at the expense of Polish elites. Kauffman contends that this undermined the legitimacy of von Beseler's putatively pro-Polish policy.

Under German rule, there was a clear appreciation of the status of the bourgeoisie. The privileges of the former aristocratic elite were also recognized albeit de facto. The applicable electoral law in the 1916 Warsaw municipal council election also reflected that configuration of forces. It was based on six curias, the order of which also reflected their hierarchy, i.e. the voting power assigned to particular categories of voters. These were: (1) property owners, (2) business owners, (3) liberal professionals,

(4) small business owners and craftsmen, (5) housing taxpayers, (6) the rest. In September 1917, the self-governing system in the occupied Polish territories was supplemented by the Regency Council. This was a semi-independent and temporarily appointed highest authority (head of state) in Partitioned Poland and formed by Imperial Germany and Austria-Hungary. The Council was meant to hold office until a new monarch or Regent was appointed, as the German plan for Poland (and other protectorates) was to transform them into constitutional monarchies with kings related and loyal to Germany. The Regency Council, as a representative of the Polish elite, ruled the *Generalgouvernement* with Governor von Beseler in Warsaw. That entity can therefore be considered a German protectorate that derived its power from the bourgeois-aristocratic elite. In this situation, the transfer of power from the Regency Council to Józef Piłsudski (1867–1935) on November 11, 1918—the anniversary of which is formally celebrated in Poland as the day of regaining independence—can also be interpreted as a significant political revolution. The bourgeois-state elite, whose interests were partly protected by the empires controlling Poland before 1918, was supplanted by an intelligentsia elite that further democratized the country and won full independence.

2.7 Economic Development of the Polish Lands During the World War I

2.7.1 Last Decades of the Nineteenth Century

First, I wish to briefly examine the history of the economic development of Poland during the nineteenth century. Initially, none of the partitions showed any particular developmental dynamics. There was growth stimulated by the exploitation of coal deposits in Upper Silesia, and some development was observed in Galicia in connection with the discovery of oil deposits and the refining and distillation methods invented by Ignacy Łukasiewicz (1822–1882). Cottage industries, small factories, and arts and craft dominated in Greater Poland. There were two phases of industrialization in the Russian partition. The first was a decidedly statist attempt undertaken under the patronage of the authorities of the

Kingdom of Poland. The second came at the end of the nineteenth century and was much less planned, being primarily based on exploiting the location of the lands annexed by Russia on the western edge of the Empire (Koryś 2018). This location gave access to the gigantic Russian market and ensured proximity to Western Europe, especially Germany, which was an important source of capital and technology. That period of dynamic economic development, especially of that part of Poland in the Russian partition, can be interpreted in several ways. One decidedly negative interpretation emphasizes the unsustainability of economic growth and its structural effects (see e.g. Kochanowicz 2014). This could be interpreted as a short-lived bonus of being on a classic semi-periphery. Poland could then be seen as a beneficiary of the exploitation by the Western core of the eastern, mainly Russian, peripheries of the world system. Economic development was largely dependent on an influx of the Western capital (primarily German, French, and Belgian). The owners of that capital expected a faster return at a higher rate than could be obtained in the core, while the level of technological sophistication it financed was much lower than in their home countries. Moreover, the industrial plants located in the Russian partition mainly exported goods to the Empire that were of inferior quality to those they sold on Western markets. Thus, the Kingdom of Poland owed its dynamic growth in large part to the customs border between Russia and Germany, the maintenance of which would have been uncertain in the long run. The internal customs borders in the Russian Empire also played an important role; in particular, they were the main source of the now somewhat forgotten economic boom in Białystok in the late nineteenth century. The same processes, however, can be viewed more sympathetically. The economic boom could be portrayed as not so much the result of temporary gains from semi-peripherality of Polish lands under Russian control, but more an expression of the structural benefits accruing from interface peripherality. In this view, the economic development of the Russian Empire, including the Polish partition, was not as dependent on external factors as, e.g. the contemporary development of Poland. Russia had a good deal of capital and controlled a significant portion of its banking sector, as even Boris Kagarlitsky (2008) admits, despite his skepticism about the potential of Russian capitalism at the time. The Bank Handlowy in Warsaw is a case in point. Established in 1870, it soon grew to become the first bank in the Russian partition of Poland (Landau and Tomaszewski 1995). And there were several others, e.g. Wawelberg Bank (*Dom Bankowy Wawelberga*). This shows that financial capital was being accumulated on a large scale in Poland. These banks financed substantial production and infrastructure projects (e.g. the rapidly expanding rail network). Bank Handlowy had a large network of branches, mainly in the Russian Empire, but also in Germany, which may indicate that the Kingdom of Poland had slowly begun to benefit from being located between the two great economies of Russia and Germany (Żor 2011).

2.7.2 Development of Bourgeoisie

Economic growth during this period was spectacular and resulted in the strengthening of the power of the bourgeoisie, which, together with the landed gentry, began to form a distinct new elite of economic capital. The attempts of landowners to become financiers and industrialists, and the attempts of the rich bourgeoisie to become landowners, mainly through the purchase of prestigious estates or matrimonial alliances, were particularly observable. The ownership of capital in Poland, however, was very diverse. Much of it came from Germany and other Western countries, although some from Russia, and many of the local owners, especially Jews, did not identify as Polish. A significant portion of the Jewish elite did, however, become wealthy and gradually Polonized, although some assimilated into German or Russian culture, which alienated them from Polish culture and was often perceived negatively. The influx of Jews from central Russia (the Litvaks) also fueled anti-Semitism. In any case, Poland witnessed an unprecedented economic boom in the final decades of Russian rule. This resulted in the Polish bourgeoisie accumulating considerable economic capital. The increasingly powerful bourgeoisie in the Russian partition largely favored conservative and liberal forces. This could be seen as the gradual emergence of the classic opposition of the Western fields of power; an opposition that pitted economic capital and its elite against cultural capital and its elite. This process split the educated class into a relatively liberal faction that favored cooperation with, or even service to, the rising economic elite, and a faction oriented toward confrontation with, or at least autonomy from that elite. This second factor included progressive circles that were unsympathetic or even hostile to the bourgeoisie and the landowning class. However, this tension can be also seen as a continuum of attitudes and adopted positions. One aspect of their differentiation was manifested in the intelligentsia-led reform of the economic-capital faction; one which would marginalize the traditional noble and aristocratic elite, and replace its feudal ethos with modern meritocracy. Be that as it may, the radical intelligentsia's political focus of interests was shifting toward the social mobilization of the broader masses—on the one hand under national (nationalistic) demands and on the other under social (leftist) slogans. The tension between these two political poles of the intelligentsia was evident during, e.g. the revolution of 1905, which saw the mass political mobilization of broad social groups under the leadership of the intelligentsia. At the level of the Russian Empire, the 1905 revolution led to a significant liberalization, which in addition to improving the political position of the intelligentsia, signaled a gradual shift in the balance of power from the state imperial elite to the bourgeoisie.

The tension between the rising bourgeoisie (economic-capital elite) and the intelligentsia (cultural capital elite) had a complex and ambiguous nature. On the one hand there was a clear dimension of confrontation, manifested for example in 1905, while on the other, there was a very clear dimension of cooperation, the strength of which prevented a fully-fledged revolution in Poland. As already mentioned, as the Polish bourgeoisie gained in strength, it devoted substantial funds to the development of Polish culture. Without this contribution, the many intellectual myths central to the modern Polish national identity could not have been popularized. One illustration of the workings of this mechanism was the financing of intellectual magazines and novels, including Henryk Sienkiewicz's famous "Trilogy" (*Trylogia*), three unconnected historical novels of great myth-making power. As mentioned above, the Polish bourgeoisie also financed a number of private institutions, and this compensated for the lack of Russian state involvement in Polish higher education. These included foundations like the Mianowski Fund, which financed the education and research activities of many Poles at the world's best universities. It can therefore be said that the spectacular rise of the Polish intelligentsia was both a byproduct of the rise of the Polish bourgeoisie, which financed its infrastructure, and a result of the bourgeoisie collapse around 1917, since this wiped the slate of the dominant sectors of the Polish field of power for the intelligentsia. Thus the Polish intelligentsia can be seen as building a dual autonomy at the turn of the twentieth century. One dimension of that autonomy was defined in relation to the economic capital and its elite. This allowed the bourgeoisie-funded institutions to be relatively autonomous tools of the intelligentsia elite. The second dimension of an autonomous intelligentsia was built on political capital, and its elite, i.e. the state elites of the three empires, especially Austria-Hungary and Russia. This allowed the intelligentsia to make use of several institutions and other social structures funded and sustained by the imperial state administration. This obviously occurred to a greater extent in the Austrian partition, but even in the Russian Empire, institutions controlled by the intelligentsia played a considerable role with the support of state financing and legitimation.

2.7.3 Nationalist Movements and the Project of *Mitteleuropa*

The outbreak of World War I led the partitioning powers to openly support the national aspirations of each other's minorities. The interface periphery mechanism now began to work at maximum strength in the region. Germany was the most active in this field for the simple reason that it occupied a large part of Central Europe during the 1915 offensive. German support was particularly evident and, it seems, decisive for the success of the creation of the new state of Lithuania, whose independence was formally recognized by Germany in March 1918. The role of Germany was less obvious in the case of Latvia, where the local Germans, who constituted a significant part of the bourgeoisie, were far less eager to support the creation of a Latvian nation-state. There was also clear German support for Belarusian and (especially) Ukrainian national movements. The latter even resulted in the short-lived institutionalization of Ukrainian statehood as the Ukrainian People's Republic, which was de facto recognized by the Central Powers within the framework of the Brest-Litovsk Peace Treaty as early as February 1918. Nor should Germany's establishment of the Kingdom of Poland (1917–1918) be forgotten. This entity was administered by the Regency Council, whose members were mostly Polish aristocrats and wealthy bourgeoisie nominated by the Austrian and German emperors. The Council and the other central institutions of the Kingdom were limited in their independence, but the Kingdom enjoyed significantly greater autonomy than its Russian predecessor had prior to 1915. In any case, both the Russian and the German Kingdoms of Poland had technically separate administrative structures that were readily accessible to the independent state that emerged at the end of 1918. Thus, it was largely on the basis of this para-state, established as a result of Russia's attempt to outbid Germany in buying Polish support, that the foundations of the administration of the Second Republic were ironically laid under German supervision.

The dominant German strategy, after realizing the impossibility of directly controlling the vast Central and Eastern European territories taken from Russia in 1915–1917, was to create a network of dependent and weak states, so as to channel national aspirations and ensure administrative order. This vision is encapsulated in the *Mitteleuropa* plan of Friedrich Naumann (Naumann 1915). Naumann defined Germany's military aims on the Eastern Front in precisely these terms. This envisaged a much more low-key domination of the region than the later radical pan-Germanic vision of the Third Reich. However, the concept is now critically evaluated—even in Germany. One reason for this is that the term conjures up a vision of German domination that is far too stark (Le Rider 2008). It has been replaced with the terms Central Europe (*Zentraleuropa*) or Eastern Europe (*Osteuropa*), although the latter is less popular in Poland and its neighbors. This type of strategic thinking had its roots in the late nineteenth century, when Germany was pursuing fairly aggressive policies toward other regions of the Russian Empire as well. These involved supporting both national movements and ideological currents, which were seen as having the potential of weakening the Russian Empire. As Sean McMeekin points out, at the beginning of WWI, these took the form of, e.g. support for Zionism, which was intended to encourage soldiers of Jewish extraction to desert (McMeekin 2011). The failure of this plan induced the authorities in Berlin to support the communist movement. Either way, however, the Bolshevik Revolution was welcomed by Germany as weakening the Russian Empire and enabling Berlin to impose the onerous Treaty of Brest-Litovsk in March 1918. As McMeekin and Aviel Roshwald (Roshwald 2001) point out, Germany made similar attempts to foment national and ideological movements in Britain's MENA colonies, especially in the hinterland of the Ottoman Empire, e.g. Egypt and the Arabian Peninsula. Curiously enough, one of the ideologies supported was jihad. Muslims would hopefully direct their holy war against the British colonists. However, these plans did not bring the expected results. The British were more successful in directing the Arab nationalist movements against Ottoman interests and thus indirectly against German interests. However, the British also supported nationalist ideologies, including Zionism, in a similarly strategic manner.

2.7.4 Fall of the Economic Elite

Poland was also at risk of becoming a German or Russian, satellite after WWI, depending on which Empire emerged victorious. For Poland to have regained its fully independent statehood in 1918 was therefore an impressive feat and a highly favorable outcome—at least in the short term. It was, however, also a very costly outcome. One of the costs was that all Poles east of the border of the Second Republic had their property expropriated. During WWI, the economic systems that had functioned in the Polish partitions had been destroyed, including the industrial infrastructure in the Russian partition. At the same time, in addition to the collapse of the financial system wherever the Bolsheviks gained control, private owners, among whom were a large number of Poles, were completely expropriated with no compensation. This brought centuries of Polish proprietorship over a vast area to an abrupt halt, and undercut the economic foundations of the largest fortunes of the Polish landed gentry. This resulted in a revolutionary turn everywhere east of

Germany. The economic crisis that followed the war, to say nothing of the destruction the war itself had wreaked, along with German looting of the industrial infrastructure, and a revolutionary wave that was upending longstanding social relations, brought about a significant crisis among the slowly emerging bourgeoisie in some regions and its complete collapse in others. The bourgeoisie simply perished anywhere under Soviet control. The Polish-Soviet Treaty of Riga (1921) not only recognized the new eastern borders of Poland, but also the expropriation of the landowners in the Soviet sphere. Many of Poland's largest landowners lost any legal basis to reclaim property expropriated by Soviet Russia. Interestingly, this settlement was primarily lobbied for by the National Democratic representatives who were opposed to having large areas where ethnic Poles would be in a clear minority incorporated into the new Polish state. This was a natural consequence of their ideal of a homogenous Polish nation-state, and was typical of an intelligentsia that was markedly indifferent to the interests of the economic elites. Poland thereby renounced its influence, especially its economic influence, over a vast swathe of territory that had belonged to the First Republic. Capitalism and private property were preserved in the Second Republic, but the landowning class and the bourgeoisie were significantly weakened (Mich 2000). Landowners not only suffered as a result of the confiscations in the east, but also as a result of the progressing agrarian reforms and the growing crisis of the manor economy in Poland itself. Industry declined significantly as a result of being denied access to Russian markets, which had driven its dynamic development prior to WWI. Finally, the Great Depression of the 1930s hit Poland especially hard.

Many of the Polish communities which remained in the Soviet Union also suffered a tragic fate. In particular, the ethnic cleansing in Soviet Ukraine and Belarus in the 1930s almost completely annihilated the Polish minority in that enormous area. It is important to bear in mind that many regions and cities in Western Russia had sizeable Polish communities prior to 1914. For instance, Poles comprised approximately 10% of Kyiv's population at the turn of the twentieth century, and many of them belonged to the elite of the city. Polish institutions were also active throughout the Russian Empire, constituting a significant, if now forgotten, part of what could be Polish economic, cultural, and institutional resources (Zarycki 2013).

The consequences of the Bolshevik Revolution are marginalized in contemporary Polish historical narratives, dominated as these are by the Polish intelligentsia. This is especially the case concerning the enormous scale of confiscation of Polish property and the collapse of the Polish bourgeoisie. What can be seen as a great triumph of the Polish intelligentsia is effectively presented as a triumph of Polishness. In the Soviet Union, by analogy, the triumph of the revolutionary factions of the intelligentsia was presented as a victory of the people of the new state (the "Soviet people"), and even of all "progressive" humanity. The Polish-Bolshevik war of 1919–1920 is likewise presented in Polish historiography as a great and largely independent triumph of Polish arms over the eastern barbarians, which threatened the whole of European civilization.

In the meantime, the liberal-bourgeois camp mentioned above were among those who suffered most in the aftermath of WWI. This was because, as the Russian Empire had ceased to exist, and its liberal and bourgeois elites (the main political partner of the Polish progressive liberals), had been annihilated (often literally), along with the foundations of their status. Some of them were killed by the Bolsheviks in the first years of the Soviet rule, and all of them were stripped of their properties. The National Democratic (*Endecja*) camp, which had primarily relied on the Imperial regime for political dialogue, was also in retreat, although it had not been completely routed, as had the liberals. For Endecja, the Tsarist authorities were only a tactical ally. At the same time, they did not represent circles so strongly connected with the existing Russian social and economic system as was in the case of liberals who were related to the wealthy bourgeoisie. At the same time, their leader, Roman Dmowski (1864–1939), gained a significant political advantage during the rebirth of the Polish state by becoming the main representative of Polish interests before the Western powers. This Western recognition, which the Polish liberal camp, more loyal to republican Russia, did not manage to obtain, was an extremely important factor that tipped the balance of power in the new Second Republic against them.

The significance of WWI, and even more so of the Bolshevik Revolution, to Poland is therefore difficult to overestimate. These cataclysmic events ended the dynamic economic development associated with the First Globalization. They also marked the end of a period of accelerated accumulation of economic capital and foreshadowed a crisis for its economic elite. One ramification of these changes was they contributed to having the role of historical figures in the field of economic capital sidelined in Poland's mainstream historical narrative. The pantheon of Polish heroes was consequently peopled almost exclusively with political and military leaders, and intellectuals and artists, i.e. leaders of the field of socio-political and cultural capital. Piłsudski and Dmowski remain the dominant actors in the contemporary memory of the political history of the late nineteenth and early twentieth centuries. It is important to note that these figures operated mainly in the dimension of political and cultural logic. The Polish liberals of the turn of the century are fading into oblivion in the same way their economic interests were neglected and then disappeared. This includes the memory of their economic resources.

2.8 The Interwar Period

2.8.1 Roots and Phases of the Interwar Period

With the restoration of Poland's independence, its field of power became formally institutionalized. Many Polish communists either became Soviet citizens and emigrated to the USSR or conducted illegal activities in Poland while remaining in the Soviet sphere of influence. These options were not necessarily mutually exclusive, especially as the Soviet Union promoted a strongly internationalist ideology under the slogan of being the "Fatherland of the Proletariat". This internationalist ideology, and the policies derived from it, were progressively dismantled under Stalin, who pushed toward a Russian-dominated unitary model of Soviet culture. In any case, as already mentioned, the Polish Second Republic can be described as an intelligentsia republic. The intelligentsia gained its key role through institutional arrangements that gave it special privileges and through the weakening of its main rivals, viz. the landed gentry and the bourgeoisie. This weakening initially resulted from the economic catastrophe of World War I, the Bolshevik Revolution and the resulting expropriations, and the post-war economic crisis. Secondly, the Second Republic was economically crippled by the tariff wars with Germany and the USSR. The intelligentsia attained a dominant position in the field of power. At the bottom of its hierarchy, the intelligentsia gained privileged access to all the elite clerical positions and a network of universities and other cultural institutions that enjoyed relative autonomy in what was otherwise a relatively poor country. These quality universities and elite secondary schools were the key engines of the reproduction and domination of the intelligentsia elite in the Second Republic. However, the state was no longer able to build or support powerful corporate entities, capital markets, or banks that could compete internationally. The elite of economic capital inherited from the Russian Empire were impoverished and some of them lost all their assets. This led to many of the intellectuals and politicians related to their circles being marginalized. Many of them could not find a place in the new post-1918 Polish field of power. At the same time, despite the difficult economic situation in which it was founded, the Second Republic inherited universities in Lviv (Lwów), Kraków, and Warsaw in addition to the two it created. Moreover, a private university was founded in Lublin in 1918. This is discussed in more detail below.

The history of Poland in the interwar period can be divided into two main periods. The first was relatively democratic, whereas the second, which began with the May Coup of 1926, was more authoritarian. The evolution of Polish politics in this respect reflected the general European trend toward "strongman rule" in which increasingly nationalistic policies and ethnic homogenization were gaining in popularity. In Poland, the military elite played a major role in building an autocratic state, especially the officer corps and other members of the intelligentsia centered around Marshal Józef Piłsudski (1867–1935). This was essentially a faction of the post-nobility intelligentsia, and it was intent on subjugating the other factions of the field of power by monopolizing the political capital of the newborn state. This political bloc, called

"Sanacja" (meaning moral healing or renewal) could be compared to Stalin's creation of the nomenklatura. The Polish case involved transforming a particular faction of the intelligentsia, viz. the elite of cultural capital, into the state elite, i.e. the elite of political capital. This project, however, was limited in scope, lacking any strong ideological basis apart from the ideals of a strong state and an ethnically homogenous nation. Most importantly, it ultimately proved unsuccessful. This was primarily due to the outbreak of World War II, as a result of which the Polish state had ceased to exist by the end of 1939. Its remnants functioned in Poland as the Underground State and the Polish government-in-exile (which operated from London) and its agencies. The Soviet Union later created a separate Polish administration in exile. However, these structures were originally weak, and did not control any Polish territory. Once more, the intelligentsia played a dominant role, although this was a very different intelligentsia from the one that formed and supported "Sanacja" rule until September 1939.

2.8.2 Poland and Soviet Union—Differences and Similarities of Fields of Power

Nevertheless, the trajectory of the field of power of Soviet Russia was on a high level of abstraction homologous to that of Poland, but the scale of its transformations was significantly greater. First, analogous processes of marginalization of the elite of economic capital took place in both Russia and Poland. Obviously, capitalists were simply eradicated in Russia, as opposed to being marginalized in the field of power, as they were in Poland. However, it is nevertheless possible to talk about the two parallel intelligentsia revolutions which produced states in which the intelligentsia, or more precisely selected factions thereof, became the dominant classes. The symbols of these elites were Józef Piłsudski in Poland and Vladimir Ulyanov (Lenin) in Russia. The two indirectly collaborated in destroying the Russian Empire and its old elite, albeit for different ends. The most notable example of this was Piłsudski's refusal to assist Denikin at the request of the Bolsheviks (Mackiewicz 2009). The new Russian state, however, was built by an intelligentsia elite with a significantly different orientation from the Polish elite. In Russia, factions of the intelligentsia which were revolutionarily oriented and ethnically diverse built a totalitarian state, the essence of which could be described in terms of the hegemony of political capital. In short order, it not only destroyed the elites of economic capital, but also a major part of the former cultural elites. After WWII, it created a new intelligentsia strongly dependent on state privileges. This supports the thesis that, since the crucial moment of the October Revolution of 1917, which should be seen as the starting point of processes that were to span an extended timeframe, the structures of the Polish and Russian fields of power have been characterized by fundamental differences from the typical fields of power of Western European societies. Their specificity can be defined as the marginalization of the logic of economic capital and its elite. In Poland, the field of power and its logic were dominated by cultural capital, as this was the only way to reproduce a relatively stable elite. In Russia, political capital, underpinned by the mechanisms of a strong state, dominated. The nomenklatura of Soviet Russia continue to have a considerable capacity for multi-generational reproduction and are still able to subordinate the elites of economic and cultural capital decades after the collapse of the system that gave rise to them. A telling example of this dependence is the fate of the oligarchs, i.e. members of the economic-capital elite who tried to gain influence in the political field of contemporary Russia. In Poland, by contrast, economic capital has been largely owned or controlled by Western interests since 1918. This does not allow for the formation of strong domestic economic-capital elites capable of maintaining their status over the long term (Łazor 2019). The factions of the intelligentsia that dominated Russia were, of course, completely different. Russia was dominated by a radical leftist intelligentsia, while Poland was dominated by a nationally oriented intelligentsia with a broad spectrum of orientations from socialist to conservative to popular. At the end of the 1920s, as part of a broader European trend, both countries took an anti-democratic and statist turn. In the Soviet Union, it took the form of building an authoritarian totalitarian state in which most of the hitherto intellectual elite was exterminated. In Poland, on the other hand, the attempt to build a new state elite under the military elite, which took over in 1926, turned out to be much less radical. Most importantly, it was short-lived. In the long run, therefore, it did not disrupt the dominance of the intelligentsia. In Germany, per contra, the new ruling elite of the totalitarian state marginalized the intelligentsia, partly by destroying it physically. In contradistinction to Poland, and especially Russia, a significant part of the German bourgeoisie, particularly those families who owned large industrial companies, was largely unaffected (the obvious exception being Jewish families). The German bourgeoisie entered into a coalition with the new authorities and many German corporations prospered as a result. Moreover, after WWII, many wealthy German bourgeois families retained their positions in West Germany, while in Poland, what was left of the bourgeoisie was being eliminated. Thus, the gap between Western Europe, whose elite was economic, and the eastern periphery, now under communist rule, where the economic elite was disappearing, widened considerably after 1945.

Poland and Soviet Russia can therefore be considered intelligentsia states from 1918, as the intelligentsia turned out to be the dominant faction of their fields of power. However, whereas there was a kind of national intelligentsia republic, as well as a relatively democratic and multi-ethnic political system, in Poland from 1919 to 1926, a dictatorship of one faction of the intelligentsia was established in Russia from the outset. Initially, it tolerated the functioning, even in the field of power, of a number of other intelligentsia circles—especially those that explicitly dissociated themselves from the pre-revolutionary order and declared loyalty to the Soviet power. Members of the intelligentsia having a variety of leftist tendencies thus occupied numerous positions in the Soviet state administration, as well as in the universities, and cultural institutions. They were also given opportunities for cultural creativity in the individual republics of the USSR, where peculiar new nation-spaces were being built in the 1920s. The young, creative intelligentsia was intensively involved in the development of these new Soviet, but nonetheless national, cultures. This period saw the development of many spectacular cultural trends, including the avant-garde and modernism, as well as unique innovations in architecture, art, and science. All this, however, was stifled in the late 1920s and ended in the 1930s with a tragic finale. A large part of the intelligentsia elite perished in the Great Purge of 1937–1938. Many others were imprisoned or marginalized. These dramatic developments offer a glimpse of an important aspect of the mutual dependence of these two fields of power. This is because the Great Purge involved the murder of the major part of the Polish intelligentsia in the Soviet Union and most of the members of the Polish Communist Party. Most of its members who survived that cruel purge only did so because they happened to be imprisoned in Poland. Ultimately, a uniform Stalinist nomenklatura class was formed, and the field of power was absorbed into the state apparatus, as no factions not aligned with the party line were tolerated.

The field of power in the sense of a differentiated social space emerged again in the USSR only after the death of Stalin. The nomenklatura became somewhat differentiated, and alongside it, the intelligentsia, understood as the elite of cultural capital functioning in relatively autonomous social spaces, began to reemerge. In the USSR, that space was predominantly created by scientific institutes, cultural institutions, and other organizations. This allowed for an exchange of ideas which was not fully controlled by the state apparatus. In this way, independent intellectual, cultural life was to some extent revived, and the intelligentsia elite, although in dominant positions, re-entered the wider field of power after Khrushchev's denunciation of Stalin in 1956. It should be noted, however, that the Soviet intelligentsia that emerged in the 1950s was a new social group, a product of Soviet universities who rediscovered the legacies of the pre-revolutionary Russian intelligentsia in a new context. In contrast, the Polish intelligentsia, despite the revolution of 1917 which also impacted Poland, and the losses inflicted by both world wars, has maintained much more continuity. This primarily involved the stable reproduction of identities and other resources.

2.8.3 Intelligentsia Domination in the Field of Power

But to return to interwar Poland. As mentioned earlier, in its field of power, the dominant positions were held by the intelligentsia of various orientations. This included factions of the petty bourgeoisie that had been partly transformed into intelligentsia. The aristocracy and the rich landed gentry were still present, however, as dominant actors who were similarly dependent on parties controlled by the intelligentsia to represent their interests. The dominant intelligentsia occupied a long continuum of political attitudes. Their main field of contention was how the country should be organized culturally. At one extreme were the communists, led by the Communist Party of Poland (KPP), for whom the future of Poland lay in joining the USSR as a constituent republic, and which was declared illegal in 1919. At the right end of the spectrum, there was National Democracy, which was inspired by fascism and whose vision for Poland was a homogenous nation-state. In this vision, the Catholic Church was to become the main pillar supporting Polish statehood and cultural homogenization. Between these two extremes, there were numerous factions of the moderate left led by the Polish Socialist Party (PPS). Closer to the right were Christian Democrats and a congeries of peasant parties, which, however, were divided internally into more conservative and progressive factions. A not insignificant force, especially in the first period of the Second Polish Republic, were parties that represented the country's many ethno-religious minorities, especially Ukrainians, Germans, and Jews. The Jewish movement was divided into a range of orientations, from conservative to socialist and Zionist. Interestingly, the coup of 1926 was supported by a number of parties, including most of the left wing (KPP and PPS) as well as some peasant parties and some national minorities. The legitimate government was defended by National Democracy, the Christian Democrats, and part of the peasant movement. A large part of the public supported Piłsudski because they were frustrated with what was a very ineffectual and highly unstable parliament, with its short-lived and often complex and unwieldy coalitions. The Sanacja governments after 1926, although supported by the left, began to slowly implement the National Democrats' program and adopt their right-wing platform. This entailed strengthening the nation-state at the expense of cultural diversity and the rights of national minorities. The left was increasingly marginalized, and the role of the military elite was increasingly strengthened and made more and more prominent. It is worth mentioning, however, that a large part of the officer corps in the interwar period came from the intelligentsia, and was often aristocratic in origin.

2.8.4 Higher Education in the Interwar Poland

I now wish to briefly discuss the dynamics of the development of the university system in the Second Republic. In 1919, two new public universities were established: the University of Poznań (now Adam Mickiewicz University) and the University of Vilnius (Wilno), which was named after Stefan Batory, the king of Poland. Both drew on the earlier traditions of higher education in their respective cities, although only Vilnius had a serious historical university heritage. The leading institution was the Vilnius Academy, which had operated between 1579 and 1832, initially as a Jesuit college, and then from 1803 as a fully-fledged Russian university. It was founded, like the University of Warsaw, by Tsar Alexander I. As already mentioned, however, it was closed in the repression that followed in the wake of the November Uprising. The University of Lublin was founded in 1918. It was renamed the Catholic University of Lublin (KUL) in 1928, and the John Paul II Catholic University of Lublin in 2005. Interestingly, it was founded on the initiative of the rector of the Catholic Clerical Academy in St. Petersburg, Father Idzi Radziszewski (1871–1922). He collected funds for the establishment of a Catholic university in Poland, primarily among the Polish community in St. Petersburg. The university was thus established partly to replace the St. Petersburg Clerical Academy, which had been liquidated by the Bolshevik authorities and whose traditions dated back to 1842. The main founder of the new university was the Polish industrialist Karol Jaroszyński (1878–1929). In 1917, Jaroszyński was an extremely wealthy financier who operated throughout the entire Russian Empire, controlling several large banks and a number of large enterprises. The Catholic University of Lublin, which has always been a quintessentially "Polish" university, not least on account of its Catholic orientation, is in a way another product of the Russian Empire, or more precisely, of the resources accumulated by its elite. For a long time, it was the only recognized private university in Poland, and like many other Polish institutions of learning, it was founded with the funds of the extremely wealthy Polish bourgeoisie operating in Russia. Jaroszyński's fate is very telling here. While he was funding the university, Jaroszyński, possibly the richest person in Poland, was also losing most of his capital. In the end, he was unable to honor his promise of financial support. He spent the interwar years in poverty and died in 1929. His fate is shared by many of the Polish bourgeoisie, and especially financiers, who only had a shadow of their former power and social status in the interwar period.

As for the evolution of university legislation during the interwar period, a very liberal law was passed in 1920 which largely sanctioned the status quo at the universities of Lviv, Kraków, and Warsaw. This included their autonomy, self-governance, and the key role of university chairs held by the all-powerful senior professor. This legislation, adopted in 1920, was drafted by professors at Jagiellonian University, for whom the Humboldtian university was the model. Professors enjoyed virtually unlimited autonomy under their new law. In 1933, another statute was passed as part of the Sanacja government's attempt to control academia. Under the new law, universities had limited authority, and the ministerial authorities had a say in the organization of internal university structures. The election of the rector and other key decisions were now subject to government approval. Not surprisingly, this law was met with resistance from the academic community and attracted protests.

2.9 The Early Post-War Period

2.9.1 The Intelligentsia as a Hegemon in the New Field of Power

The Second World War was an extremely bloody and tragic episode in Polish history, but in some dimensions, the changes it brought were not as revolutionary for the Polish field of power as those of the First World War. Admittedly, the Polish state had been completely and brutally destroyed in 1939–1945, and the country had never witnessed death, destruction, repression, plunder, and loss of territory on anything like this scale. However, as the Second Republic, for all Sanacja's ambitions, was not all that strong, this was not a period of complete discontinuity. Poland's pre-war economic system had all but stopped functioning and its economic elite were almost completely eliminated. Moreover, this process did not cease with the armistice, but continued well into the

1950s until it was completed. It can be argued, however, that this was not as revolutionary a change as that which occurred in 1917–1921, as the pre-war Polish economy was weak and its economic elite were not all that prominent in the Polish field of power. The Polish intelligentsia were mercilessly persecuted during WWII and its aftermath. Some were executed *en masse*. Katyń (near Smolensk), Palmiry (near Warsaw), and Ponary (near Vilnius) are among the most symbolic places of mass killings of the Polish intelligentsia. The bulk of the Jewish intelligentsia in Poland was also exterminated, although some survived in the USSR. However, these persecutions were unable to weaken the structurally dominant position of the intelligentsia vis-à-vis other elites, which suffered even more severely in the structural sense. In particular, WWII deprived the already weak landowning and bourgeois elites of whatever material resources and political influence they still had. The extermination of Polish Jewry was another blow to the Polish bourgeoisie, as the Jewish elite constituted a sizeable portion of the capital-owning class prior to 1939. These brutal killings of the Polish intelligentsia, with the Warsaw Uprising of 1944 at the forefront, have become milestones in its legendary victimization and have served to legitimize its symbolic domination. In the conservative narrative, this decimation of the intelligentsia elite constituted a major crisis. Some have gone so far as to argue that it marked its downfall (e.g. Gella 2001). It can be argued, however, that in the long run, these tragedies did not decisively weaken the intelligentsia, although they obviously significantly altered the power relations within its elite and opened new channels of inclusion. The Stalinist period saw the formation of a new party-state elite; one which could be seen as a key opponent of the intelligentsia, viz. the nomenklatura. However, it too drew much of its membership from the intelligentsia, especially initially. This is particularly true of those factions which mostly comprised previously marginalized leftist and communist circles. Although the stated aim of this new elite was to create a "state of workers and peasants", the status of the intelligentsia was elevated as a result of having the category of "working intelligentsia" incorporated in the official social structure of the People's Republic of Poland (as the new incarnation of the Polish statehood was officially named in 1952). Over time, part of this elite evolved in a technocratic direction, and had become the dominant elite of the political field of power by the 1970s, although its position in the wider field of power was far from hegemonic. Some remained a purely political elite (mainly party "apparatchiks" who did not have excess cultural capital), while others left the nomenklatura, particularly after 1968, and moved toward an autonomous intelligentsia field, which over time shifted toward a dominant position within the broader field of power. A significant strengthening of the left wing or liberal intelligentsia faction coincided with the emergence of the "Solidarity" trade union in 1980 and the growing influence of the Catholic Church over the course of 1970s. By the end of the 1970s, a coalition of liberal and conservative members of the intelligentsia elite had formed, and by 1989, had attained both formal and (especially) symbolic and structural domination.

Thus, it can be argued that with the collapse of the People's Republic of Poland (PRL), the intelligentsia not only became a hegemonic class but for a certain period, mostly in the 1990s, a significant portion of political power was transferred to some of its elite members. This period was short-lived, but very important. The marginalization of the heirs of the nomenklatura by the intelligentsia elite ensured that no clearly formed and self-reproducing political class of the nomenklatura type endured in Poland. The role of the political elite has instead been taken, especially after 1989, by actors from different sectors of the field of power, but they have not usually retained their positions for long. Polish political elites can therefore be seen to have a strongly dynamic nature but a low degree of self-reproduction. This stands in stark contrast to the situation in the Soviet Union and now in Russia. There, the political elite, i.e. the nomenklatura, stabilized and became structurally entrenched after the Stalinist period, and within a well-defined institutional structure, managed to activate effective mechanisms of multi-generational reproduction (Kryshtanovskaya and White 1996; Wasilewski 1995). This elite did not subordinate itself to the new economic elite (oligarchs) or the cultural elite (intelligentsia) after 1991. The intelligentsia was given a few limited political roles in the late perestroika period of the 1980s, but by the mid-1990s, it had been almost completely ousted from them. Similarly, the slowly consolidating group of Russian oligarchs aspiring to political independence in the 1990s was broken up, and their successors were subordinated to the political elite.

Thus, as discussed earlier, Poland and Russia exhibited two very different modes of peripherally compensating (cultural and political, respectively) for their dependence on the core of the world system. Consequently, the intelligentsia elite enjoys a dominant role in Poland, while the field of power in Russia is dominated by the political elite, i.e. the redefined heirs of the old nomenklatura. The two countries have also adopted very different strategies to cope with the economic domination of the Western core. Poland has completely opened up its economy to Western capital while preserving the intelligentsia's sovereignty over the cultural and political spheres. In Russia, by contrast, the political elite has retained full control of a relatively strong state as well as the most profitable and strategic sectors of the economy. These sectors underpin the strength of the dominant political elite and are its primary resource, compensating for the country's peripherality in relation to the Western core.

2.9.2 Political Landscape of the Communist Poland

I now wish to draw a more detailed picture of the configuration of the field of politics, as well as the broader field of power, in communist and post-communist Poland, and link them to the configuration of the social sciences field. The communist order, established in Poland in 1944–1945, was firmly consolidated around 1948. The unification of the Polish Socialist Party (*Polska Partia Socjalistyczna*, PPS) and the Polish Workers Party (*Polska Partia Robotnicza*, PPR) to form the Polish United Workers Party (*Polska Zjednoczona Partia Robotnicza*, PZPR), which was the Polish equivalent of the communist party in other Soviet bloc countries, was a defining moment. This marked the beginning of Stalinism in Poland, which lasted until the "thaw" that began in October 1956 (hence the name "Polish October"). The change was symbolized by replacing Bolesław Bierut (1892–1956), a staunch Stalinist, with Władysław Gomułka (1905–1982), a more independent and nationally oriented communist, as the First Secretary of the PZPR (i.e. the facto head of state). The first months of the thaw brought an unprecedented degree of liberalization as the communist party remained preoccupied with its own factional power plays. Some restrictions were reintroduced in 1957, after the situation had stabilized, although in a relatively liberal configuration by communist standards. However, this belied increasing tension in the Polish field of power, which finally erupted in a massive power struggle in March 1968. The outcome proved to be the turning point for communist Poland. Gomułka was replaced as first secretary by Edward Gierek (1913–2001), who ruled the country until he was brought down by a wave of strikes in 1980. Internal conflicts within the communist party (mostly related to the economic crisis) during Stanisław Kania's (1927–2020) brief term as first secretary gave rise to a second brief period of liberalization, sometimes known as the "Solidarity Carnival", due to the independent Solidarity trade union being the key legal institution opposing communist rule. This liberalization came to an abrupt halt when Gen. Wojciech Jaruzelski (1923–2014), replaced Kania and imposed martial law in December 1981. Solidarity and its elite now constituted an informal and partly underground opposition. Jaruzelski presided over a period of hardline Party rule, but was unable to reverse the continuing decline of the system, which began to be liberalized as a desperate last resort around 1988, before it permanently collapsed in 1989. A degree of cyclical dynamics can therefore be observed in the communist system, although there is also a noticeable economic and social developmental pattern from 1945 until 1968, followed by a steady decline in most respects from the late 1970s until 1989. The post-communist period can similarly be divided into two periods. The first, between 1989 and 2005, can be seen as a period of enthusiastically embracing the Western economic and political systems. This culminated with Poland joining the EU in 2004. The second period began with the first accession to power of the conservative and Euroskeptical elite, which continues to rule the country. This second phase of post-communism is marked by a division of the political field (and other fields) into opponents and supporters of liberal European integration.

2.9.3 The Stalinist Period

The early phase of the development of the system, i.e. the late 1940s, is now examined in more detail. By 1948, all political opposition to communist rule had been crushed and the regime was firmly entrenched.

The main political opposition in the early post-war years was the Polish Peasants' Party (*Polskie Stronictwo Ludowe*, PSL). However, it was quickly marginalized by having election and referenda results falsified and its members and supporters arrested, intimidated, and harassed. Repressions against the anti-communist intelligentsia were also carried directly by the Soviet special forces, which operated in Poland at the time. Conservative (i.e. non-communist) politicians and activists were persecuted. Some were imprisoned and even killed. As far as the academic field was concerned, anyone whose loyalty to the Communist Party was questionable was marginalized or even dismissed. In short, all pre-war non-communist political parties were dissolved and their social elites marginalized and dispossessed. Nationalizing large estates, industry, and other economic and financial assets (without compensating the owners) was a particularly important part of the communist program. This resulted in the annihilation of the landowner elite and the bourgeoisie (wealthy and petty). Even Western capitalists had their property nationalized, so they and their managers also disappeared from the social landscape of the Polish elite. The remnants of these classes and strata joined the ranks of the intelligentsia, which was now considerably widened and differentiated as a result of the diverse groups it had co-opted. The political scene was dominated by hardline Stalinists, while academic institutions were, at least institutionally, controlled by party loyalists.

One of the key cleavages within the intelligentsia was between its "older" and better-educated members and new, freshly educated members with working-class and peasant backgrounds. Another was between those who supported the new regime, based as it was on Soviet domination, and those who opposed it. The latter group was further split between radicals, who joined the underground anti-communist guerillas, and moderates, i.e. members who tacitly accepted the new order, but were reluctant to join the PZPR. During the entire communist period, the Catholic Church, which remained relatively independent of the party-state, was one of the main institutions that sided with those factions of the intelligentsia that had reservations about fully embracing the communist order. These two dimensions remained partly independent, even if the official ideology of the ruling party held that the pre-war elite had been replaced by a new elite of predominantly working-class and peasant origins. In reality, the new elite was composed of both old and newly educated intelligentsia members, as well as freshly promoted party apparatchiks.

2.9.4 Higher Education in Stalinist Poland

Of crucial importance here is that communist Poland began producing newly educated cadres immediately after it was founded. First, it quickly revived the universities it had inherited from pre-war Poland, viz. those of Warsaw, Kraków, and Poznań. The University of Warsaw was the most difficult to revive, as the city had been completely destroyed. Many of its faculties and staff were initially moved to Łódź, but the University of Warsaw was eventually revived and has since remained the largest university in Poland. Four new universities were additionally established at the end of 1944. The first was the University of Łódź, which was to become a model "people's" university. Agata Zysiak describes the history of the University of Łódź, which thwarted the political plans laid out for it by becoming a typical intelligentsia-dominated university (Zysiak 2016). In other words, it soon fell under the sway of the old intelligentsia academics who had educated successive generations of intelligentsia elite. The other two new universities, in Toruń and Wrocław (pre-war Breslau, Lower Silesia, Germany), replaced the universities of Vilnius and Lviv (Polish: Wilno and Lwów), which lay in territory annexed by the Soviet Union in 1939 (with the new border being formally recognized by Poland in 1945). Many of the staff of these former Polish universities moved to Toruń and Wrocław, where they found employment at the newly established universities. In Wrocław, whose population was composed almost entirely of immigrants after 1945, there was considerable continuity with the heritage of the University of Lwów, not only by virtue of the staff, but of the entire intelligentsia elite. In 1945, a second university was established in Lublin. It was named after Maria Curie-Skłodowska (Uniwersytet Marii Curie-Skłodowskiej, UMCS). The main motivation for its establishment was to neutralize the influence of the Catholic University of Lublin (KUL). Although Lublin is only a medium-sized Polish city, it was the only one that had two universities during the entire communist period.

At the same time, several new institutions of higher education were established, including some directly subordinated to the communist party. One of the most prominent of these was the Scientist Training Institute (*Instytut Kształcenia Kadr Naukowych*, IKKN), established in 1950 by Adam Schaff (1913–2006) on the example of the Moscow Institute of "Red Professors". Schaff also served as rector. Its teachers included such notable figures as economists Włodzimierz Brus (1921–2007), Bronisław Minc (1913–2004), and Edward Lipiński (1888–1986). The biography of Adam Schaff is definitely worth a brief digression. Schaff was born into a gradually Polonizing Jewish bourgeois family in Lviv in 1913. He graduated from the University of Lwów in 1935 with majors in law and political economy. He was a member of the illegal Communist Party of Poland (KPP), and in 1941–1948, lived in the Soviet Union, where he defended his doctoral dissertation and habilitation in philosophy at the University of Moscow. When he returned to Poland, he joined the central committee of the PZPR and became prominent in Stalinizing Polish academia. Among his numerous posts was the chair of the Institute of Philosophy at the University of Warsaw (1952–1953). Several IKKN students later became leading economists, e.g. Tadeusz Kowalik (1926–2012) and Kazimierz Łaski (1921–2015). The Institute was also supposed to train lawyers, as well as loyal officials, and experts in other disciplines (Dębska 2019: 101). The Institute became less prominent after 1956 (Czarny 2015). In 1954, the Institute was transformed into the Institute of Social Sciences of the PZPR Central Committee (*Instytut Nauk Społecznych przy KC PZPR* or INS), and in 1959, into the Higher School of Social Sciences of the PZPR Central Committee (*Wyższa Szkoła Nauk Społecznych przy KC PZPR,* or WSNS), before being closed in May 1968. This decision signaled a change in the system of educating party cadres. There would no longer be an institution dedicated to training party cadres and administrators run directly by the PZPR. In 1968–1971, WSNS functions were taken over by the Central Party School at the PZPR Central Committee. In 1971, this was replaced by the Higher School of Social Sciences of the Central Committee of the PZPR (WSNS). In 1974, the Institute of Fundamental Problems of Marxism-Leninism (*Instytut Podstawowych Problemów Marksizmu Lenininzmu*, IPPM-L) was established as a department of the Central Committee. In 1984, the WSNS and IPPM-L were merged to form the Academy of Social Sciences of the PZPR (*Akademia Nauk Społecznych*, ANS). This was abolished in 1989, and most of its staff were transferred to conventional universities (Degen and Hübner 2006). The creation of a "party science", one directly subordinate to the authorities, was thus neither easy nor successful. The academic prestige of institutions under party tutelage was low, especially after 1956, and their political subordination exposed them to considerable instability, best exemplified by the history of institutional change described above. It is therefore safe to say that relatively autonomous learning, fostered by classical universities, proved indispensable to the communist authorities. This autonomy, especially of the social sciences and the humanities, was obviously a necessary precondition for the reproduction of the intelligentsia. It could even be argued that the intelligentsia succeeded in using the communist state for its own advancement, particularly when viewed from the perspective of its long-term reproduction. This thesis seems particularly defensible with respect to the 1960s, when the institutional system of science and higher education created by the communist state gained significant autonomy and the political elite's control over it weakened. This was due to the liberalization of the regime, the internal divisions within its power elite (which lasted until 1968), and to some very senior PZPR positions being occupied by prominent members of the intelligentsia. The best examples of this are Adam Schaff (discussed above) and Stefan Żółkiewski, who thereby raised the status of the academic field.

Far-reaching changes were nevertheless made to the academic system during the Stalinist period. For one thing, academic autonomy was severely restricted, especially in fields that were of potential ideological importance to the communist authorities. Academia and its elites were assumed to be subordinate to the communist state and its elite, i.e. the nomenklatura. The reforms of the social sciences that were made from 1949 onward were therefore quite radical and involved significant personnel changes. Within particular scientific institutions, politically loyal persons were promoted, even though many of them had modest qualifications and experience. This was especially true of managerial positions. Politically suspect persons were dismissed or marginalized, although they were often allowed to conduct research out of the public eye. Sometimes such people were removed from teaching for fear that they might have a negative influence on students. Many renowned scholars whose views and backgrounds were considered conservative, e.g. the philosophers Władysław Tatarkiewicz (1886–1980), Henryk Elzenberg (1887–1967), and the sociologist Jan Stanisław Bystroń (1892–1964), were dismissed. The famous philosopher Roman Ingarden (1893–1970) was banned from teaching, as were literary scholars such as Juliusz Kleiner (1886–1957), Stanisław Pigoń (1885–1968), and Stefania Skwarczyńska (1902–1988). Nothing less than strict adherence to the PZPR approach to philosophy and science was demanded. As for institutional reforms, there were only three types of philosophy chairs at universities in 1950: history of philosophy; logic; and dialectical materialism, which meant, among other things, the formal liquidation of sociology. A symbolic moment of consolidation of the communist grip on Polish science[3] was the First Congress of Polish Science, held at the end of June 1951. Its final resolution called on the government to establish the Polish Academy of Sciences. The Polish Academy of Sciences (*Polska Akademia Nauk*, PAN) was duly founded the same year. The Academy has proved to be a stable institution and it is still in operation. It was established on the basis of two institutions with an impressive historical heritage: the Polish Academy of Learning (Polska Akademia Umiejętności, PAU); and the Warsaw Scientific Society (*Towarzystwo Naukowe Warszawskie*, TNW). Both contributed not only their scientific traditions, but also their property. The Polish Academy of Sciences was modeled on the Academy of Sciences of the USSR. As such, it was structured so as to allow for more effective state control of the academic field. The years 1951–1952 also saw major changes in the organization of science and teaching. The degrees of doctor and assistant professor were abolished, and, following the Soviet model, the degrees of candidate of sciences, which qualified graduates for associate professorships,

[3] Unless otherwise indicated, this work uses the term "science" in its broad sense to refer to the study and/or advancement of a body or knowledge. It does not denote the physical sciences.

and of doctor of sciences were introduced. Three-year courses were introduced to accelerate the education of primary school and junior high school teachers, and five-year courses (leading to a master's degree) were introduced for high school teachers.

2.9.5 Duality of Views on Stalinism

In this context, it is worth noting at least two perspectives from which the communist period, and Stalinism in particular, is viewed today. On the one hand, there is a very critical narrative that presents the Stalinist reforms of science as extremely harmful and as having destroyed much of the legacy of Polish culture and broken up the scientific community. An example of this is the interpretation of Dorota Degen and Piotr Hübner. As they put it, the authorities "in the sphere of humanism, the humanities, and the humanistic sciences were actually heading backwards towards a new barbarism" (Degen and Hübner 2006: 15). Degen and Hubner characteristically go on to stress that the new system of science did not allow for "a humanistic foundation, i.e. a system of values". In its place were "Marxist-Leninist dogmas" (Degen and Hübner 2006: 15). Opposed to this is the image of the communist period, including the Stalinist years, as a heavy-handed and ham-fisted—but nevertheless important—time, as it accelerated the modernization of Polish academia and more fully developed its potential. The reform and expansion of the university system, together with the construction of new institutions, with the Polish Academy of Sciences being the largest, all contributed to the modernization of Polish science. These measures constituted a significant institutional strengthening, and provided for universal access to higher education. As is shown in the chapter devoted to Polish linguistics, the Stalinist period was not necessarily a time during which Polish culture was destroyed, as many conservatives would have us believe. It can alternatively be seen as a period of reconstructing its paradigms and attempting to use them to legitimize the communist project.

This duality of views on the communist period also concerns the extent to which it changed the position of the intelligentsia. First, there is an anti-communist position, which posits that the communist period, especially its initial phase, which included the Stalinist reforms of the scientific system, was a period of destruction and persecution of the intelligentsia. The communist system is anti-intelligentsia in this perspective, as many members of the intelligentsia, especially those with anti-communist leanings, were imprisoned or persecuted. As mentioned above, some even lost their lives. Moreover the communists imposed their ideology and priorities on the intelligentsia elite. This anti-intelligentsia nature of the communist power was almost certainly a major part of its early plans. It was certainly the case with Stalin in the USSR, where there was a great purge, followed by the domination of the Party elite, which later stabilized in multi-generational reproduction as the nomenklatura. The opposing view sees the communists, and more especially the new elites they educated, as assuming intelligentsia identities, pursuing the intelligentsia calling, and adopting intelligentsia priorities by developing science. This second perspective, which recognizes the strength of the intelligentsia and the structural benefits it derived from the communist science reforms, obviously does not claim that these results were intentional when the communist system was first imposed on Poland.

As several authors have pointed out, the communists undeniably intended to radically reform, or even partly reinvent, higher education and research institutions in Poland when they came to power in 1945 (e.g. Connelly 2000; Behr 2017, 2021; Zysiak 2016). The new model of academia was designed to educate a new loyal intellectual elite, who would study, research and teach all academic disciplines through the lens of Marxism–Leninism. However, this project proved difficult to implement. First, there was the hegemony of the old intelligentsia elite, which could not be eliminated completely as it was in neighboring countries. The recruitment and promotion policies of the communists gave preference to those whose positions were avowedly left-wing (and preferably communist), but even among those academics mostly educated in pre-war Poland, class identities proved stronger than political views. Most social science disciplines were dominated, at least intellectually, by members of the old intelligentsia families, which had been respected for generations for their cultural capital assets. Even young students with peasant or working-class backgrounds were usually inculcated with the traditional viewpoints and worldviews of the intelligentsia elite at university.

2.10 The Liberal Period (1956–1968)

2.10.1 The Thaw (October 1956)

The Stalinist order ended suddenly in 1956. The communist regime in Poland liberalized radically in stark contrast to the trajectory of Hungary. Soviet military intervention was also being considered for Poland, but for reasons that are difficult to fathom, this option was not pursued. Possible reasons include Chinese threats against the Soviets and the preparedness of Polish army factions to fight Soviet troops. In any case, in October 1956, Władysław Gomułka was elected PZPR secretary, while Marshal Konstantin Rokossovsky (1896–1968), a Warsaw-born Soviet WWII hero of Polish descent who had served as minister of defense since 1949, was dismissed and repatriated. Khrushchev traveled to Warsaw, and after tense talks, approved the new PZPR Central Committee. This allowed for a relaxation of Soviet control of the party. Rapid de-Stalinization followed, and a significant group of Soviet officers and "advisers", who had been performing managerial functions in every government department, as well as the armed forces, left Poland. In addition, about 30,000 Poles who had been detained in the USSR after the war were able to return to Poland.

It was crucial from the standpoint of the field of power that an "enlightened", more liberal faction of the communist elite had gained the ascendant in the political field and the broader field of power. This faction ruled the country until 1968, although it was confronted by an increasingly assertive internal opposition. Numerous liberal leftist intelligentsia members rose to senior party positions, often becoming key figures in the academic field as well. In addition to Adam Schaff (discussed above), literary scholar Stefan Żółkiewski (1911–1991) and sociologist Julian Hochfeld (1911–1966), were members of the PZPR Central Committee. Schaff and Żółkiewski were heads of institutes at the Polish Academy of Sciences, while Hochfeld was the director of the Polish Institute of International Affairs (1957–1960). Stanisław Leszczycki (1907–1996) enjoyed a similarly privileged position in geography and was a director of the Institute of Geography of the Polish Academy of Sciences, although he was never a member of the PZPR Central Committee (Zarycki 2020b). Interestingly, the liberal intelligentsia camp within the field of power was largely formed by people who had only recently been ardent Stalinists. However, given the context of the political change, they quickly moved into liberal positions and were joined by many other intelligentsia members who had thitherto not been politically involved. Thus, for example, there was a huge staff turnover at the Scientist Training Institute after 1956, with Adam Schaff and Włodzimierz Brus (1921–2007) being among those dismissed (but eventually moved into "revisionist" positions). The academic elites, especially those who had belonged to the pre-war intelligentsia (and had therefore had an academically rigorous and non-ideological education, even if they had abided by the rules of the game imposed by the ruling Communist Party during the Stalinist period, availed themselves of the opportunities that opened up after 1956 to adopt a more liberal and cosmopolitan stance. In the academic field, this often meant repudiating Soviet science as a point of reference and embracing Western science as a source of inspiration and new ideas. Initially, this inspiration was mostly drawn from the left of the Western intellectual field, but it later drew on other Western intellectual currents. Social science fields not only gained relative intellectual autonomy after 1956, but also became relatively (by Soviet bloc standards) internationalized. This allowed many in the intellectual elite to gain foreign experience. Several members of this rising faction of the academic elite also inherited considerable cultural capital assets, not least proficiency in foreign languages and manners. These proved useful in forging friendships with Western scholars.

The thaw of October 1956 can be considered a turning point, as all social fields began to achieve at least partial autonomy from the apparatus of political power. This was made possible by the erosion of party structures and competition for dominance within the party. The two camps that emerged, viz. "Natolin" or the Natolinians (*Natolińczycy*) and "Puławy" or the Pulawians (*Puławianie*), represented different interests and had different views on how the socialist state should develop. Their names were derived from their favorite meeting places, viz. Natolin Palace and a luxury apartment blocks in Puławska St., Warsaw. This rivalry marks the beginning of the formation of a proper political field. To use Bourdieu's language, this moment can be seen as a transition from a centralized "apparatus" to a meta-field of power understood as a space of competition between various factions of the political elite, each representing different interest groups. It is worth mentioning some of the reforms of the day. Economic reforms included abandoning collective farming. Passport policy was liberalized, so there was more freedom to travel—even to the West. Poland became more active in the international arena (e.g. by participating in UN General Assembly sessions). A fragile compromise was reached with the Catholic Church and the Primate of Poland, Card. Stefan Wyszyński (1901–1981) was released from prison. The Catholic catechism was even reinstated as a school subject for a few years. Censorship was relaxed. Several previously blacklisted authors could be now published, as could literature from non-communist countries. There were also changes in the field of art. Pre-war professors who had been banned from teaching were now allowed to return to universities. These included Tadeusz Kotarbiński (1886–1981), Władysław Tatarkiewicz (1886–1980), Edward Taylor (1884–1964), Józef Kostrzewski (1885–1969), and Konrad Górski (1895–1990). Promotions that had been withheld for political reasons were now approved. Academic exchanges with Western countries were resumed. Noteworthy here was the Ford Foundation, which offered research and training scholarships to young academics (for social scientists and artists in particular) in the United States and some Western European states. In the first five years of the program (1957–1961) alone, 330 Poles were granted scholarships. The British Council and Germany's Alexander von Humboldt Foundation offered fellowships on a more limited scale (Kilias 2017).

2.10.2 New Sector of the Field of Power Emerging in 1956

Of crucial importance from the point of view of the transformation of the field of power were freedom of association and the ability to engage in public debate. Discussion clubs began to appear. The first, and longest-lasting (until February 1962), was the Crooked Circle Club (*Klub Krzywego Koła*). Catholic Intelligentsia Clubs (*Kluby Inteligencji Katolickiej*, KIKs) also sprouted up, especially in Kraków, Poznań, Toruń, Warsaw, and Wrocław. By the spring of 1956, there were some 130 discussion clubs. A year later, clubs had to be registered. As a result, most of them vanished overnight. However, in a new wave of activism the "Inter-school Discussion Club, Contradiction Seekers Club" (*Międzyszkolny Klub Dyskusyjny, Klub Poszukiwaczy Sprzeczności*) was set up in 1962 by Adam Michnik (b. 1946) and his friends, many of whom, e.g. Jan Józef Lipski (1926–1991), went on to become well-known public figures.

Michink and Lipski are central figures in the post-1956 history of Poland. Adam Michnik was born to Polonized Jews. Both parents were active communist party members in the interwar period. After the war, they belonged to the middle-range nomenklatura, as did many in Michnik's social circle in the 1960s. Michnik enrolled at the University of Warsaw in 1964 and studied history, but was expelled and imprisoned after the March Events of 1968. In the 1970s, he worked as a welder in a factory, then focused on independent intellectual and opposition activities. By that time, he had become one of the leaders of the anti-communist opposition and its key ideologist. In 1980, he became one of the main advisors to the Solidarity movement and after the imposition of martial law in 1981, spent several years in prison as one of the main opponents of the communist authorities. In 1988, he became one of the main opposition representatives in the Round Table Talks with the authorities. These resulted in parliamentary elections in 1989 and Michnik was elected to the lower house of parliament (the Sejm), where he served one term. In the same year, he co-founded the daily newspaper *Gazeta Wyborcza*. Since then, he has been its editor-in-chief and one of the key intellectual leaders of the liberal intelligentsia. Michnik is a highly respected columnist and intellectual with a worldwide reputation.

Jan Józef Lipski was born into a petty intelligentsia family in Warsaw in 1926. He fought in the Warsaw Uprising of 1944, which attracted the attention of the communist security services after the war. Nevertheless, he was able to enroll in the University of Warsaw, where he graduated with a major in Polish philology in 1953. He was employed part-time at the Institute of Literary Research of the Polish Academy of Sciences (IBL) in the Dictionary of Literary Theory Concepts Lab (*Słownik Terminów Literackich*) section from 1950. In 1952–1959, he was employed at the prestigious PIW literary publishers. He was also a journalist at *Po Prostu* weekly for a brief but crucial period (1956–1957). Lipski found permanent employment at IBL in 1961. He signed the "Letter of 34" in 1964 and the "Letter of 59" in 1975. He defended his doctoral thesis in literary studies in 1965 and his habilitation thesis in 1975, both at IBL. The latter was withheld by the Central Qualification Commission (CKK), the central state body that approved academic degrees. His habilitation was only approved in 1981. By that time, Lipski had been an honorary leader of the liberal opposition since the early 1970s.

The Contradiction Seekers Club's speakers included such prominent intellectuals as Leszek Kołakowski (1927–2009), Bronisław Baczko (1924–2016), Zygmunt Bauman (1925–2017), Włodzimierz Brus (1921–2007), Janusz Kuczyński (1930–2017), Witold Dąbrowski (1933–1978), Karol Modzelewski (1937–2019), and Janusz Zabłocki (1926–2014). Although it folded in 1963, it played an important role in integrating the younger generation of the intelligentsia. Interestingly, the club was founded on the recommendation of Adam Schaff, who was a member of the Central Committee of the PZPR at the time. The club played a crucial role in forming an independent sector of the field of power. It was a formative experience for Adam Michnik. Its members included such prominent future politicians and intellectuals as Jan Lityński (1946–2021), Jan Gross (b. 1947), Aleksander Perski (b. 1947), Włodzimierz Kofman (b. 1945), Andrzej Titkow (b. 1946), and Marek Borowski (b. 1946).

The Crooked Circle Club played an even greater role in integrating the liberal intelligentsia elite, especially its inner elite circle, which included social scientists Włodzimierz Brus, Józef Chałasińki

(1904–1979), Stanisław Ehrlich (1907–1997), Aleksander Gella (1922–2014), Julian Hochfeld, Paweł Jasienica (1909–1970), Jerzy Jedlicki (1930–2018) and Witold Jedlicki (1929–1995), Jakub Karpiński (1940–2003), Tadeusz Kotarbiński, Stanisław Manturzewski (1928–2014), Karol Modzelewski, Maria (1896–1974) and Stanisław Ossowski (1897–1963), and Jan Strzelecki (1919–1988), with Jan Józef Lipski serving as president in 1957–1959. As Osęka (2008: 58) reports, there was also contact between the club and homologous young Soviet intelligentsia circles. For example, when the poet Witold Dąbrowski spoke of a new poetic movement in the USSR, Yevtushenko and Voznesensky came to be read in Poland. The Crooked Circle Club also established contacts with the Catholic Intelligentsia Club, in a way anticipating the alliance that eventuated in the second half of the 70s. Significantly, most members came from intelligentsia families, were often highly placed, and had usually attended prestigious secondary schools.

2.10.3 "*Środowisko*" as a Basic Unit of the Intelligentsia

The field of power was thus expanded to include a large swathe of the intelligentsia after 1956. A dense network of formal and informal intelligentsia networks emerged and was partly institutionalized—largely through the Catholic Intelligentsia Club and the Crooked Circle Club, although there were others, e.g. the Contradiction Seekers Clubs. At the same time, individual Catholic MPs entered parliament, thereby providing an additional link between the informal intelligentsia network and the formal institutions of power. In this context, it is worth emphasizing the importance of the Polish notion of *środowisko* (social circle or milieu; plural: *środowiska*). This denotes a smaller, more close-knit, and usually more elite group of the intelligentsia. *Środowiska* are usually centered around their leaders or institutions, although groups built around friends and family members might also may be significant, especially among the descendants of former landowners and aristocrats (Smoczyński and Zarycki 2017, 2021). Intelligentsia groups built around families constituted an important social structure within the *środowiska*.

Their significance has been exhaustively documented by Magdalena Bajer (e.g. Bajer 2013), who compiled histories of multi-generational families boasting several prominent academics. These histories of prominent families, most with aristocratic or bourgeois (including Jewish) roots, have been published with the support of the Foundation for Polish Science (*Fundacja na rzecz Nauki Polskiej,* FNP). As the FNP is a government agency, these historical elites could be seen to have been "canonized" in the current academic field. In any case, *środowiska* were the main vehicle by which the elite participated in social life, and can therefore be seen as fundamental building blocks of the field of power. The crucial role of this social structure for the Polish elite was highlighted by Janine Wedel in her first book, "Private Poland" (Wedel 1986). Wedel claims that *środowisko* cannot be accurately translated into English because of its special connotations. Her observations led her to conclude that affiliations with social circles is often a more important social distinction than any other measure of social status or affiliation, including religious and political ones. *Środowiska* usually have an important aspect of direct social interaction and other informal contacts. This aspect is best encapsulated in social events that bring the members of a given *środowisko* together. Interestingly, at an elite level, these sorts of gatherings sometimes acquire the status of historical events, and accounts of them often dominate the memoirs of the intelligentsia elite. A good example of this were the name-day meetings at Jan Józef Lipski's house. These were dubbed "the President's ball". Lipski's nickname "president" was born of discussions to make him the President of the Polish government-in-exile in London. However, it was mainly a term that ironically reflected his undisputed authority among the liberal intelligentsia. The notion of the "President's Ball" was immortalized in a mocking poem ("satirical opera") by Janusz Szpotański (1929–2001), entitled "*Cisi i gęgacze, czyli bal u Prezydenta*" (The Laconic and the Prolix, or the President's Ball). In 1967, the author was sentenced to three years in prison for the poem, as it alluded to Władysław Gomułka. Another mention of "the President's ball" is in the volume of Jacek Kuroń's (1934–2004) memoirs titled "Star Time" (*Gwiezdy Czas*) whose eponymous first chapter describes one of these gatherings at Lipski's apartment (Kuroń

1991). It is worth noting that, apart from lists of names, the expression "everybody was there" features in descriptions of these meetings, regardless of whether they were purely social or, e.g. intellectual or scientific. The notion of "everybody" most often refers to the elite of a given milieu. The frequent use of this expression indicates that the elites of a given *środowisko* were usually rather small and easily identifiable. This language is even adopted by historians and biographers. For example, Cyril Bouyeure, a French author writing about Adam Michnik, mentions that "all of intellectual Warsaw" was at a particular Crooked Circle Club meeting (Bouyeure 2007: 64).

2.10.4 Higher Education and Research Since 1956

The thaw also brought significant changes to academia. Three-year courses and candidate and doctor of science degrees were abolished in 1957. The Marxism-Leninism Graduate School was liquidated and a considerable number of its faculty dismissed. The crisis that afflicted the ideological system of the state after 1956 is interesting in this respect. In particular, compulsory university subjects such as "Fundamentals of Marxism-Leninism" were discontinued after 1957, as orthodox Marxist–Leninist teaching had lost its legitimacy and fallen out of favor amidst the general ideological change. Political education was not reintroduced until 1963, when "Foundations of Political Science" was made compulsory at all Polish universities (Warczok and Zarycki 2018). This lack of ideological courses can be interpreted as yet another aspect of the autonomy of the academic field and an aspect of the power of the intelligentsia. Curiously, the theoretical framework for the new courses was not borrowed from the USSR, as was the case with earlier Marxist–Leninist education, but from the West. The academic framework used was virtually non-existent in communist countries. As there was an acute shortage of staff to teach these new courses, the Central Methodological Unit was set up under the auspices of the Ministry of Higher Education (COM SNP) in 1967. COM was tasked with educating lecturers in political science, and its foundation can be viewed as a major institutional reinforcement of the emerging field. The first regular political science program was inaugurated at the University of Warsaw the same year. The emergence of "political science" in Poland, first at the University of Warsaw and later at other universities, was clear evidence of the Westernization of the social sciences. This can be interpreted as an attempt to modernize them and to conform to changing global ideological tendencies. But it can alternatively be seen as a sign of weakness on the part of the PZPR, as it was something of a compromise. Although Polish political science remained strongly subordinated to political directives and was relatively heteronomous in the broader context of the social sciences at the time, moving away from the Soviet academic model was a bold move. The general weakening of research and education in the spirit of political economy and class analysis was an obvious aspect of the replacement of Marxist–Leninist study centers with such a hybrid version of political science. This was the beginning of a trend that continued throughout the 1970s and 1980s as Polish social science underwent a process of "de-economization", by which is meant the marginalization of class-based, materialistic analyses and their gradual replacement with political, cultural, or eventually psychological analyses. It is instructive to compare this trend with what was happening in the USSR. According to Derluguian (2005), the new discipline of Soviet social studies was instituted, under the name of "scientific communism" in the mid-1960s. Its introduction seems to have had (at least in part) a similar motivation as in Poland, viz. the intractable difficulties involved in legitimizing Marxist–Leninist studies. However, whereas in Poland sociology was fully restored in 1956 and political science introduced in 1963, these disciplines were absent in the USSR until the perestroika movement of the late 1980s. Derluguian (2005) suggests that the introduction of "scientific communism" partly compensated for the absence of political science and sociology in USSR. The Soviet nature of "scientific communism" as a scientific discipline severely limited the compatibility of Soviet and Western (and even Polish) social sciences. For this reason, the works of Polish social scientists served as important mediation channels between Soviet and Western scientists. The works of Jerzy J. Wiatr (b. 1931) are a case in point, especially his numerous books on the sociology of politics. These were translated into Russian and attracted considerable interest in USSR. The situation was similar in Czechoslovakia, particularly after 1968. Czech and Slovak translations of Wiatr's works were an important window for scholars in that country to access new Western trends in sociology and political science. This was possible because Wiatr benefited considerably from exchange programs, especially Ford Foundation scholarships, and had the opportunity to follow intellectual developments in Western universities on a regular basis. Such tendencies could also be observed in linguistics, where Russian translations of Polish works gave Soviet scientists access to new Western trends. This is exemplified by, e.g. the works of Michał Głowiński (b. 1934) and Janusz Sławiński (1934–2014) on structuralism.

2.10.5 Institutional Changes After 1956

New institutions were founded on the wave of the thaw, e.g. the Institute of Philosophy and Sociology of the Polish Academy of Sciences (*Instytut Filozofii i Socjologii*, IFiS PAN), which was established in 1956. It was headed by Adam Schaff, who was now leaning toward a much more liberal position in the field of power. The new institute employed several prominent intellectuals, usually from intelligentsia families and of a more or less liberal bent. These included philosophers known for their work during the interwar period, e.g. Kazimierz Ajdukiewicz (1890–1963) and Tadeusz Kotarbiński. Stanisław and Maria Ossowski, Józef Chałasiński and Jan Szczepański (1913–2004) were among the sociologists who helped found the institute. The thaw also saw the launch of several periodicals that were to play a pivotal role in building the autonomy of the social sciences, humanities, and cultural fields. *Po prostu* [Simply] was the periodical of choice among the critical liberal intelligentsia, especially its younger members. Although founded in 1947, it changed its formula and editorial team in 1955 and went on to play a symbolic role. The weekly "*Polityka*" was founded in early 1957 as the official organ of the PZPR. Stefan Żółkiewski was the first editor-in-chief, and Jerzy Putrament (1910–1986), Andrzej Werblan (b. 1924), Mieczysław F. Rakowski (1926–2008), and Adam Schaff were on the editorial team. Polityka set out to be a magazine of the "center" and to channel the intellectual energy of the new intelligentsia that the thaw had unleashed. The monthly *Kultura i Społeczeństwo* (Culture and Society) was founded in 1957, with Józef Chałasiński as its first editor-in-chief (Stefan Żółkiewski was editor-in-chief in 1960–1966) and became an important academic periodical. In the first issue, Chałasiński served notice of its editorial direction: "We do not want to eliminate politics and economics, class issues and class struggles from our interests. However, we want to treat them in a scientific manner and from a spiritual and cultural perspective, which has been neglected in our writings" (Chałasiński 1957). This declaration reflects the zeitgeist of the period, which can be interpreted as a shift in the dominant ideology toward "humanism" and "humanistic values". This was a crucial change in outlook and "ethos" for the traditional intelligentsia. Chałasińki's declaration hints at a departure from class analysis, particularly in the traditional Marxist–Leninist vein. It is worth noting that journals such as *Kultura i Społeczeństwo*, although strictly academic, were read by a wide circle of the intelligentsia, especially in the 1960s. This demonstrates the importance of the intelligentsia elite and its intellectual life to the field of power. The field of social sciences assumed unprecedented significance, in particular in terms of its position in the broader field of power during this period. Academic disputes were related to political change. They were followed by the general public and attracted a fair degree of commentary. However, academic scholarship was not overly politicized. Relations between the purely political field and the field of social sciences seem to have been balanced, with the latter enjoying considerable autonomy and maintaining high professional standards. Although the political field had a formal advantage, it had to cede considerable autonomy to the academic field for the sake of its own legitimacy.

The end of thaw was marked by the closing of *Po prostu* in October 1957. Censorship was also tightened. The first wave of resignations from the PZPR followed. These included Stanisław Dygat (1914–1978), Paweł Hertz (1918–2001), Mieczysław Jastrun (1903–1983), Jan Kott (1914–2001), and Adam Ważyk (1905–1982). The field of power was reconfigured. According to Andrzej Werblan, the Natolin vs. Puławy (*Natolińczycy vs. Puławianie*) rivalry was a cleavage that emerged in 1956 and had a clear international aspect, viz. an attitude toward the

Soviet Union as an external center (Stasiński and Werblan 2018). The Natolin faction was mostly neo-Stalinist, and mostly consisted of former members of the pre-war Communist Party of Poland. Their position was strong so long as Soviet military intervention remained a possibility. However, it effectively disintegrated when Khrushchev publicly declared his support for Gomułka during his 1959 visit to Poland. Some members, resentful of what they saw as an act of betrayal, declared their loyalty to the communist parties of China or Albania. Another cleavage emerged in the early 1960s; one which had a much clearer internal aspect, in particular opposition between political and cultural capital. Cultural capital was represented by the old intelligentsia faction and political capital by the newly educated faction in the field of power, mostly nomenklatura members. Werblan emphasizes the differences between these two cleavages, which are often perceived as one. The "Partisans", i.e. the political capital camp, differed from the Natolin faction in eschewing too close a dependence on the Soviet Union. Nationalism lay at the core of the Partisan ideology. They were amenable to working with a pragmatic faction of the pre-war nationalists and were receptive to anti-Semitic slogans. By contrast, many members of the Natolin camp were of Jewish origin. The March events of 1968 were formally a victory for the "Partisans", but Werblan contends that a new pragmatic faction took over in 1970. This group was clearly technocratic in its orientation (Stasiński and Werblan 2018).

2.10.6 The Field of Power in the 1960s

The field of power was organized around at least two axes in the 1960s. First, there was a cleavage that cut through the party elite. The opposing camps were the Partisans, who had replaced the former Natolin faction, and the Liberals, or Revisionists, who were the heirs of the Puławy faction. The latter group was strongly associated with activists of Jewish origin and were generally well-educated. During the Stalinist period, many of them had been loyalists, but after the thaw of 1956, they moved to relatively democratic positions, and distanced themselves from Marxist orthodoxy. The second axis was connected to the field of power in a much broader sense, i.e. not only the political elites, but also the para-political ones. Here, there was a cleavage that counterposed the left wing (in the broad sense of the term), which accepted the Polish-Soviet alliance as permanent and unassailable and even as the basis for the functioning of Poland, and the right wing, whose main point of reference was pre-war Poland, which they positively perceived as a sovereign country. The right's chief ally was the Catholic Church. The Church was growing in strength as an independent political actor and had gained considerable freedom to act after 1956. In this respect, the uniqueness of Poland in the communist bloc is worth noting. It was the only country where the Catholic Church, despite attempts by the authorities to subordinate it, retained considerable institutional independence, and was a de facto important political actor in the field of power. The Church in Poland partly owed its strength to there being very few collective farms—another peculiarity of communist Poland. This was not only due to the resistance of the Polish peasantry, which had the support of the Church, but also to Gomułka's reluctance to collectivize. In any case, the conservative right wing worked closely with the Catholic Church and was able to rely on the support of a large part of the Polish diaspora in the West, especially those who had remained there after the war, having served under General Władysław Anders (1892–1970). In Poland, many former Home Army (*Armia Krajowa*, AK) soldiers had similar right-wing views. Despite having fought against the occupying Germans, they were severely persecuted during the Stalinist period. Some were even murdered by the Polish and Soviet secret services. Their military deeds could not be honored until after the fall of communism. However, from 1956, several representatives of the moderate Catholic right were active politically, especially as deputies in the Polish parliament. In particular, the "Znak" parliamentary group, which was founded in 1957 with only a few members, played a symbolically significant role, despite being a political lightweight. This was mostly because its MPs abstained from voting on several key bills, most notably the incorporation of the Polish-Soviet alliance and the leading role of the PZPR in the Constitution of the People's Republic of Poland (in 1976) for which it was disbanded. Znak had earlier attracted attention for submitting an interpellation on police violence during the events of March 1968. Several members had high public profiles, e.g. Stanisław Stomma (1908–2005), Stefan Kisielewski (1911–1991), Konstanty Łubieński (1910–1977), and Tadeusz Mazowiecki (1927–2013), who was elected the country's first non-communist prime minister in 1989. The circle's leading journals were the intellectual monthlies *Więź* and *Znak* and the weekly *Tygodnik Powszechny*. This last was published by PAX, a pro-communist Catholic association, during the Stalinist period. Association PAX (*Stowarzyszenie Pax*) was founded by Bolesław Piasecki (1915–1979), who had been a right-wing activist before the war. Piasecki was arrested by the NKVD at the end of the war and subsequently declared his loyalty to the new system. Piasecki was given free rein of the association and its parliamentary representatives. He also was allowed to set up a large company in Poland that financed the organization's activities, as well as a high school and the only publishing house in Poland that was permitted to publish the memoirs of former Home Army soldiers, religious books, etc. Piasecki remained loyal to the communists and increasingly to the fundamentalists. During the Stalinist period, he broke ranks with the Church by not protesting the arrest of Card. Wyszyński in 1953. This provoked a split within PAX that led to the resignation or dismissal of its influential young intellectual leaders, most notably Janusz Zabłocki and Tadeusz Mazowiecki. These disaffected former members remained under the influence of French Catholic philosophers, such as Jacques Maritain (1882–1973) and Emmanuel Mounier (1905–1950), founder of the *Esprit* journal (which was overtly political and flirted with the leftist causes, including communism). Zabłocki later became more conservative and Mazowiecki more liberal (Kosicki 2016).

I would like now to return to the first axis of the division of the field of power in 1956–1968, i.e. the opposition between "Partisans" and Liberals. The liberal camp attracted much of the intelligentsia, including many outstanding intellectuals and scholars. Many, like their counterparts in Czechoslovakia, were attracted by the slogan of "socialism with a human face" and many were involved in international, Western intellectual circuits and were fascinated by the successive waves of ideological transformation of the Western Left. It was not uncommon for them to take advantage of international travel opportunities and foreign scholarships, which allowed them to meet many prominent Western intellectuals. The trajectory of Adam Michnik is a case in point. Michnik made his first trip abroad in 1964. He visited the Polish embassy in Vienna and met with Radio Free Europe journalists in Munich and in Paris through the mediation of the émigré *Kultura* publishing house. He also met Zbigniew Brzeziński and others. Michnik later met a number of well-known Western intellectuals on subsequent trips (Bouyeure 2007). This "internationalization" of the liberal camp, partly driven by a fascination with Western intellectual currents, aroused increasing resentment among the Partisan faction, which moved closer to the Polish right-wing ideological canon, which they pragmatically combined with their adherence to a fairly orthodox, but de facto and largely declarative, Marxism. The Partisans, whose leaders included Interior Minister Mieczysław Moczar (1913–1986), effected a *sui generis* synthesis of nationalism and communism in which antipathy toward Germans and Jews (albeit less overt) played an important role. The historical successes of the Polish armed forces were also foregrounded.

Tensions on this central axis of political division became publicly visible through open letters to the authorities signed by intellectuals and scholars. For example, in 1964, the "Letter of 34" appeared, protesting censorship and the reduction of paper allocations for books important to culture and learning. Many prominent authors signed it, including writers and literary scholars, even some associated with the Institute of Literary Research of the Polish Academy of Sciences (*Instytut Badań Literackich*, IBL) (Eisler 2018). Kazimierz Wyka, Mieczysław Jastrun, Jan Kott, Aleksander Gieysztor (1916–1999), Adam Ważyk, Melchior Wańkowicz (1892–1974), Jerzy Turowicz (1912–1999), and Stanisław Dygat were among the signatories. The authorities composed a counter-letter, which was signed by Stefan Żółkiewski, Jerzy Putrament, and Jarosław Iwaszkiewicz (1894–1980). Wańkowicz, the respected writer, was most severely punished for signing the Letter. He was tried and sentenced to three years in prison in 1965, although he only spent a brief period in pre-trial detention, as the party leadership feared loss of national and international reputation. His case illustrates the constraints on persecuting the intellectual elite after the Stalinist period. One interesting aspect of the trial is that Wańkowicz demanded—and was given—a personal meeting with Secretary Gomułka and Minister Moczar. After these meetings, the charges against him were dropped. These contacts were an interesting aspect of the power field at the time, which, as can be seen from this example, had a clear bodily aspect—direct contacts between many of its members, including the party leadership and the liberal intelligentsia. Other signatories of the Letter were punished with sanctions of various kinds, mostly bans on publication and foreign travel. It is worth noting that both the party and the secret services maintained detailed records of everyone who signed an open protest letter, and these records were used to impose further restrictions on them (or to ease them). Another symbolic event that took place in 1965 was the trial of Jacek Kuroń (1934–2004) and Karol Modzelewski, who wrote a "revisionist" letter to the party. An interesting aspect of this trial, from the point of view of the logic of the field of power and its corporeal aspect, was the physical presence of many active or future oppositionists, whose milieu was slowly consolidating and whose role in the field of power was growing, at the court hearings. As Andrzej Friszke wrote, the trial "was one of the most important events in Poland during those years and contributed to the creation of a *środowisko* that, coming from the party, criticized the existing system, demanded democratization, freedom of speech, and showed great determination in its clashes with the authorities" (Friszke 2018: 8).

Another publicly visible moment in this confrontation was a seminar held in October 1966 to mark the 10th anniversary of the "October 1956" breakthrough. Leszek Kołakowski, who was already a well-known philosopher, presented a paper entitled "Polish Culture in the Last Decade", as did Krzysztof Pomian (b. 1934). Both were expelled from the PZPR after the event as part of the party's purge of "revisionists", as the liberal camp was increasingly called. The term "revisionism" originally denoted a "revision" of the orthodox interpretation of Marxism (i.e. the PZPR's interpretation). Revisionists were even accused of wielding the sword of Marxist criticism against the party. Such a "provocative" or "subversive" use of Marxism was also associated with a *środowisko* which was now termed "commandos". The name came from their perception as youth who "raided" party meetings, which they disrupted by asking difficult questions aimed at exposing the hypocrisy of the party in relation to Marxist ideals. As scholars such as Eisler (2018), Osęka (2008), and Jakub Karpiński (2001) point out, the definition of "revisionism" was always contextual, and should therefore be considered relational. It mostly referred to the formal position of the Party at any given moment (i.e. the "Party line"). At the same time, this changing point of reference constantly rearranged the positions of individual actors in the field of power. In particular, more and more of the intelligentsia, especially those from nomenklatura families with old intelligentsia roots, including several of Jewish origin, moved to "revisionist" positions, as they were often directly attacked by the increasingly influential Partisan camp.

The 1960s can be thus seen as a period of relative intellectual freedom and considerable openness of Poland toward Western and, more generally, global academic and cultural trends. International contacts were quite intensive. Previously marginalized academics and intellectuals were now free to pursue their intellectual passions and many of them were even allowed to travel abroad—on the proviso that they did not question the authority of the PZPR or impugn "socialist values". In fine, the period from 1956 to 1968 saw the intelligentsia, which included many representatives of its old elite, intensely involved in the mainstream state-controlled fields of the economy, academia, and culture. Meanwhile, former Stalinists, whose cultural capital was wanting, were relegated to the internal and informal opposition within the field of power. This started to change during the second half of the 1960s, when that internal opposition, mostly from officials who failed to qualify as intelligentsia, began to challenge the status quo.

The 1960s, in particular, the period between 1956 and 1964, can therefore be seen as a period where the intelligentsia dominated, at least in the fields of the arts, academia, and education. This eventuated despite the omnipresent socialist slogans trumpeting the official party ideology and the relatively egalitarian policies of the communist state. Nevertheless, a considerable segment of the intelligentsia elite entered the field of power and enjoyed numerous privileges. Many members of the party elite, not only its dominated elite (i.e. its academic or artistic elite), were of intelligentsia origin. Numerous representatives of the intelligentsia, particularly those whose sympathies lay with the political left, had joined the new state elites during the first phase of their formation, which began in 1944. During the communist period, different factions of the intelligentsia were engaged in politics, academia, and culture with varying degrees of intensity. Many became disillusioned with communism sooner (in the mid-1950s) or later (in the late 1970s), and by the mid-1980s, most of them were critical of the communist government. Once communism had collapsed, these same elites acquired dominant positions in their respective fields, which can be seen as part of a broader process of consolidation and full institutionalization of what I have previously termed "the intelligentsia hegemony" (Zarycki 2009; Zarycki et al. 2017). As for the field of social sciences, this group was able to take full advantage of the opportunities to participate in academic exchange programs with Western countries as of around 1956 (Kilias 2017; Sułek 2010).

2.10.7 March 1968

The events of March 1968 were the culmination of the clash between the Partisans (hardline communists) and the Revisionists (liberals). This was a watershed in the history of Poland's communist period (Osęka 2019). It was then that the Partisans, who possessed political rather than cultural capital, and who relied on overtly nationalistic discourse for their legitimacy, gained control. The liberals (the faction party leadership possessing higher cultural capital assets and its rank and file, along with other beneficiaries of the liberal climate) were marginalized, or forced out of, the field of institutionalized politics, and even the field of power, even as its very definition was being contested. In any case, many of the revisionist elite lost their political influence. Several lost not only their privileged positions, but also their jobs. Some left the country. While not formally forced to emigrate, they were given informal incentives to do so, while their employment situations were made intolerable. Some were even dismissed from their posts and prevented from obtaining alternative employment. Some moved to internal exile (opposition), and others were relegated to less prominent positions. The sociologist Zygmunt Bauman is probably the best known of those who were forced out. Adam Michnik, who was a student at the time, is the most prominent of those who were forced into internal exile. Only about 15,000 people emigrated in the years following the events of March 1968. As Piotr Osęka points out (Osęka 2008: 266), the emigration was mostly non-Zionist, and was not particularly directed toward Israel. Most of the country's remaining Jews had already left in 1945–1951 and 1956–1960. Those emigrations combined had involved more than 140,000 people of Jewish origin. This is an interesting contrast: a wave of Jewish emigration that was only about a tenth the size of the previous two, resonated much stronger—not only in the Polish field of power, but also on a global scale. One reason for this, apart from the pressure exerted on the individuals concerned in 1968, was that major part of those leaving Poland used to be members of the wider field of power or had friends or family in it.

It is vitally important to understand the other structural conditions of March 1968 and the mechanisms that triggered the events in question. The logic of the clash between the two factions has to be factored into any account of that period. As Osęka puts it, March 1968 can be seen as a revolution of quadragenarian apparatchiks (Osęka 2008). For this reason, psychological interpretations of the March events are best avoided. Among the structural determinants of the events of March 1968 was the increase in the number of university students and recent graduates during the communist period (Warczok 2022). By the end of the 1960s, a lot of first-generation educated specialists in various fields had emerged in social life. Many of them manifested significant political, as well as professional, ambitions. These people had been partially socialized into the intelligentsia ethos at universities run by the "old" intelligentsia. This put them at odds with the old intelligentsia elite in three dimensions. The first was the political field, especially the competition among the PZPR elite. The second was professional competition in various social fields, including the field of science, where the inflation of degrees and diplomas was beginning to be felt. The third was competition in the intelligentsia field. This was manifested in the struggle to be recognized as a "real" member of the intelligentsia". Tension in this last dimension began to be evident as early as the late 1950s. For example, Mieczysław F. Rakowski (1999), a journalist for the weekly *Polityka*, in a diary entry dated February 18, 1959, noted that there was a discussion on the intelligentsia in *Przegląd Kulturalny*, a weekly founded on the initiative of the sociologist Jan Szczepański and published in 1952–1963 (Rakowski 1999: 89). Szczepański wrote that "the old intelligentsia is melting away" and being replaced by "not a new intelligentsia, but a new social creation, a stratum of white-collar workers". He described this process negatively, using the term "antipathetic cancer", and claiming that its symptoms included indifference to the fate of others and a tendency to aggression in social relations. Andrzej Krzysztof Wróblewski (1935–2012), among others, polemicized with Szczepański, who, as Rakowski recorded, claimed that "white-collar workers", of whom there were already 2.1 million, could not all be counted among the intelligentsia.

I now wish to briefly discuss the growth dynamics of the number of students in post-war Poland compared with the interwar period. Some 40,000 students were enrolled annually in Polish universities in the 1920s and about 50,000 in the 1930s. This figure remained relatively stable until the outbreak of WWII in 1939. About 56,000 students were enrolled in the first post-war academic year (1945–1946). By 1950, that number had more than doubled to 125,000. This growth then decelerated, with 165,000 in 1960, 330,000 in 1970 and 454,000 in 1980. After that peak, a slight drop was observed with 340,000 students enrolled in 1985 (Popiński 2018).

Not surprisingly, tensions similar to those that triggered the explosion in the field of power in 1968, when the new, advancing intelligentsia clashed with the established elite and the up-and-coming generation had no guarantee of inheriting their parents' social positions, can be observed in Poland today. Disputes as to who qualifies as "a real member of the intelligentsia" are just as pronounced as ever. After 1968, tensions of this nature did not lead to a conflict on anything like the same scale, but as I argued in a recent text, this is how the rise to power of the conservative right, usually called "populist", can be interpreted. In particular, its anti-elitist slogans can be interpreted as being partly the result of the frustration of successive generations of university graduates who have been disillusioned by the lack of access to prestigious jobs due (at least in part) to diploma inflation (Zarycki 2020a).

It is also interesting to note that although the Soviet field of power did not undergo a structural shift analogous to those of Poland or Czechoslovakia, similar steps were nevertheless taken. One striking example was the new editor-in-chief of *Novy Mir*, an internationally respected literary magazine. Alexander Tvardovsky (1910–1971) lost the editorship in February 1970. This is considered a symbolic moment in the marginalization of the liberal faction in the Soviet field of power. It was in *Novy Mir* that Alexander Solzhenitsyn's novella "One Day in the Life of Ivan Denisovich" was first published in 1962. Moreover, the first wave of Jewish emigration from the Soviet Union began in 1970. As Bogusław Żyłko (Żyłko 2009) reports, many of the semioticians associated with the Tartu-Moscow school were driven out of the Soviet Union in the 1970s. In particular, Alexander Piatigorsky (1929–2009) went to the United Kingdom, while Aleksander Zholkowsky (b. 1937) and Yuri Shcheglov (1937–2009) went to the United States.

2.11 The Last Decades of the Communist Period

2.11.1 Toward Technocratic 1970s

One of the key differences between the Soviet Union's field of power and the Polish field of power appears to be that part of the Polish intelligentsia elite actually participated in the field of power. The key tension in its structure, particularly in the 1960s, was between the old, multi-generational intelligentsia, which was possessed of a solid education and whose cultural capital was of a high standard, and on the other, the new intelligentsia, which was more of a nomenklatura elite primarily based on political capital. Derluguian argues that the USSR field of power was dominated by a factionally divided nomenklatura after 1953. The conservative faction sought to limit changes to the country's political structure to those necessary to ensure the security of the nomenklatura and preserve its privileges. The reformist faction strove to modernize the system, and called for greater liberalization and openness to the West. The divisions within the intelligentsia could be seen as homologous to the opposition between these factions. The conservative mouthpiece was the journal *Okriabr*, and that of the liberals was the journal *Novy Mir*. The liberals were also known as the "Sixers". They were most active during the Khrushchev Thaw, i.e. from the 20th CPSU Congress in February 1956 until Khrushchev was ousted in October 1964. The conservative faction was obsessed with its own social stability, and opposed any calls for mobility within the field of power, as well any but the most cosmetic changes to policies or ideology. As already mentioned, the nomenklatura never managed to cut itself off from the intelligentsia or reproduce itself multi-generationally in Poland. The boundaries between the nomenklatura and the higher intelligentsia remained permeable, even after 1968, when most of the multi-generational intelligentsia resigned or were dismissed from the party leadership. Over time, however, new members of the intelligentsia joined the upper echelons of the party. Gen. Wojciech Jaruzelski (1923–2014), who became First Secretary of the PZPR in 1981, although he did not complete his university studies due to the war, was the son of a typical Polish intelligentsia family that had descended from the petty nobility. His successor as First Secretary, Mieczysław F. Rakowski, appointed to the post in July 1989, was a first-generation member of the intelligentsia, but at the same time deeply connected to its identity and *środowisko*, e.g. through his marriages. He was also an active intellectual who helped create the intelligentsia discourse in Poland (Przeperski 2021). In this context, it is worth reexamining the differences between the Polish intelligentsia and the Soviet intelligentsia. Not only was the Soviet intelligentsia subordinated, and even made subservient, to the nomenklatura, its identity was largely reconstructed after Stalin's death. In Poland, by contrast, the intelligentsia was mostly unaffected. Its main dimension of reproduction was constituted by close-knit circles and often comprised multi-generational families that transmit traditional identities and ethos. This difference was also noticed by Derluguian, who pointed to the fact that "dissident democratic mobilizations in the socialist states of Central Europe drew strength from the survival of latent oppositional traditions within the networks of middle-class families that dated back to the petty nobility, bourgeoisie, and intelligentsia of pre-socialist formation. Inside the USSR, similar conditions were found only in several smaller national republics, mainly Estonia, Latvia, and Lithuania, which had been independent states during 1918–1940, but also in Armenia and Georgia, to a still lesser degree in Moldavia and the western part of Ukraine. The fact that the national intelligentsias in these regions traced their lineages to the states that at some point had been outside Soviet control determined the fusion of nationalism and the intelligentsia's opposition to the rule of communist officials" (Derluguian 2005: 108). He goes on to assert that "The actual number of old-high status families could be quite small, because communist repressions took a heavy toll and Soviet industrialization brought its own specialists in vast numbers. Nonetheless, the families of old prestige tended to monopolize the gate-keeper functions in the formation of high-culture fields, and this exerted a disproportionate influence. The members of such families served as an attractive example for Soviet-era newcomers who were seeking admission into the intellectual and professional elite. Thus, the old families could preserve their status despite the loss of political and economic power in socialist times" (Derluguian 2005: 148).

The events of March 1968 also had significant repercussions for the core of the field of power, which was mostly constituted by the political field. The PZPR expelled many prominent intelligentsia members it deemed too liberal. These included Stefan Żółkiewski and Adam Schaff, who were expelled from the Central Committee at the November 1968 plenum. Żółkiewski had provoked the Party to expel him by publicly supporting students who had been beaten and expelled from universities in March. Paradoxically, however, March 1968 was not an unalloyed victory for the Partisan faction. For one thing, its leader, Moczar, was replaced as Minister of the Interior. Admittedly, he remained in the Politburo, but was demoted to the rank of deputy member. It was not long before the entire team, including Gomułka, was replaced. In December 1970, with Moscow's backing, Edward Gierek, often described as a technocrat, became First Secretary of the PZPR. "Technocratization" in the ideological dimension weakened the nationalist line by softening the pathos of the Partisan discourse. Nevertheless, the national and patriotic narratives assumed an increasingly dominant role, while Marxist and other communist ideological narratives were firmly ritualized.

In more general terms, the events of 1968 brought about a considerable reconfiguration of the field of power. The dominant opposition was now between the hardliners (or orthodox communists) and their technocratic allies (younger, recently educated pragmatists) (Eyal et al. 1998) and the former communist "liberals", many of whom gradually assumed anti-system or anti-communist stances from the mid-1970s. Most of them left the institutionalized political field and eventually joined forces with the Catholic Church and anti-communist conservatives. This division was reinforced with the emergence of the independent Solidarity trade union in 1980 and the imposition of martial law in 1981. An important aspect of the formation of this structure within the field of power was that the anti-communists, which included a major part of the 1960s liberals, repudiated their former leftist affiliation. This included renouncing all references to Marxist theory and disavowing all criticism of capitalism. Western free-market capitalism and liberal democracy now became ideals to emulate. After the fall of communism, this central cleavage assumed the form of a confrontation between post-communists and radical liberal reformers (Zarycki 2011).

2.11.2 Academic Field in the 1970s

The field of the social sciences and humanities was structured in a homological manner during the communist period. The more "conservative" scholars usually professed greater loyalty to the PZPR and at least paid lip service to Marxist ideology. They generally had less cultural capital and tended to be party members. The more "liberal" scholars, per contra, stood aloof from communist ideology and displayed greater openness toward Western ideas. They generally belonged to the intelligentsia, and some came from families with several generations of intelligentsia traditions. In some fields, and during some periods, particularly the 1980s, the most radical representatives of the "liberal" pole could even assume an anti-communist, i.e. openly oppositional, political stance. This was the case with, e.g. sociology (Bucholc 2016) and history toward the end of the communist period, although most overtly anti-regime historical works were published outside the formal academic system, in the *samizdat* (underground publishing system), or *tamizdat* (diaspora publishers in the West) (Behr 2017, 2021). This opposition was much less visible in the case of the political sciences, a field that was far less autonomous from the PZPR and which was dominated by Party loyalists (Warczok and Zarycki 2018). Geography, especially social and economic geography (which also includes regional planning in Poland), fell somewhere in between. Its "liberal" pole was much weaker and far less radical when compared to the liberal poles of the fields of sociology or history. Be that as it may, the field of geography nevertheless enjoyed a great deal more autonomy from the PZPR than did the political sciences, which mostly performed an overtly legitimizing function. It thus clearly developed the essential opposition described above, albeit not in the most radical or strongly politicized manner.

As mentioned above, the declining of the importance of Marxist–Leninist ideology in legitimizing the communist authorities in the 1970s was an important aspect of technocratization. This also resulted from the 1968 "socialism with a human face" movement, which delegitimized the nomenklatura and confronted conservative bureaucrats with moral and ideological arguments that they could not answer (Derluguian 2005: 108). Marxist–Leninist ideology was sidelined in favor of social and economic development. A scientific rationality that relied on optimization, calculation, and cybernetics was the centerpiece of this approach. As was the case with the Communist Party in the USSR, the PZPR sought to compensate for its waning ideological legitimacy by raising living standards and improving general wellbeing (Derlugian 2005: 108). In the USSR, these economic initiatives were financed by increased gas and oil revenues after 1973. For their part, Poland and Hungary had to resort to Western loans. Such credit was readily available on account of the capital surplus that had accumulated in the Western financial system after the massive increase in oil prices.

There was now less emphasis on social theory (including political economy), classical class analysis, and the humanities, and more on "practical" disciplines, especially management, organization theory, and obviously STEM disciplines. As far as the social sciences were concerned, the main beneficiaries of this shift in priorities were social psychology and political science, which were expected to provide modern solutions for human resource management and social control. The newly installed Party leaders were hopeful that these methods would resolve social tensions, including those previously analyzed in terms of class frictions. This change of direction might have been disappointing for some humanists, as shown by Degen and Hübner, who critically cite a speech by the sociologist Władysław Markiewicz (1920–2017). In the view of the authors, Markiewicz pushed for the social sciences at the expense of the humanities. In the speech, titled "The State and the Prospects for the Development of the Social Sciences in Poland" (*Stan i perspektywy rozwoju Nauk społecznych w Polsce*) and delivered at the General Assembly of PAN in May 1975, Markiewicz said, *inter alia*, "We give priority to research that serves the accelerated social and economic development of our country and the realization of the great ambitious development programs associated with the scientific and technological revolution and the transition to the construction phase of a developed socialist society" (Degen and Hübner 2006: 36). The departure of a large part of the liberal intelligentsia, including scholars and academics, from the political field after March 1968 indicated that the scientific field had lost its key representatives in the political field, thereby giving the latter a numerical advantage over the former. However, the intelligentsia that had left the political field began to reorganize itself outside it, as is discussed below. This caused a new pole to appear within the field of power; one that soon began to strongly impact the scientific field as well.

Emigration abroad continued into the 1970s, as did "internal emigration", especially toward underground or semi-legal activities that more or less directly undermined the legitimacy of the communist authorities. It can therefore be said that a field of independent circles, movements, and even organizations, whose main engine was the intelligentsia, began to take shape as a fully autonomous independent social space from the beginning of the 1970s. The papal visit of Pope John Paul II in 1979, and the subsequent establishment of the independent Solidarity trade union, combined to consolidate the entire field. At the level of the field of power, it became a strong pole of opposition led by the intelligentsia elite.

2.11.3 Consolidation of the Anti-Communist Opposition

But to return to the first half of the 1970s, minor radical organizations challenging communist rule emerged at one pole of the independent field. One such was *Ruch*, which was secret, but others openly demanded democratization and decried Poland's dependence on the Soviet Union. The Polish Independence Alliance (*Polskie Porozumienie Niepodległościowe*, PPN) was founded in 1976. It mostly published analytical texts penned by, e.g. Zdzisław Najder (1930–1921), Andrzej Kijowski (1928–1985), Jan Olszewski (1930–2019), and Jan Józef Szczepański (1919–2003). The Alliance played an important symbolic role, despite operating on a small scale, as it was the first organization to openly call for Poland to be fully independent of the Soviet Union. This made it instrumental in institutionalizing a new cleavage in the Polish field of power—one based on attitude toward the Soviet Union. Less radical and institutionalized were the open letters to the authorities that subsequently appeared in the public space. As mentioned above, the intelligentsia elite had already been writing such letters in the 1960s. However, they became more intrepid and exigent in the 1970s. The best known was "the Letter of 59" (also known as the Memorial or Memorandum of 59) of 1975, which protested the Party's proposed changes to the Constitution of the People's Republic of Poland. It was initially signed by 59 Polish intellectuals with others signing in January 1976.

The Workers' Defense Committee (*Komitet Obrony Robotników*, KOR; transformed in 1977 into the Social Self-Help Committee KSS KOR) was crucial to consolidating the opposition sector of the power field under intellectual leadership. This organization provided assistance to workers persecuted by the authorities after the wave of strikes which swept through Poland in 1976. The Committee broke a major barrier of fear by entering the public space in a visible way—primarily by publishing the addresses of most of its members. Moreover, "discussion clubs" began to emerge in the second part of the 1970s, although most, in contrast to the 1960s, ignored the official sphere. The new clubs were completely informal, i.e. illegal, in contradistinction to the 1960s, when clubs sought formal registration. New clubs included "Free

Discussion Clubs" (*Kluby Swobodnej Dyskusji*) run by the Movement for Defense of Human and Civil Rights (*Ruch Obrony Praw Człowieka i Obywatela*, ROPCIO) and the "Independent Discussion Club" in Łódź. Another discussion club was the Walendowski Salon, which operated from November 1976 until 1979 in an apartment on Puławska street in Warsaw. The apartment belonged to Anna Erdman (1945–2004), wife of and Tadeusz Walendowski (1944–2004) the granddaughter of Melchior Wańkowicz, which demonstrates the role of inherited family symbolic capital in intelligentsia circles. The secret services took a close interest in these clubs. After the fall of communism, their activity was documented in numerous publications, as a good many of their participants went on to become prominent public figures. One such publication is a volume devoted to the Walendowski Salon, which lists all its meetings and the names of those who attended, using both secret services documents and accounts (historical and current) of participants (Gluza 2016). Informal parties at private apartments can therefore be seen as part of the formal political history of Poland and as symbolic capital for many current members of its field of power. A closer look at the records of those who attended these meetings shows that they included representatives of various fields, including academia, culture, and politics. Many were already established scholars, writers, or poets. But representatives of the younger generation were also present, with many of them becoming prominent intellectuals a decade or two later. Most of the younger attendees at the Walendowski salon meetings were active in radical underground anti-communist activities, including samizdat, e.g. Mirosław Chojecki (b. 1949), who founded the independent publishing house NOWa in 1977. There was a clear overrepresentation of sociologists, psychologists, and historians. At least two permanent employees of the Institute of Literary Research (IBL) were Walendowski Salon regulars, viz. Tomasz Burek (1938–2017) and Jan Józef Lipski.

The development of this organizational infrastructure, independent of the authorities, significantly changed the geometry of the field of power in Poland. The earlier dispute between the liberal leftist intelligentsia and the new intelligentsia with more authoritarian and nationalist leanings, which dominated the second part of the 1960s, was replaced by a new axis of division in the 1970s. One pole was the anti-communist opposition circles that openly challenged the authorities. The consolidation of the anti-communist (and implicitly anti-Soviet) camp in the field of power reached its apogee with the creation of the Independent Self-Governing Solidarity Trade Union (*NSZZ Solidarność*) in September 1980, following a wave of large-scale, protracted strikes throughout the country. Intelligentsia opposition circles became formally institutionalized at the same time. Their status, viewed from the perspective of the field of power, might not have been clear earlier, but as of September 1980, they also entered the field of power by participating in talks with the government. Their most visible media presence was the committee of advisors to the strike committee in Gdańsk in August 1980. These processes intensified after martial law was imposed in December 1981, and acquired a geopolitical dimension of Soviet-Western confrontation in which Solidarity was increasingly involved. Anti-communist circles received a great deal of moral and material support from the West. First, the Polish diaspora had its own channels of support and political influence. One of the crucial coordination centers in this respect was the Paris headquarters of *Kultura* monthly. Second, governmental and public organizations from the United States and other Western countries became increasingly committed to supporting the anti-communist opposition. Finally, the Western European trade union movement offered its assistance, as did numerous churches. There was a large stream of financial aid, in-kind donations, and publications. At the same time, the opposition elite built up a vast network of international contacts. In a structural sense, there was a widening division of the field of power according to geopolitical criteria. Thus, the opposition was not only becoming increasingly pro-capitalist, and pro-West generally, but its links to the Western world and its elites were becoming stronger and more numerous through networks of contacts.

The foundation of the Association of the Academic Courses (*Towarzystwo Kursów Naukowych*, TKN) on January 22, 1978 was of crucial importance to the academic field. The name alluded to the self-educational institutions of the nineteenth century and was proposed by Edward Lipiński (1888–1986). This was just one example of a wider tendency to frame the anti-communist movement within the historical legacy of nineteenth-century organizations led by the intelligentsia. TKN's predecessor was the KOR's "Flying University", founded in October 1977. Interestingly, Adam Michnik was one of the lecturers. Although the Association was persecuted by the Security Services (*Służba Bezpieczeństwa*, SB) and those who attended its illegal lectures were sometimes dispersed and even beaten, it steadily grew in strength and numbers. In addition to giving lectures, TKN published scholarly works—mostly its "course materials". From the standpoint of the present work, the booklet entitled "The Language of Propaganda", which was the first in a series of lectures published by TKN in 1979, is especially significant. It was edited by Stefan Amsterdamski (1929–2005), Aldona Jawłowska (1934–2010), and Tadeusz Kowalik (1926–2012). Subsequent volumes in the series included: "Which Literary History Do We Need Today", edited by Tomasz Burek; *After the Great Leap* by Waldemar Kuczyński (b. 1939); *The Polish Underground State* by Władysław Bartoszewski (1922–2015); and selected writings by Tomáš Masaryk (Kaminski and Waligóra 2014). These initiatives received foreign support, e.g. in January 1980, *The New York Review of Books* published information on the establishment of the International Committee of Support for Freedom of Teaching in Poland. There were 65 signatures, including 3 Noble Prize laureates, as well as such Polish intellectuals in exile as Bronisław Baczko, Zygmunt Bauman, Włodzimierz Brus, Maria Hirszowicz (1925–2007), Leszek Kołakowski, and Krzysztof Pomian. However, TKN was closed when Martial Law was imposed in December 1981 (Kamiński and Waligóra 2014). Another parallel initiative was the "Experience and Future Seminar" (*Konwersatorium Doświadczenie i Przyszłość,* DIP), which was something of an independent think tank *avant la lettre*. It was initiated by Stefan Bratkowski (1934–2021) in 1978. Its "Service Team" published several analytical reports on the state of economy and society, most notably "A Report on the State of the Republic and Ways to Repair it" (*Raport o stanie Rzeczypospolitej i drogach wiodących do jej naprawy*). DIP was formally dissolved in 1983. The Independent Publishing House NOWA (*Niezależna Oficyna Wydawnicza* "NOWA") was the best known publishing house of the anti-communist camp, and the scale of its activity was impressive. Between 1977 and 1989, it managed to publish around 300 books, as well as a number of underground magazines (such as *Zapis*, *Krytyka*, and *Tygodnik Mazowsze*). The establishment of these institutions in the second half of the 1970s signaled the emergence of a clear dualism in the intellectual field. This was reflected in the nature of the publications of that period, e.g. "official circuit" and "independent" (or "second "circuit"), and can be seen as strengthening the existing dualism, especially in publishing and media institutions operating in exile (including Radio Free Europe). The 1970s also saw the deepening of dualism in other spheres, especially the economy, where the distinction between "official" and "black market" was increasingly pronounced and partly institutionalized. This was reflected by the legalization of foreign currency possession (including private bank accounts in the state-owned Pekao SA bank) and foreign currency shops, in particular the Pewex chain.

By way of comparison, it can be instructive to examine what the intelligentsia was doing in the USSR at the time. Organized intellectual dissident activity had a longer tradition in the USSR than in Poland, e.g. the Initiative Group for the Defense of Human Rights was established as early as 1968. The Soviet dissident movement developed a network of human rights defenders and published underground books, magazines, etc. Poland even borrowed the concept of "samizdat" from the USSR. Polish dissidents thus lagged behind their Soviet peers, as their movement only gained momentum in the second half of the 1970s. However, the scale of opposition in Poland was incomparably greater than that in the USSR. There were two reasons for this: the Polish intelligentsia elite headed initiatives independent of the authorities; and the Polish dissident movement, which soon became a broad opposition movement, encompassed a broad social spectrum. The intelligentsia, particularly through its assistance to workers from 1976 onward (and especially during the strikes of 1980), managed to win the support of entire social sectors, including the crucial group of industrial workers and a wide variety of professional circles (Kennedy 1991). The opposition elite of the Soviet intelligentsia never enjoyed anything like this level of support—not even when the USSR was imploding. Prior to perestroika, most of the Soviet intelligentsia elite—even its most liberal members—lived partly in conditions of internal emigration. They were not involved in any unsanctioned social activities and limited their unofficial activities to narrow circles of friends. They typically conjured up an image of a small kitchen in a Soviet apartment where their "kitchen conversations" were held. The social space of the kitchen can be related to what Ingrid Oswald and Victor Voronkov called the public–private sphere of the late Soviet Union (Oswald and Voronkov 2004). Even public issues remained private when discussed there. Meanwhile, there were two public spheres in Poland, viz. formal/official and informal/unofficial. Media and public intellectuals were active in either or both.

2.11.4 The Re-traditionalization

Re-traditionalization was another process that encompassed most of the elite, including the field of power, and especially its crystallizing opposition pole in the 1970s. This differs from the trajectory of cultural change that took place in most Western countries, where secularization and a "counter culture" was adopted by the main social and political forces after 1968. This was accompanied by a tendency in the social sciences toward approaches deconstructing traditional cultural norms and historically conditioned customs and beliefs (e.g. post-structuralism). Poland, however, exhibited the opposite tendency after 1968. The victorious camp in the field of power, i.e. the dominant faction of the Communist Party, adopted moderately nationalist rhetoric combined with increasingly ritualized references to Marxism. A Western ideology of soft individualism compatible with the dominant technocratic framework of PZPR ideology was gradually adopted. All this boded well for standard of living and quality of life, particularly for the traditional family, and promised modernity in various dimensions and meanings, but importantly, without undermining traditional mores. Western counter-cultural movements were viewed with suspicion by the communist authorities. The other side, however, was significantly transformed, primarily as a result of the events of March 1968. The demotions of 1968 and the PZPR's turn to nationalist legitimacy came as a great shock to many liberals. Some not only lost faith in communism, but in all left-wing ideas.

A key aspect of this transformation was that the Soviet Union and Marxism–Leninism were no longer seen as the wellspring of Poland's modernization. This disillusionment was global, especially after the Warsaw Pact crushing of the "Prague Spring". This intervention cost the Soviet Union its presumed leadership of an alternative modernization project. Many people no longer saw communism as a modernizing project. Interestingly, a variety of sources, including personal memoirs, reveal that this was even true of the Soviet Union. It is intriguing that the much bloodier Soviet intervention in Hungary in 1956 did not similarly alienate people, which indicates that the conditions for significant homologous effects were lacking. The lack of synchronization of the transformations occurring throughout the Soviet bloc was probably due to liberalization and the thaw, which attracted crowds of young intelligentsia who were nevertheless disillusioned with what was happening in Hungary. But to return to the aftermath of 1968 in Poland, the key circle of left-wing intelligentsia liberals—former revisionists whom Adam Michnik called "the secular left"—found themselves in a difficult position. Having lost their privileged positions in the field of power, while losing their faith in communism or socialism under Soviet patronage as the most rational framework for modernization (with themselves at the helm), they had no option but to find new allies and formulate new ideas. An alliance with the politically marginalized but numerically strong right wing, with the Catholic Church as their main institutional patron, was now in the offing. The best known and detailed outline of such an alliance on an ideological level is Adam Michnik's seminal "The Church and the Left". Published simultaneously in Paris and (underground) in Poland in 1977, the book focuses on the search for the values common to leftists and conservatives. Michnik enumerated democracy, individual freedom, freedom of expression, and, more broadly, human rights, all of which had been trampled underfoot by the ruling Communists. However, behind the references to these common values, there was a clear fascination with many elements of the conservative cultural tradition. In particular, Michnik pointed to the contribution of conservative authors and thinkers to Polish culture, perceiving in their works a depth which had often been overlooked and which most purely Marxist authors lacked. This fascination turned out to be part of a more serious intellectual trend. It included a multidimensional rapprochement of the previously liberal intelligentsia with the Church and religiosity. For many individuals, these dimensions were separate. Some took up active religious practices, seeing them primarily as a form of independent civic activity, without necessarily undergoing a deeper spiritual transformation, while others underwent profound religious conversions without converting (or reverting) to any institutional religion. The latter group declared their interest in the absolute, transcendent, and spiritual values of the culture. Yet others either converted to other religious traditions, e.g. Hinduism, or to the faiths of their forebears, e.g. Judaism (Krajewski 2019). Stanisław Krajewski (b. 1950) coined the phrase "de-emancipation of the Jews" to describe the period. It is interesting to note that similar discoveries of conservative traditions as well as a fascination with religion, were beginning to appear among the previously liberal intelligentsia in the USSR, albeit less visibly and on a much smaller scale (Arkhangelskiy and Laskari 2011). This wave of re-traditionalization coincided with similar tendencies in several MENA countries, from Iran to Egypt, which were swept by religious revival with clear political overtones in the 1970s. The failure of most developmentalist projects in peripheral third world countries, to which Derluguian (2005) attributed the collapse of the Soviet project, probably accounts for this. Although these failures did not cause deep and prolonged economic recessions in these countries, they made their economies dependent on the Western core and led to growing inequalities. Their counter-elites, similarly to their Polish counterparts, turned toward national cultures and religious values as crucial symbolic capital which allowed for compensatory strategies in both the national field of power and global politics. In Poland, where the role of cultural capital had been of crucial political importance since at least the turn of the twentieth century, compensatory strategies proved to be extremely effective for the opposition elite. This was also possible because of the role of the Roman Catholic Church. The second half of the 1970s, and in particular 1978 when Karol Wojtyła was elected pope and took the name John Paul II, was their period of public triumph. The massive attendances at the masses Pope John Paul II celebrated during his 1979 pastoral visit to Poland demonstrated just how religious Poles had become. In this context, it is worth mentioning

Stefan Nowak's study of Warsaw students over several decades of communist Poland. It shows that, after a period of fairly low student religiosity in the 1960s, there was a gradual increase to a very high level during the 1970s, which was maintained throughout most of the 1980s, only beginning to decline after the fall of communism. In the USSR, this growth was much smaller and only accelerated during perestroika and then in the 1990s, i.e. there was no synchronization with Poland (Nowak 1991).

Coming back to the 1970s, it is interesting to note that the transformations discussed above, which also reoriented the geometry of the field of power, were clearly reflected in what could be called a neo-conservative turn in various fields of culture (within the broad meaning of the term). Most affected were the cultural field, the social sciences, and the humanities. The evolution of the creative work of the renowned composer Krzysztof Penderecki (1933–2020) provides a spectacular example. At the dawn of the 1970s, Penderecki, who had been known for his avant-garde, experimental stylistics, increasingly composed tonal works and orchestrated in the style of late nineteenth-century German symphonic music (e.g. his Symphony No. 2 "Christmas", composed in 1980). The critics accused him of betraying the avant-garde, but his new style found favor among the general public. Works alluding to political events, such as his Polish Requiem (1980–1984), which was classified as liturgical social realism, attracted particular interest. The *New York Times* obituary read as follows: "Mr. Penderecki was most widely known for choral compositions evoking Poland's ardent Catholicism and history of foreign domination, and for his early experimental works, with their massive tone clusters and disregard for melody and harmony. Those ideas would reverberate for decades after he himself had pronounced them 'more destructive than constructive' and changed course toward neo-Romanticism. (His decision to move on was partly political: The Polish avant-garde movement had created an unhealthy illusion of freedom in a country living under Communism, he said. But it was also artistic: Experimentation had reached an impasse, he told a Canadian interviewer in 1998, because 'we discovered everything!')" (Lewis 2020). Similar transformations can be observed in a number of other fields and subfields of culture and academia. Earlier formalists, Marxists, agnostics, and their ilk, discovered tradition, religion, and Polish history in its more traditional, conservative readings in the 1970s.

2.11.5 The Collapse of Communism

The communist system gradually collapsed during the 1980s. The failure of the communist project, first in Poland and then in other countries in the region, can be seen as part of a broader context in which peripheral and semi-peripheral developmental projects whose modernization had been driven by the state collapsed (Leszczyński 2017). Derulgian notes that the developmentalist project, which collapsed between the 1970s and 1980s, included several Arab and African states, as well as the Soviet bloc. They all harbored nationalist aspirations, they all implemented socialist policies of various kinds and in varying degrees, and they all emulated each other. Derluguian argues that the "similarities between developmentalist states were both structural-positional and dynamically acquired - a fact which helps to account for the broad similarity of the ways in which the majority of these state regimes were to come to an end in the 1980s" (Derluguian 2005: 79). According to this view, the consequences of the collapse of former developmentalist states were interrelated, and re-traditionalization was most likely one of them. Derluguian further argues: "Where the hegemonic imagery of globalization presents the worldwide liberalization of economic and political controls as a historical new beginning, we might rather discover that these trends represent the breakdown of previous world order and its attendant regime of accumulation" (Derluguian 2005: 79).

The significance of the economic failure of communist countries globally cannot be understated. That Poland and Hungary championed "democratization" at the end of the 1980s can be explained not only by political or cultural factors, as is usually done, but also by economic factors. These included their indebtedness, which had been increasing since the mid-1970s and was now unmanageable (Gerőcs and Pinkasz 2018). US sanctions on Poland after the imposition of martial law in

1981 further contributed to the deterioration of the economic situation. The transfer of power from the communist government to the opposition therefore involved handing over the management of a bankrupt country. This left no option but to seek Western capital on any terms and privatize, i.e. sell off whatever public property could be sold. Not surprisingly, the Polish nomenklatura was left thoroughly debilitated. Most of its erstwhile power was now in the hands of intelligentsia opposition elites, who embarked on a program of rapid and wholesale privatization targeted at Western corporations. This effectively eliminated the economic power of the former political elites and narrowed the window of opportunity for any new economic elites (e.g. oligarchs) to emerge. As a result, much of the Polish economy is now in the hands of multinationals, although some of it remains under state control. Most businesses owned by Poles are SMEs. No oligarch class on the scale of Ukraine or even the Czech Republic has therefore emerged. The absence of a powerful domestic economic elite appears to be conducive to the persistent dominant role of the intelligentsia elite, which wields considerable influence on the economy through managerial or political control, e.g. as public servants or as local agents or managers of multinational corporations. Thus, the current configuration of Poland's external dependencies can mostly be explained by the country being radically opened up to international capital and rapidly integrated with Western structures in all dimensions in an extremely asymmetric manner immediately after the collapse of communism (Hardy 2007; Shields 2008). An important component of this strategy was the more or less conscious decision of the intelligentsia elite (primarily anti-communist in orientation) to prevent the formation of an economic oligarchy in Poland at all costs. This was done primarily out of fear that such an economic elite would be dominated by the former PZPR elite (Staniszkis 1999). The sale of a significant portion of the national wealth to Western corporations prevented such a scenario. At the same time, it allowed a significant part of the intelligentsia elite to take up privileged positions in privatized enterprises and numerous economic, political, and above all, cultural institutions operating in the style of global capital and the local environment (Drahokoupil 2008). Meanwhile, in Russia, key sectors of the economy, including the extremely profitable mining industry, were kept under state control. Foreign capital has also been prevented from dominating many sectors, including the banking sector. This gives the ruling Russian political elite a significant degree of autonomy on the international scene and a hegemonic position on the domestic scene.

These observations justify the conclusion that Poland has, in effect, become an internal periphery of the Western core, while Russia has become an external periphery, retaining the character of an autonomous regional power pole (Kagarlitsky 2008). This gives an ambivalent character to contemporary Russia, as discussed by Viacheslav Morozov (Morozov 2015), who calls it a "subaltern empire". Moreover, the Russian social hierarchy is structured differently from both the Western and Polish models. The dominant capital in Western societies (and due to the dominance of the Western core, the entire world) is economic capital, to which both political and cultural capital are subordinated. Among the key ideologies of this system are supply and demand, competition, limited government, economic efficiency, freedom to contract, freedom of association, freedom of choice, and freedom of expression. In Russia, the dominating principle is political capital, whose central ideology is a strong and even an authoritarian state. This is even more pronounced in some of Russia's neighbors (and former Soviet republics), e.g. Belarus or Kazakhstan. By emphasizing state sovereignty vis-à-vis the Western core, this ideology is the principle means of compensating for the country's overall weakness vis-à-vis the global center. As mentioned above, it is cultural capital, with its diverse ideologies, that dominates in Poland. This is manifested in national identity and cultural heritage (the conservative variant of which assigns a significant role to the Catholic Church), which are considered the most important resources in compensating for Poland's economic and political weaknesses. Elements of the dominant legitimization narratives in Poland include the aforementioned ideology of post-communism, the Russian threat, and the key role of intelligentsia's heroes in ensuring the survival of Polishness as a primarily cultural game. Most of these narratives and ideologies have been reinforced, and some even developed, with the wave of re-traditionalization which started in the 1970s. Similar ideologies have a subordinate and often niche status in Russia and Belarus, where the post-imperial ideology of the "great victory" in the "Great Patriotic War"

(1941–1945), as WWII is known there, dominates. Also important are the motifs of the Western, and especially US, threat, and a widespread belief that a technocratic political elite is the most effective guardian of independence and the surest guarantor of modernization.

Another important difference between Russia and Poland is that the pro-core/Western (Euro-enthusiastic) option is incomparably stronger and may even be dominant in Poland. It prevailed throughout most of the post-communist period (especially until 2005), and continues to dominate in key intellectual circles (and many others). In Russia, the anti-core/Western option is close to hegemonic, while the liberal intelligentsia, which is the main Euro-enthusiastic force, is strongly dominated. In addition, the main source of capital in Russia is the state-controlled sale of mineral resources, while in Poland, it is the provision of cheap labor (primarily in assembly plants and call centers) and money earned abroad. The sale of property was an important component of the state budget in previous decades.

2.12 The Post-Communist Period

2.12.1 Post-Communist Transformations of the Field of Power

As far as the political dimension is concerned, the initial phase of the post-communist period (approx. 1989–2005) was organized around the main political cleavage that emerged in the late 1970s. It obviously evolved and adopted new meanings after the collapse of communism, but the two opposing camps that define its essence remained, as did the semantic opposition they engendered. Thus, on the one hand, there was the former anti-communist opposition (a highly diverse group), along with a motley of allies of various persuasions, united under the banners of a "return to normality" or a "return to Europe". They found common cause in their support for capitalism and condemnation of Poland's communist past and its evil legacy. Their best known leader was indisputably Lech Wałęsa (b. 1943), the first head of the Solidarity trade union, who was elected president in 1990. Arrayed against them were the "post-communists" whose core consisted of former PZPR activists. This group reinvented itself as a social-democratic party called The Democratic Left Alliance (*Sojusz Lewicy Demokratycznej*, SLD). Its leader, Aleksander Kwaśniewski (b. 1954), was elected president in 1995 and served until 2005. This camp set out to preserve the material and symbolic interests of various circles connected with the former system. The SLD declared its support for social justice, it implemented radical neo-liberal market reforms just as expeditiously and vigorously as its opponents (sometimes referred to as "post-Solidarity" forces). The two camps were therefore mainly differentiated by a historical dispute, albeit one that was nevertheless pragmatic. This was because the dominant historical narrative was contingent on the outcome and would be instrumental in legitimizing the ideology and policies of one camp while delegitimizing those of the other. Eventually, the dominant historical narrative came to be that of the post-Solidarity camp. The post-communist camp was left with no choice but to recognize its symbolic, and to a large extent, political, defeat. Anti-communism has thus become the dominant ideology of contemporary Poland. A largely naturalized symbolic frame contrasting generalized communism (considered evil) with generalized anti-communism (considered the essence of goodness) became and continues to be a doxa in the Polish public sphere. As a result, political rhetoric frequently invokes the bugbear of the communist as the personification of social evil. Political disputants invariably label their opponents neo-communists and their ideology neo-communist, neo-totalitarian, and anti-democratic. This rhetoric is employed by all intelligentsia circles toward their opponents, be they conservative, liberal, or social-democratic (Zarycki 2020a). Opponents are derided as "Homo Sovieticus" and their discourse ridiculed as "newspeak". This role of the communist past as a naturalized component of political discourse (a negative frame of reference) has strengthened the role of historians in Poland and has proved conducive to increases of funding of historical research, especially of the communist period. A special state agency called the Institute of National Remembrance (*Instytut Pamięci Narodowej*, IPN) was established to build an anti-communist historical framework. The Institute has both a research and an administrative function (e.g. it manages the archives of the communist secret services and issues certificates of non-cooperation) (Behr 2015). The liberal faction of the former anti-communist opposition sat somewhere between the two camps on either side of the key cleavage that characterized the first 15 years of post-1989 Poland. Tadeusz Mazowiecki, the first non-communist prime minister after 1989, was one of its leaders. Adam Michnik, and his *Gazeta Wyborcza* could be seen as the key intellectual or ideological forces behind this faction of the intelligentsia in the field of power. At times, the liberals were closer to the post-communists; at others, they were closer to the radical post-communists, especially when a dialogue between Michnik and post-communist prime minister Leszek Miller (b. 1946) ended in an open confrontation known as the Rywin Affair (Zarycki 2009). This clash resulted in the political marginalization of the post-communists and the SLD. This moved the conflict between the liberals and the radical anti-communists (as well as many of the conservatives) to center stage and reinforced the naturalization of the anti-communist symbolic frame of public discourse. The general context, including the geopolitical one, was however changing and the tension between liberals and conservatives, who had been close allies since about 1977 (as described in Michnik's programmatic manifesto "The Church and the Left"), erupted in a new cleavage which was to acquire a dominant role and have strong homological effects.

2.12.2 The Second Phase of the Post-Communist Transformation

After 2005, the political field, and to a large extent the field of power, was divided according to a new criterion. Generally speaking, this was the attitude toward the West, or more precisely, toward the core of the European Union. This change can be interpreted as a reflection of the geopolitical change in the field of power with a twenty-year time lag (Zarycki 2011). When Poland joined the EU, conflicts over its current dependence on the Western core became far more important than conflicts over its past dependence on the Soviet Union. That such historical interpretation has nevertheless remained as prominent as it is can be viewed as the result of the inertia of the structure of the political scene and the related semantic structures it produced, as well as the persistence of historically defined interest groups (in particular the post-communists). However, the main cleavage of both the political scene and the broader field of power has been evolving since about 2005. On the one hand, there are the Euro-enthusiasts, who generally support political, economic, and cultural integration with Europe. These forces in the field of power see European integration as the best remedy for most of the country's woes, including its economic backwardness and cultural traditionalism. They number liberals of various stripes, former moderate anti-communists, and former communists. Euroskeptics and conservatives, by contrast, tend to resist Western influences, especially liberal tendencies in the cultural and political dimensions. Interestingly, the economic divide has not become dominant, even though social inequality has widened considerably since the fall of communism. Conservative Euroskeptics, however, have been conducting a relatively pro-social policy since they came to power in 2015, so they can count on the support of a large part of the lower classes. The pro-European side is mostly (except for the minority camp of the young left) neoliberal in its economic outlook and mostly represents the interests of the more affluent middle and upper classes. Its best known political leader is former PM Donald Tusk (b. 1957), who was also the leader of the Civic Platform party (*Platforma Obywatelska*, PO). Tusk graduated from the University of Gdańsk in 1980 with a major in history. This is a common educational background among many of the political leaders of his generation who were active in the anti-communist movement. We can also speak of a relatively close synchronization between the Euro-enthusiastic camp and the representatives of international capital, as well as attempts to establish a dialogue between the Euro-skeptic camp and domestic business, which usually feels marginalized and under the heel of global corporations—even in Poland (Drahokoupil 2009). The core of this camp is made up of the former radical anti-communists. The current leader of this camp Jarosław Kaczyński (b. 1949) is also the leader of the ruling Law and Justice party (*Prawo i Sprawiedliwość*, PiS). Kaczyński was prime minister in the first PiS government (2006–2007) and served as a deputy prime minister between 2020 and 2022.

Born into a typical intelligentsia family, he graduated from the faculty of law at the University of Warsaw in 1971 and defended his Ph.D. at the same faculty in 1974. His mother, a long-time employee of the Institute for Literary Research (IBL) at the Polish Academy of Sciences, helped him get a job there. The fact that he was briefly a librarian at IBL would seem to illustrate the role of intelligentsia milieus in the biographies of members of the Polish field of power. However, recent research on the behavior of the banking elite in Poland (Naczyk 2021) reveals an interesting tendency within that part of the field of power. This is generally seen as an effect of the 2008 global financial crisis, which made those Polish bankers dependent on the Western core, i.e. those who managed the subsidiaries of global financial corporations, as well as those of other central European states, painfully aware of the vulnerability of the national economies of the region to global systemic shocks (Nelson 2020). Marek Naczyk (2021) argues that it prompted these bankers, who could be seen as a "comprador economic elite" (Drahokoupil 2008), to push to have foreign-owned banks nationalized (or, more precisely, to have the state buy them out), and development institutions reformed so as to better support domestic firms. This goal was largely achieved by state institutions, which, as Naczyk argues, were used after some influential public actors had been co-opted. This campaign was launched under liberal governments by Jan Krzysztof Bielecki (b. 1951), a former prime minister and former president of the then Italian-owned Bank Pekao SA. It was continued under conservative rule by the current PM, Mateusz Morawiecki (b. 1968), onetime head of the Spanish and Irish-owned Santander Bank (formerly BZ WBK). Morawiecki, who graduated from the Faculty of History of the University of Wrocław, was active in the anti-communist underground, and his father, Kornel Morawiecki (1941–2019), was one of its best known leaders. As prime minister, he has been overseeing an attempt to build a new developmental model in the Central European periphery; one based on the use of state institutions to support local accumulation. This mechanism has parallels to what can be observed in other countries in the region, notably Hungary. In Poland, however, as Naczyk points out, it was not so much populist politicians or the local proto-bourgeoisie, but former comprador bankers who were the main promoters of the campaign. Still, the case of Morawiecki seems to suggest that they largely remain elite members of the intelligentsia, and that their ability to implement the abovementioned national developmental project is contingent on their ability to form diverse alliances and utilize state institutions. Currently, the most important group is the Euroskeptic faction of the intelligentsia elite, whose legitimacy is derived from conservative (or nationalist/populist) rhetoric. It is telling that such crucial economic transformations were not initiated by the local economic elite, which seems too weak in the field of power, but managers of foreign-owned banks, who owe their positions to their presence in intelligentsia elite networks in the days of the anti-communist opposition.

The post-2005 dominant cleavage of the field of power is homologically reflected in most of the social sciences and humanities, albeit in varying degrees. The Euroskeptic, conservative side has a tendency to value Polish traditions, is skeptical toward liberalism, especially as manifested in diverse, progressive cultural reforms, and views anything related to communism with abhorrence. Opposing it is an anti-populist ideology that tends to idealize all things Western and is suspicious of Polish traditionalism, nationalism, and often the Catholic Church, which openly supports the ruling party. These conservatives are also often seen by liberals as the "heirs" of communism (e.g. they supposedly evince a "communist mentality"). Thus, a significant part of this sector of the social sciences also shares the basic doxa of anti-communism, and sees most of its opponents as infected with the evil heritage of communism, usually imagined in its nationalist version (e.g. the Partisan group in the 1960s). This may be related to a certain cyclicality in the evolution of the central cleavages of the Polish field of power. Even the cleavages of 1944–1957 and 1970–2005 bore some resemblance, as the hegemony of the Soviet Union was the main point of reference in both cases. Similarly, the dominant cleavages of 1958–1969, and well as the post-2005 cleavage, can be seen as homological in certain respects. Here, the main point of reference in both cases was/is the West (real or imagined). Relatively cosmopolitan liberals were confronted with locally oriented and nationalist (at least liberals viewed them as such) conservatives. For this reason, Euro-enthusiasts sometimes compare Jarosław Kaczyński to Władysław Gomułka, as an archetypal communist and nationally oriented leader. Valentin Behr (2017) shows how this current cleavage translates by refraction into two threads in the field of historiography. The conservative side is manifested by a "patriotic" (i.e. military, heroic and martyrological) history; one strongly supported by the IPN. Other historians, however, have been investigating the narratives of groups that have been overlooked or marginalized in this mainstream "national pride" narrative, e.g. Jews, the working class, the peasantry, and women. The two opposing sides also differ in their methodological approach. While the conservative camp tends to rely on conventional historical methods, the liberal camp has been increasingly turning to ethnographic and anthropological methodology in order to "give a voice" to the marginalized narratives (Gospodarczyk and Kożuchowski 2021). This duality is also visible in several other social science and humanities disciplines, although it varies in extent. At times, even the same theoretical schools, e.g. post-colonial studies, are adopted from the Western core and interpreted in two parallel ways: one conservative, attempting to present Poland as an undeserving victim (both currently and historically); and one left-liberal, attempting to present Poland and its elite as an oppressor of both its neighbors (e.g. Ukrainians) and its own marginalized groups (Zarycki 2014).

The change in the organization of the field of power which occurred between 2005 and 2015 can therefore be seen as a reorientation from a configuration based on differences in attitudes toward Moscow, which emerged at the end of the 1970s, to one based on differences in attitudes toward the generalized West. Thus, these are two center-periphery cleavages, although they are oriented toward different centers. It is also worth noting that within the field of power, these cleavages translate into a division between factions based on the ability to benefit from relations with an external center. This implies possession (or lack thereof) of networks of direct social contacts in the fields of power of respective centers (countries of the core). In the communist period, some of the higher party elite enjoyed the privilege of having direct relations with members of the Soviet communist elite in Moscow. At the same time, some of the liberal intelligentsia had contacts with marginalized dissident intelligentsia circles in USSR. The liberals in the Polish field of power have a clear advantage over their opponents in terms of direct contacts in most Western fields of power. This gives them considerable political and symbolic assets, especially in influencing the international perception of Polish society and politics.

2.12.3 The Fate of the Intelligentsia in the Post-Communist Period

It is worth noting that there are two opposing views of the fall of communism and the post-communist period from the perspective of the intelligentsia. On the one hand, the fall of communism is associated with the final collapse of the intelligentsia (its key crises had previously been seen as occurring during WWII and the Stalinist period). From this perspective, the intelligentsia is seen as destroyed primarily by capitalism and the free market, in which it does not know how to find its place. In effect it has been replaced by a Western-style middle class. One argument in support of this thesis cites the pauperization of part of the intelligentsia, especially its lower ranks (e.g. teachers). Another cites the decline of important intellectual institutions, including intellectual journals. The monthly *Res Publica*, published underground in 1979–1981 on the initiative of Marcin Król (1944–2020) and Wojciech Karpiński, makes for an interesting case study. The magazine was relaunched as a legal publication in 1987, which aroused controversy, as it had the approval of the authorities, leading it to be condemned by a large part of the opposition intelligentsia. It initially had a large readership, but circulation fell immediately after the fall of communism. In its place, the journal *Res Publica Nowa* appeared in 1992. Neither it nor any other intellectual journal regained their previous popularity during the post-communist period. Another example is the weekly *Po prostu*, which was closed in 1957, relaunched in 1990, and closed again after a year due to lack of readership. This can be seen as part of a more general tendency, viz. the decreasing social significance of intellectual debates and the diminishing role of intellectuals, who have always been not only members of the intelligentsia, but also its public face. Many of them have even been living legends (Turkowski 2018). These debates are gradually giving way to purely political discussions in which intellectual rigor is no longer relevant, and where academic credentials count for nothing. This demonstrates that the former intelligentsia has been replaced by a political class and a new business elite.

On the other hand, however, a number of arguments can be made to support the thesis that the intelligentsia has not been marginalized in the field of power, although its roles have changed significantly. Although the former role of the intelligentsia elite, viz. as intellectuals and scholars, has diminished in importance, it has acquired other significant roles, most notably those of politicians, journalists, and managers. I should point out that I understand the term "intelligentsia" in a much more technical way than do most of its Polish members. Identity and ethos are usually stressed as crucial membership criteria, whereas I emphasize the technical criterion of cultural capital. More precisely, I see the role of cultural capital as the dominant resource in determining social status in relation to other types of capital. Thus, an intelligentsia member who happens to be a manager and who makes his living primarily from employing his cultural capital (e.g. selling his knowledge and expertise) rather than drawing on his accumulated economic capital remains an intelligentsia member provided his accumulated economic capital becomes a more important determinant of his social status than of his cultural status. Therefore, although many intellectual journals have folded or declined in circulation, and the status of the social sciences and humanities in relation to several other fields have dropped, since 1989, a number of intelligentsia circles managed to retain their key positions in the field of power. They rely on new types of organizations, e.g. foundations NGOs, political parties, or businesses. They have also gained access to media, including television and high-circulation dailies. The best example of this is the daily newspaper *Gazeta Wyborcza*, edited by Adam Michnik since 1989. As Poland's largest circulation daily, it is a very important tool for the broader intelligentsia faction to influence the Polish field of power and Polish social life (Zarycki 2009). As already mentioned, the post-communist period can therefore be seen as being dominated by the intelligentsia, although it no longer principally operates in the academic and humanistic fields, but rather the political and economic fields. In the latter, the intelligentsia appears primarily as managers working for foreign companies, managers of state-owned companies, and expert advisers and consultants. Crucially, the new bourgeoisie, i.e. the landowning class for whom economic rather than cultural capital is the main resource for building social status, remains relatively weak. At the same time, no political elite has been able to self-reproduce in the long term in Poland, in contrast to the nomenklatura in Russia and post-Soviet countries such as Belarus.

2.12.4 The Field of Higher Education and Its Autonomy

To conclude this overview of post-communist developments, I would now like to focus on the transformations in the field of higher education in Poland, which was expanded considerably after 1989. First, the number of students increased several times, as already mentioned. This was the result of facilitating access to higher education, e.g. by radically expanding number of students at public universities, with most of the additional places available only for tuition fees-paying students. At the same time, many private universities offering fee-only degrees were established. Most employed lecturers from public universities for whom working at private universities provided additional income. This spectacular expansion of the university system has resulted in several significant changes. First, the increase in the number of graduates has been met with degree inflation in many disciplines. Some 378,000 students were reported to be enrolled at Polish universities in 1989. This figure increased to 795,000 in 1995 and to over a million in 1997, mainly due to the mushrooming of private universities that mostly offered social science and humanities courses. Numbers peaked at 1,950,000 in 2005 and have been slowly declining since (Popiński 2018, and Statistical Yearbooks of Poland). Second, the quality of research declined, as academics overburdened with teaching did not have time to conduct it. Third, many academics, particularly professors in the social sciences, saw their incomes increase significantly. This obviously made them economic beneficiaries of the new system. The downside was a weakening of the potential of Polish research and scholarship, especially in the social sciences. Its position repeatedly fell in the global rankings during the 1990s (Warczok and Zarycki 2018). In a symbolic sense, then, the debasement of the social sciences and humanities, which had been ongoing since the 1970s, continued unabated. For many of the intelligentsia, academia became a means to earn a living rather than an opportunity for creative development. Lecturing at public universities was just one of many occupations, which often included lecturing at private colleges and activities in other spheres. This could be seen as the peak of the multi-positionality of the intelligentsia, which was later legally restricted. Nevertheless, multi-positionality remains one of the characteristic traits of the intelligentsia elite and is one of its key strengths. For one thing, it enables, them to easily adapt to changing economic and political conditions, and play the role of middle-man in a variety of fields, including the field of power. This disposition to activity in diverse spheres and tendency to acquire more than one position among the Polish intelligentsia had been already noticed by Wedel in the 1970s (Wedel 1986). The ensuing crisis in the social sciences and humanities in Poland is often interpreted as a crisis of the intelligentsia. I see it rather as an aspect of the transformation of the intelligentsia and its main areas of activity. In particular, it evidences an increase in the importance of the political and economic fields, in which many of the intelligentsia retained their basic social status and identity as defined by cultural capital. Thus, the intelligentsia elite is not threatened either by a strong native economic elite or by a multi-generational political elite (nomenklatura). The field of power continues to be dominated by internal struggles between various factions of the intelligentsia.

It should be recalled that Polish science was granted a great deal of autonomy and self-governance on many levels after the fall of communism. As a result, power in public scientific and learning institutions, especially those subordinated to the Ministry of Science and Higher Learning, passed almost exclusively into the hands of their academic and institutional elite. Universities and the institutes of the Polish Academy of Sciences elect their own leadership at various levels. It is also significant that the Council for Scientific Excellence (until recently the Central Commission for the Academic Degrees and Titles which was a heir to Central Qualification Commission), a key state institution, is elected by Polish academics in possession of a habilitation degree. This body is responsible for approving scientific promotions, awarding degrees and conferring titles. During the communist period, appointments to this central institution, and consequently its decisions, were largely controlled by the authorities. This is still the case today in many of the former Soviet Republics, e.g. Belarus and Russia, which have the same central institutions (known there as Higher Attestation Commission, VAK). As a result of this autonomization of the field of science, which has mainly become a tool to further the interests of its elite, it has also become largely autonomous from the system of world science. This tendency can be seen as a part of a broader process, viz. the autonomization of most key social fields after, or even during, the fall of communism. That fall brought about several major changes. First, there was a loosening of the administrative grip of the state, especially on the part of its central actors, e.g. the communist party, on these fields. Second, there was a deep and persistent economic recession throughout the 1980s, which continued well into the 1990s. This forced funding cuts to many public institutions, and their staff additionally saw the purchasing power of their salaries eroded by high inflation. The management of many such fields, which include not only academia but also health services, law enforcement, and the judiciary, was largely taken over by their elites. While this implies increased autonomy, even in matters such as strategic planning and structural reforms, it induced managers to find creative ways of improving the material situation of both their institutions and their staff. This resulted in diverse forms of privatization, semi-privatization (including privatizing part of the revenue earned through the provision of public services, and outsourcing to private institutions run by the elite of public agencies), and corruption.

The multi-positionality of the elites of these fields was an important aspect of these processes, as it allowed for a combination of activities in the public and private spheres. One example of such multi-positional individuals were the "institutional nomads" (Ozieranski and King 2016), but overall, these processes seem to have reinforced the role of the intelligentsia as a flexible elite possessed of sufficient cultural capital to be simultaneously active in several fields. The prime ministers Jan Krzysztof Bielecki and Mateusz Morawiecki are prime examples of this type of intelligentsia elite. Interestingly, corruption has been considerably reduced in Poland, while in Russia, it remains a serious concern. One possible explanation for this is that, in Poland, most "additional" incomes have been institutionalized through a creative and complex system of public and private institutional channels, whereas in Russia, what Derluguian (2005) terms "nomenklatura's rent" (a hangover from Soviet times), has been preserved in some form. Given its purely informal and partly illegal nature, it gives additional leverage to the Kremlin to arbitrarily punish insubordinate officials. These time-honored techniques, combined with a strong executive in place of the former communist party apparatus, and powerful law enforcement institutions (including the secret services), enabled the state to regain control over most of the fields related to public sphere by the end of the twentieth century. The role of the Higher Attestation Commission (VAK) is but one small example of a structural outcome colloquially known in Russia as the "vertical" or backbone" of power. In Poland, by contrast, the state, despite the best efforts of consecutive governments, remains relatively impotent when faced with the autonomy of several fields, which largely ignore its demands. This is particularly noticeable when compared to Russia, and even to Western countries in some respects. For example, Artur Wołek notes that a peculiar weakness of Poland's central government is that it directly controls only a minor part of the state budget (Wołek 2014). Most public spending is fixed and most of the remainder is directed through diverse semi-autonomous agencies, often partly self-financing and thus not visible in the central budgetary listings. The genesis of the proliferation of these agencies in post-1989 Poland is complex. First, it can be related to the important trend of "decentralization", which was a central plank of the structural reforms of the 1990s. This movement away from a centralized, party-controlled authoritarian state could be seen either as a positive aspect of "democratization" or as the precipitate dismantling of stable state structures, resulting in such negative consequences as the almost complete disappearance of the welfare state infrastructure they had supported. Derluguian contends that this process had two basic dimensions: territorial, with national separatism as the main driving force; and sectoral, with capitalist privatization as the main driving force. In the Soviet Union, the territorial dimension of the fall of communism was one of the most dramatic, while in Poland, it was rather smooth and led to the more or less successful reform of the regional organization of the country (Zarycki and Kolankiewicz 2003).

Another reason for the proliferation of state agencies after 1989 was that there was little, if any, structural reform in most Polish government institutions. This might have resulted from the relatively strong influence of the post-communists as a political force in the 1990s. As they had a vested interest in preserving existing communist institutional arrangements, they resisted structural reforms, especially of the state administration. Nor had the elites of many of the fields discussed above, including academia, medical services, and the judiciary, any interest in structural reforms, as they found the autonomy they had acquired with the fall of communism most agreeable. In this context, the reforms introduced at the beginning of the 1990s, especially the radical economic reforms, were partly implemented by circumventing the institutional structures inherited from communist times. This was achieved by establishing a number of semi-autonomous agencies that were partly exempted from ministerial oversight and control, and which were assigned new tasks and charged with overseeing crucial areas of economic and social reforms. Poland stands apart from other CE European post-communist countries in that the scope of institutional reform was minimal after 1989. This may be due to the gradual and negotiated nature of the fall of communism in Poland. As mentioned above, it was based on a consensus of the former communist elite with part of the anti-communist, mostly intelligentsia, elite. Some of the most striking contrasts are provided by Lithuania, Latvia, Estonia, and Czechoslovakia. These countries had much deeper structural reforms than Poland in the early 1990s. They also had a much weaker traditional intelligentsia elite than Poland. The reforms of legal institutions in Estonia and Poland, as Christopher Hartwell and Mateusz Urban point out, followed completely different paths (Hartwell and Urban 2021).

The genesis and mechanisms of this kind of autonomy were analyzed in a case study of the Polish field of political science conducted by Tomasz Warczok and myself (Warczok and Zarycki 2018). Again, the minimal scope of academic reform in Poland from 1989 stands in stark contrast with Estonia, where particularly ideologically sensitive fields, e.g. political science, were reconstructed from scratch (or more precisely built from scratch, as there was no formal political science in the Soviet Union before 1991). In Poland, by contrast, political science had begun to be institutionalized as early as the late 1960s. Thus, in Poland, most of the institutional infrastructure of the field was preserved, and most of the staff from PZPR-affiliated institutions moved to state universities. This was accompanied by an ideological transformation of those academics who had worked on areas such as Marxism–Leninism, and who now generally turned to, e.g. European integration, state security, and Euro-Atlantic military integration. In this context, it is worth emphasizing that the field currently has considerable autonomy from the global field of science (more specifically, the global sub-fields of particular disciplines) and the bureaucratic field (more specifically, the Polish state). However, this does not mean, in most cases, especially as our study of political science has shown, complete autonomy toward the field of power. Political science remains a strongly heteronomous discipline in Poland, but its dependence on the field of power mainly takes the form of the academic elite being involved in other sectors of the field of power, and even other fields, especially the political and administrative fields. Thus, as with many other fields in academia and the public sector generally, political science is not formally or institutionally subordinated, but it is used by individual academics, many from the intelligentsia or their circles, to advance the interests of other sectors of the power field. As a result, the divisions of the field of power have been homologically transferred to the structures of most academic fields, of which political science is a good example.

The autonomy of the Polish social sciences and humanities from the system of global science, combined with their informal subordination to the interests of the national intellectual elite has eroded the international competitiveness of most Polish science disciplines. This international degradation of Polish science in the first two decades after the fall of communism prompted reforms to ameliorate the situation. These were initiated by Science Minister Barbara Kudrycka (b. 1956) in 2009–2011 (Kwiek 2021). The National Science Centre (NCN) was established to fund internationally competitive research. The ministry also started using international citation databases to evaluate the scientific output of institutions. These reforms aroused considerable resistance from many in the scientific community. Every discipline had a critical contingent that usually declared that any attempt to exert pressure on internationalization would endanger fundamental intellectual values and principles and threaten to commercialize research, which, they claimed, should serve the national public and not international "customers". The division into supporters and opponents of internationalization has become an important axis dividing most social science and humanities fields in Poland.

The cyclical nature of state policy toward the scientific field in Poland is also worth noting here. By this I mean the constant fluctuating between autonomy and the restriction thereof. The periods during which autonomy was curtailed have usually been portrayed extremely critically, if not dramatically, by the intelligentsia. It is worth noting, however, that autonomy has never been abolished. This independence of the political elite (at least to some degree) has allowed for the long-term development of Polish science, and particularly its intelligentsia elite. As already mentioned, even in the post-war period, there were cyclical changes in higher education legislation, which were largely dictated by the dominant relations between the forces in the field of power. However, even before WWII, the Sanacja government, at the beginning of its autocratic rule in 1933, amended the law on higher education to limit the autonomy of universities. A similar cycle, only longer and more complex, began soon after the war. The first declarations of the authorities in 1945 and 1946 proclaimed that the subjection of science to the influence of political parties was harmful. They defended the pre-war condition of Polish science and warned against arbitrary measures. Beginning in 1947, however, the emphasis shifted to a need for change and a strengthening of ties between science and the state. This tendency was reversed on the wave of the reforms that followed 1956. A law adopted in 1958 significantly liberalized the system of higher education, mainly by introducing the election of rectors and granting faculties broad autonomy. The Act on Higher Education was amended in 1958 to partially decentralize universities, restored them to the rank of research centers, and increase their independence by reintroducing elections for their authorities. As a result, the rectors, vice-rectors, and deans became non-partisan individuals who enjoyed respect among the academic community. The new law also abandoned the Soviet model and reinstituted the traditional system of academic degrees. These liberal laws were replaced in turn after the wave of events of March 1968. Gomułka criticized the lack of respect of academic elites toward the authorities and opined that the 1958 Act was the work of "revisionist forces".[4] Elections of rectors were suspended, universities were subordinated to ministerial fiat, and dozens of "March docents" (assistant professors without habilitation) were appointed, and many departments were either liquidated or merged into institutes. This state of affairs lasted until the "Solidarity" period, i.e. the end of 1980, when, as a result of liberalization, a new law was drafted to give universities back much of their lost autonomy. It was only implemented in 1982, as the country had been under martial law since December 1981, and the authorities had tightened their control over many areas of social life. This paradoxical state of affairs lasted until 1985, when another law was introduced, limiting the autonomy of universities. Higher education was liberalized soon after the fall of communism. A 1990 statute granted full autonomy to HEIs with most of their decision-making authority vested in their senior academic staff. This situation prevailed until 2005 when the Polish education system was changed to comply with EU standards (Poland joined in 2004). Unfortunately, space does not permit a detailed discussion of the changes this entailed.

Similar cycles can be observed in the management of the Polish Academy of Sciences. As already mentioned, PAN was established to replace two independent institutions with historical traditions, viz. the Polish Academy of Learning (PAU) and the Scientific Society of Warsaw (TNW), which had been liquidated by the communist authorities. This "reform" was obviously intended to limit the autonomy of researchers that the historical institutions had provided. However, PAN assumed a dual role from the outset, as it combined the features of an autonomous corporation of scholars with those of a government system of research institutes. Its statist role was dominant during the Stalinist period,

[4] These changes were discussed by Jarosław Kaczyński in his doctoral dissertation (Kaczyński 1976), which was devoted to university management in post-war Poland.

when its management became a conduit for government pressure on its functioning. PAN was partially decentralized in 1956. The Program Committee, chaired by the renowned physicist Leopold Infeld (1898–1968), prepared a paper with the significant title: "The Polish Academy of Sciences - an autonomous organization of scholars". In January 1957, the General Assembly elected Tadeusz Kotarbiński President. A new PAN Act adopted in 1960 divided management into an elected Presidium (including the President and Vice-Presidents) and a Scientific Secretariat appointed by the authorities. A 1970 amendment stripped these corporations of their authority to directly manage PAN's establishments. Finally, in 1997, a new PAN Act provided for significant self-government of scientists (Pleskot and Rutkowski 2009).

2.13 The Polish Fields of Social Sciences and Humanities After WWII—A Synthesis

2.13.1 Interface Periphery Effects and the Polish Academic System

I would now like to summarize the foregoing analysis, briefly reconstruct the basic structures and evolution of the academic field in the communist period, and summarize its key homologies with the Polish field of power and other fields. During the Stalinist period, the main opposition in the academic fields of the social sciences and humanities was homological to the one in the field of power. Some academics were strongly relying on Marxist approaches, including class analysis and anti-imperialist and anti-bourgeois (in short, anti-Western) assumptions. Others, in politically marginalized positions, tended to defend, or at least implicitly retain, positivist values and stand on a strong empirical platform defined in a rather traditional way. From their point of view, those loyal to the partly line were ideologists rather than scientists. In the humanities, the marginalized camp defended humanistic, national, and other values which could be considered traditional. Some of its members lived in exile in Western countries. In any case, this division of academic fields was clearly homological. The main political cleavage of the period pitted supporters and activists of the Stalinist reforms against a variety of opponents whose positions ranged from neutral to vehemently anti-communist.

In his excellent study of the field of Polish historiography, Valentin Behr (2017, 2021) reconstructed a configuration of that specific realm during the Stalinist period. Whether it was fully representative is debatable, but it definitely shared several basic structural characteristics with most of the other social science and humanities fields at the time. Thus, as Behr argues, at the advent of Stalinism, many historians focused on "politically safe" subjects and periods (invariably prior to the twentieth century). Modern history, as strictly politicized, became the exclusive domain of a faction of "party historians" (*historycy partyjni*). These scholars were mostly promoted on political grounds. They were not always young, but were seldom overly distinguished, and usually lacked top-level international experience, especially in the West. They were invariably PZPR members and devoted much of their time and energy to the Party, which made it near impossible to enhance their knowledge and expertise or engage in serious research. This resulted in "dominant" historians having far less impressive academic and professional credentials (according to academic criteria rather than Party loyalty) than their dominated counterparts, who were focused on accumulating scientific capital. By drawing on their solid pre-war education and international experience, they worked on in-depth studies, many of which concerned distant epochs. These works were often methodologically innovative. Many of them synthesized traditional approaches with relatively unorthodox, yet original interpretations of Marxism. This was one way traditional historians played the political game. They referred to Marxist works partly to establish their ideological legitimacy, as this was expected by the authorities. However, they often did so in a sophisticated manner, which led to the methodological innovations mentioned above. The authorities usually turned a blind eye to these departures from orthodoxy as these works were only of interest to a narrow circle of specialists and only published in small editions. For many of these scholars, however, adopting Marxist approaches was more than simply a matter of tactics. Some had been interested in such methodologies before the war and were sometimes associated with the leftist movement. A similar milieu of young leftist scholars, who had begun their careers before the war, entered the new academic institutions in the fields of linguistics and literary studies, as is described in the next chapter. Both cases involved circles that had been marginalized before the war. Many of these scholars had worked as archivists or university tutors prior to 1939. Their status changed after WWII as they gained access to much better working conditions, and they applied research paradigms they had known before, consciously adapting them to the field context. John Connelly (2000) notes that "ironically, it was those who took Marx most seriously, like historians Witold Kula, Marian Małowist, or Stefan Kieniewicz, who proved most resistant to treating Marxism as an orthodoxy" (Connelly 2000: 177).

As Behr shows, however, the situation in the field of Polish historiography changed radically after 1956, when "Party" historians lost their previous political support and had to fall back on their paltry scientific capital. This allowed previously marginalized classical historians to move into the mainstream and assume dominant positions in the field. They brought with them the innovative methodologies they had developed during the Stalinist period, including those that utilized unorthodox approaches to Marxism and related schools. Among the best known works in this stream are those of Marian Małowist (1909–1988), who made important contributions to dependency theory. In particular, his study of Poland's dependence on the Western core in the sixteenth and seventeenth centuries provided an important inspiration and empirical contribution to Immanuel Wallerstein's world-system theory (Kola 2018; Sosnowska 2018). The confrontation between these two circles has left a strong positivist trait in the dominant Polish historiography. This mainly concerns the conscientious commitment to sources and classical archival work as the core of the historical workshop sustained by traditional historians. Attachment to such positivist values was supposed to emphasize fidelity to the norms of scientific rigor (which was of much less concern to Party historians), and to provide a bulwark against political interference in scholarship and research (more broadly, against ideologization). Thus, since the end of the 1950s, the identity of the discipline has been mostly defined by the use of sources and the adoption of an empirical approach to limit ideological interference in research work. These values permanently formed the professional ethos of the Polish historian. "Work with sources" underpinned the discipline's autonomy. Only the most recent history was politicized, and traditional historians left it to "Party historians". Although the tradition of unorthodox experimentation with Marxist approaches came to an end in the 1980s, the strong attachment to the traditional, archival determinants of scientific rigor has remained in the field of Polish historiography. There are clear parallels here with the scrupulous attention paid to survey data and quantitative methods in Polish sociology. As in historiography, the dominant social scientists of the 1950s and 1960s attempted to build their capital of scientism on the "objectivity" of quantitative methods, particularly those taken directly from the United States. This positivism placed them at odds with researchers loyal to the Party, who usually resorted to undisguised, sloganeering references to Marxism. This has led liberal sociologists to dismiss them as ideologues. They responded by taking the positivists to task for their excessive attachment to surveys as the main source of knowledge about society. In the liberal camp, the defense of quantitative methods, similar to the defense of archival work among independent historians, was defined as defending the fundamental value of truth. Implicitly, the Communists stand accused of, inter alia, falsifying statistics and other sources and are generally portrayed as enemies of the truth. Those scholars who resisted political pressure mostly attributed their positivist stance to the Warsaw-Lviv school with its devotion to truth and logical or ethical reasoning.

The 1960s were a period of transformation. Many former Stalinists adopted much more liberal positions. While both radical camps of the previous era retained some say, in the centers of most of the fields of social sciences, a moderate center emerged. Its members frequently tried to combine Western-inspired theoretical ideas, including those of a more positivist bent, with Marxist approaches and values, e.g. central planning, and socialist principles. These attempts by Polish scholars to develop original variants of social theory, including what could be seen as critical theory—partly inspired by creatively interpreted Marxist tradition and new Western intellectual trends—were considered innovative and attracted interest from all over the world. These scholars included economists Michał Kalecki (1899–1970) and Oskar Lange (1904–1965). Revisionism resulted in several interesting critical and often reflexive studies of Marxism and Marxist-inspired approaches in several fields. These were arguably the most creative years for Polish scholarship and research since WWII. However, due to the marginalization of the former revisionist camp, and the general transformation of the field of power, which strongly impacted academia, these unique conditions disappeared in 1968. By the end of the 1970s, native Polish schools of thought had noticeably lost much of their dynamics and attractiveness. This was the case with, e.g. the Polish school of planning (e.g. Secomski 1974), the Polish school of social stratification research in sociology (e.g. Ossowski 1963; Wesołowski 1979), and the field of Polish economic history, which flourished in the 1960s and strongly relied on dependency theory and even contributed to the development of world-system theory, as Anna Sosnowska has shown (Sosnowska 2018). Besides Marian Małowist who has been discussed above, one could mention in this context such leading figures of that field as Witold Kula (1916–1988) or Jerzy Topolski (1928–1998). The unique configuration of the fields of the social sciences and the humanities in the 1960s was conducive to the creative processes mentioned above and could be interpreted in interface periphery terms. The 1960s was also a period of relative autonomy from the USSR, which communist Poland gained after 1956, as well as more openness in contacts with Western and global academia. Communism, and especially the Soviet Bloc, to which Poland belonged, still enjoyed considerable international prestige. For many around the globe, communism represented an alternative model of modernization. The USSR made considerable scientific progress in many fields in the 1960s. This allowed Polish scholars to benefit from the prestige of the communist bloc and socialist ideas, and from being able to work directly with Western scholars and gain privileged access to Western academic centers (compared to scholars from other communist countries). This also made Poland attractive to Western scholars, who were interested in establishing such contacts and who followed the work of many Polish researchers, whose originality was prized not only on account of their talents, but also

Poland's unique geopolitical condition. The flourishing of international contacts in the 1960s can thus be seen as a distinct aspect of Poland's double peripherality during this period. There are at least two dimensions to this. The first is political. Poland was the most liberal country of the Soviet bloc during this period. This allowed for visitors from the West, including academics, scholars, and researchers, and for Poles to travel to the West in fairly large numbers (especially those belonging to elite circles). This aroused interest in Poland in other communist countries, especially among the intellectual elites of, e.g. the USSR, Czechoslovakia, and Hungary. Polish institutions, including the mass media, granted access to Western culture, which was almost completely inaccessible in other Soviet bloc countries. This, in turn, led to an interest in the Polish language as a medium to access that culture. The second dimension of Poland's double peripherality was science and culture. Here, Eastern and Western influences began to cross-fertilize in a most creative manner. Poland was still influenced by Russian culture, through institutional pressure to propagate Soviet culture, informal cultural transfers built on direct contacts between members of the intelligentsia, and interest in the legal and informal cultural output of the USSR (which had also been slightly liberalized) on the part of individuals. At the same time, however, more ideas were being imported from the West. Additionally, Westerners and Easterners could meet face-to-face in a fairly comfortable manner in Poland. For this reason, a number of important conferences were held in Poland in the 1960s.

Consequently, it should come as no surprise that many older Polish scholars speak of "the Golden 60s" (e.g. Sułek 2016), especially in connection with Poland's academic and cultural fields. This positive perception of the 1960s can therefore be related to the interface periphery status of the day, which optimally balanced the relevant geopolitical forces so as to bestow considerable autonomy on Poland and its social sciences and cultural fields. After 1968, there was a weakening of Soviet symbolic and academic capital and a rise of Western power in the region. These processes eventually dismantled the unique configuration of global forces of the 1960s. Thus the status of a double (interface) periphery began to diminish in the 1970s.

The emergence of a new configuration of the field of power in the 1970s also changed the dominant oppositions in many academic fields. The remaining revisionists were not only abandoning Marxism–Leninism, but gravitating toward anti-communist positions. Many of them embraced conservative, and even spiritual philosophies, e.g. Leszek Kołakowski. Although Kołakowski was forced into exile in 1968, he retained a strong influence on Polish social sciences and humanities. Positivism was also popular in that sector of the social sciences that was homological to the increasingly oppositional pole of the field of power. This rejection of Marxism was accompanied by an abandonment of class analysis, which was gradually replaced by modernization theories. The Warsaw-Lviv school, discussed in more detail in the following chapter, proved to be increasingly popular in that sector of the academic field. What particularly made it attractive was that it combined traditional positivism with strong normative cultural assumptions. It sometimes even included spiritual aspects, as in, e.g. the works of Roman Ingarden. Logic, which was at the center of the Warsaw-Lviv School, resonated very well with traditional intelligentsia values, in particular that of truth. Its insistence on "empirical" evidence stood in stark contrast to the supposed "ideologization" of Marxist, and more generally, "leftist" approaches. In this way, the liberals made archival work and its techniques central to their historiography, while in sociology, it was the social survey and its methodology which they regarded as the indicia of "true science". The national, native roots of the Warsaw-Lviv school also accorded with the general tendency toward re-traditionalization discussed above. The development of the Warsaw School of the history of ideas can be interpreted similarly. Richard De George, writing in the late 1960s, noted that it was the liberal camp, with its strong reliance on the heritage of the Warsaw-Lviv school, that dominated the Polish social sciences and humanities fields after 1956 (De George 1967). He went on to point out that historical materialism had been largely discredited by Polish philosophers trained in logic, analysis, and related philosophical streams. By contrast, he observed that the "Soviet Union does not have the tradition of logic and analysis which existed in Poland. Second, the Soviet analysis [of basic categories of historical materialism] which is taking place is being done not by outsiders who hope to challenge the theory but by dedicated Marxist-Leninists who hope to strengthen it" (De George 1967: 79).

The evolution of the field of philosophy in the 1960s and 1970s perfectly illustrates the tendencies under discussion. A large part of its elite were systematically heading in a revisionist or liberal direction. The evolution of its various branches, e.g. the Warsaw School of the History of Ideas, can be interpreted the same way. It was clearly moving away from its initial basis of Marxist historicism and the reductionist assumption of philosophy as a worldview. This is the evolution of, e.g. Andrzej Walicki and Barbara Skarga (1919–2009), who, as one of the first in the philosophical field, strived for the complete rehabilitation of metaphysical research (Borzym 2006: 78). Another example is the evolution of the materialist Władysław Krajewski (1919–2006), who attempted to revise and enrich the Marxist program with non-materialist aspects. After being expelled from the PZPR in 1968, he began to cite the work of Karl Popper in the 1980s (Borzym 2006: 84). Marxist orthodoxy was subsequently abandoned by a succession of important figures in the field, e.g. Stefan Amsterdamski (1929–2005) and Leszek Nowak (1943–2009), who began to build up non-Marxist historical materialism in the late 1970s. This brought him into conflict with the authorities, and when martial law was imposed in 1981, he was interned, along with many other intellectuals associated with the opposition. Adam Schaff, by contrast, moved closer to the Eurocommunist movement (viewed with suspicion by the authorities) after being marginalized in the field of power. Andrzej Walicki increasingly focused on studying the great Russian thinkers and came closer to the leaders of the revisionist movement, viz. Kołakowski and Baczko. As a result of this strong tilt in the balance of forces in the field, philosophy studies were suspended in Warsaw in 1968. Enrolments had been suspended in 1966 and were not resumed until 1973 (Borzym 2006). In 1979, Paweł Śpiewak (b. 1951), then a young sociologist, presented the Warsaw School of the History of Ideas at the Walendowski Salon (Gluza 2016). Although he declined to analyze its political aspects, he was inscribing it into the context of the anti-communist intellectual mobilization of the late 1970s by the very nature of his lecture.

On the other side of the academic space stood those academics who maintained their loyalty to the communist state, even though it was now acquiring a new technocratic identity. Most of them relied on political capital, i.e. they were PZPR members, and tended to lean toward the "heteronomous" pole in their respective disciplines. While that pole had been overtly politicized, it was now becoming increasingly "neutral". It was generally characterized by a strong reliance on conventional, often "outdated", approaches in given disciplines, weaker internationalization, and avoidance of politically controversial issues. There were still references to Marxism–Leninism, but they were becoming less common and increasingly ritualistic. The importation of ideas from the Soviet Union was clearly decreasing, while from the West, it was increasing exponentially—even among this heteronomous group. However, both Soviet and (especially) Western approaches were devoid of overtly political or ideological elements. The dominant trend was to neutralize them in the technocratic spirit of modernization theory. This camp was obviously losing members and influence by the 1980s, while the opposite camp of increasingly pro-Western and liberal scholars was gaining both, despite the best efforts of the PZPR. The key determinants of this shift were the force of attraction of the West and Poland's deepening economic crisis. Most Western-oriented and cultural-capital-equipped scholars were increasingly distancing themselves from both the PZPR and anything related to communist ideology. That pole of the respective academic disciplines might have been autonomous from the PZPR, but it was increasingly heteronomous toward an idealized West and the global capitalist system. Beginning in the late 1980s, most scholars, academics, and researchers eventually became involved in the transformation of post-communist Poland and the introduction of a market economy and democratization.

2.13.2 The Homology of Structures and Disappearance of Critical Approaches

The structural configuration presented above can explain the role of the Polish social sciences in the transformation period, and in particular, the legitimization of the neo-liberal reforms introduced after the fall of communism. The increasingly dominant (especially as of the late 1980s) liberal faction saw its vocation as challenging its opponents, whom it labeled as "communist" or "post-communist", and attempted to portray as both morally and professionally inferior. As its opponents often referred to Marxist approaches, albeit in a superficial, rhetorical way, or had a history of using Marxist language, the identity of the liberal camp was built on opposition to Marxism, communism, and other "leftist" approaches. Participating in international exchanges, which many members of the camp had done in the 1960s and 1970s, provided valuable assets of a different kind, particularly after 1989. The liberal (non-Marxist, Western-oriented) camp in most social science disciplines were poised to take on leading roles when communism fell. This had already occurred much earlier in many cases. Western experience gave those who had participated in academic exchanges additional credentials, as well as the self-assurance that comes with having forged personal relationships with internationally famous academics. After the fall of communism, the basic message of the social sciences in Poland and other countries of the region was that the generalized West represented an ideal social model to follow, and the local academic elite was the most competent teacher of Western standards (Böröcz 2006). This increasingly dominant liberal pole in most of Poland's social sciences and humanities fields—not only sociology, economics, history, and legal studies, but also geography, psychology, and linguistics—started to provide the field of power with both crucial ideological legitimization of the new order and the first post-communist government and state administration with fresh representatives. Many of the old intelligentsia joined forces to form the first post-communist government and parliament in 1989 and 1990,[5] or in fact, a transitory political elite. This could be seen as a return to power of the old intelligentsia, who later partly retreated from the most visible political roles. However, they are still the country's dominant intellectual elite, and continue to shape national ideology, including the meaning of Europeanization (Zarycki 2009).

At the same time, many Polish social scientists, especially the leaders of the liberal, pro-Western camp, which became hegemonic in 1989, abandoned their critical edge. This process had already begun in the late 1970s and was accompanied by a clear shift from the critical study of economic processes toward a focus on culturalist approaches and others that had been inspired by social psychology, which was deemed the perfect tool to manage "social change". This was probably part of a wider global tendency, but was very visible in Poland given the political transformation of an entire generation of the liberal intelligentsia, which gradually moved from a dominated to a dominant position in, or in relation to, the field of power. The turning point of this process came in 1989, when a major part of the intellectual elite, who would have remained within "the dominant sector of the dominating class" in most Western countries, were catapulted to the pinnacle of the social hierarchy. This transformation reversed what was considered best for the country from the long-term planning of its social and economic development to the dictates of market forces. The focus was now on "catching up with the West", first "culturally" and "psychologically", and then by following new intellectual trends such as globalization. This trend was most visible in disciplines such as economics, sociology, and psychology, but was also noticeable in geography. This transformation in the dominant modes of thinking eliminated all traces of dependency theories, both in their early Latin American form and their later incarnation as world-system theory, in the Polish social sciences. It also eliminated notions such as "exploitation" and "political economy", which were now considered non-scientific Marxist newspeak. From the 1980s (at least), these dominant actors in

[5] The genealogist Marek Jerzy Minakowski, has estimated that the first post-war, non-communist parliament, elected in partly free elections in 1989, had a spectacularly high number of descendants of noble and old intelligentsia families, comparable only to the pre-war Polish parliaments (Minakowski 2014).

the Polish social sciences had been able to redefine Western critical social theory in ways that seemed to render it compatible with modernization theory, which became the hegemonic approach during the transformation period. In an earlier work I wrote in collaboration with Tomasz Warczok, I demonstrated this mechanism on the example of the theory of Pierre Bourdieu, elements of which have been imported to Poland and modified from being critical tools to become normative models (Warczok and Zarycki 2014). Thus, the relational apparatus built by Bourdieu to deconstruct inequalities has been reinvented in Poland as a normative model of what modern, i.e. Western, society looks like. The intelligentsia elite then use these redefined, neutralized Western theoretical imports to legitimize the neo-liberal reforms it oversees.

The specific structure of the social science fields in Poland, with the dominant role of the old intelligentsia, might explain why so many Polish academics, especially those from upper-class families and those who had had international experience during the communist period, were so strongly committed to legitimizing the post-communist transformation, whereas no influential critical schools of social thought emerged after Marxism lost its appeal in the 1970s. A broad faction of the intelligentsia provided the expertise needed to liberalize the Polish economy and reconstruct the country's public administration along Western, mostly neo-liberal, lines. This was possible due to the competence and social status of these scholars, including their family cultural capital and the experience they had accumulated during their careers in communist Poland, which often included long periods of international (in particular US) fellowships from the Ford and Rockefeller Foundations. Johanna Bockman and Gil Eyal show that these visits also involved building collaborative networks between Polish (and Hungarian) scholars with liberal inclinations and the elites of the US neo-liberal schools of economics and related fields (Bockman and Eyal 2002). Poles and Hungarians provided the empirical data needed to support analyses aimed at demonstrating the inefficiency of socialist economics, and received new ideological and theoretical models in exchange. These models helped legitimize the market reforms that began in the late 1980s. In this way, the social sciences also helped legitimize and naturalize the disadvantages of the transformation and European integration, e.g. a growing dependency on the Western core, rising inequalities, and, last but not least, the hegemonic position of the intelligentsia as the dominant social stratum in Poland.

The beginning of the 1990s can be interpreted as a certain renaissance of interface peripherality. Poland was the fastest liberalizing country of the former communist bloc. Thus, it still retained a certain potential, although it was more of an "exotic" capital, as a former communist state, dependent on the USSR, and at the same time a space now open in every respect to Western influences, including the inflow of Western capital. This special role of Poland, where extremely radical economic reforms were introduced ahead of other countries of the region in the early 1990s, also attracted the attention of Russia, particularly its liberal elites, who had recently assumed power. It was not long, however, before Poland lost its privileged status of neo-liberal reform champion, and became just another country in the "new" CE Europe, which in turn became just another internal periphery of the Western core. Thus, Poland's dependence on the West increased in tandem with its isolation from the East. One sign of this growing isolation was the end of visa-free travel to Belarus, Russia, and Ukraine and the closing of the borders with those countries. This was done in 2004, when Poland joined the Schengen Area. These geopolitical shifts also affected the geometry of the field of power, and consequently, the fields of the social sciences.

References

Antonowicz, Dominik. 2016. *Między siłą globalnych procesów a lokalną tradycją. Polskie szkolnictwo wyższe w dobie przemian.* Toruń: Wydawnictwo Naukowe Uniwersytetu Mikołaja Kopernika.

Arkhangelskiy, Aleksandr, and Yelena Laskari. 2011. Zhara. Tsikl dokumental'nykh filmov Aleksandra Arkhangel's kogo. *Telekanal «Kultura»*, 4 filma. https://tvkultura.ru/brand/show/brand_id/32570.

Bajer, Magdalena. 2013. *Rody Uczone. Kreski Do Szkicu.* Toruń: Wydawnictwa UMK.

Bazylow, Ludwik. 1984. *Polacy w Petersburgu.* Wrocław: Zakład Narodowy im. Ossolińskich.

Behr, Valentin. 2015. Genèse et usages d'une politique publique de l'histoire. La «politique historique» en Pologne. *Revue detudes comparatives Est-Ouest* 3: 21–48.
Behr, Valentin. 2017. *Science du passé et politique du présent en Pologne: L'histoire du temps présent (1939–1989), de la genèse à l'Institut de la Mémoire Nationale*. Strasbourg: Thèse de doctorat pour obtenir le grade de Docteur de l'Université de Strasbourg, soutenue le 18 octobre 2017.
Behr, Valentin. 2021. *Powojenna historiografia polska jako pole walki: studium z socjologii wiedzy i polityki*. Warszawa: Wydawnictwa Uniwersytetu Warszawskiego.
Bockman, Johanna K., and Gil Eyal. 2002. Eastern Europe as a Laboratory for Economic Knowledge: The Transnational Roots of Neoliberalism. *American Journal of Sociology* 108 (2): 310–352.
Böröcz, József. 2006. Goodness Is Elsewhere: The Rule of European Difference. *Comparative Studies in Society and History* 48 (1): 110–138.
Borzym, Stanisław. 2006. Filozofia w Polsce w latach 1945–1990. In *Humanistyka polska w latach 1945–1990*, ed. U. Jakubowska and J. Myśliński, 67–95. Warszawa: Fundacja Akademia Humanistyczna; Instytut Badań Literackich PAN.
Bouyeure, Cyril. 2007. *L'Invention du politique: Une biographie d'Adam Michnik*. Paris: Éditions Noir sur Blanc.
Brubaker, Rogers. 1992. *Citizenship and Nationhood in France and Germany*. Cambridge: Harvard University Press.
Bucholc, Marta. 2016. *Sociology in Poland: To Be Continued?* London: Palgrave Macmillan.
Chałasiński, Józef. 1957. Drogi i bezdroża socjalizmu w nauce polskiej (1949–1954). *Kultura i Społeczeństwo* 1 (1): 7–42.
Chwalba, Andrzej. 1999. *Polacy w służbie Moskali*. Warszawa, Kraków: Wydawnictwo Naukowe PWN.
Connelly, John. 2000. *Captive University: The Sovietization of East German, Czech, and Polish Higher Education, 1945–1956*. Chapel Hill and London: The University of North Carolina Press.
Czarny, Bogusław. 2015. Wpływ aspirantów Katedry Ekonomii Politycznej w Instytucie Kształcenia Kadr Naukowych w Warszawie na polską ekonomię po II wojnie światowej. *Ekonomista* 41/2015.
Dębska, Hanna. 2019. Między autonomią a heteronomią. Trajektoria polskiej filozofii prawa w latach 1945–1956. *Państwo i Prawo* 11: 94–113.
Degen, Dorota, and Hübner, Piotr. 2006. Polityka naukowa władz Polski Ludowej w zakresie humanistyki (1945–1990), In *Humanistyka polska w*

latach 1945–1990, ed. Urszula Jakubowska and Jerzy Myśliński, 11–38. Warszawa: Fundacja Akademia Humanistyczna; Instytut Badań Literackich PAN.

De George, Richard T. 1967. Philosophy. In *Science and Ideology in Soviet Society*, ed. G. Fischer, 47–82. New York: Atherton.

Derluguian, Georgi M. 2005. *Bourdieu's Secret Admirer in the Caucasus: A World-System Biography*. Chicago: University of Chicago Press.

Drahokoupil, Jan. 2008. The Rise of the Comprador Service Sector: The Politics of State Transformation in Central and Eastern Europe. *Polish Sociological Review* 162: 175–189.

Drahokoupil, Jan. 2009. *Globalization and the State in Central and Eastern Europe: The Politics of Foreign Direct Investment*. London: Routledge.

Eisler, Jerzy. 2018. *Czterdzieści pięć lat, które wstrząsnęły Polską: Historia polityczna 1944–1989*. Warszawa: Czerwone & Czarne.

Eyal, Gil, Iván Szelényi, and Eleanor R. Townsley. 1998. *Making Capitalism Without Capitalists: Class Formation and Elite Struggles in Post-Communist Central Europe*. London and New York: Verso.

Friszke, Andrzej. 2018. Karol Modzelewski. *Nauka* 2019: 7–14.

Gella, Alexander. 1976. An Introduction to the Sociology of the Intelligentsia. In *The Intelligentsia and the Intellectuals: Theory, Method, and Case Study*, ed. A. Gella, 9–34. London, Beverly Hills: Sage.

Gella, Aleksander. 2001. Inteligencja polska i siodło po niej. *Arcana* 37 (1): 20–38.

Gerasimov, Ilya, Serguei Glebov, Alexander Kaplunovski, Marina Mogilner, and Alexander Semyonov. 2005. In Search of a New Imperial History. *Ab Imperio* 1: 33–56.

Gerőcs, Tamás, and András Pinkasz. 2018. Debt-Ridden Development on Europe's Eastern Periphery. In *Global Inequalities in World-Systems Perspective: Theoretical Debates and Methodological Innovations*, ed. Manuela Boatcă, Andrea Komlosy, and Hans-Heinrich. Nolte, 131–153. New York: Routledge.

Gluza, Zbigniew, ed. 2016. *Salon: Niezależni w „świetlicy" Anny Erdman & Tadeusza Walendowskiego 1976–79*. Warszawa: Ośrodek Karta.

Gospodarczyk, Marta, and Łukasz Kożuchowski. 2021. Nowa ludowa historia: Charakterystyka i społeczno-polityczne korzenie współczesnych narracji o historii chłopów polskich. *Studia Socjologiczne* 241: 177–198.

Grabski, Władysław. 2016. *Idea Polski: wybór pism. Ośrodek Myśli Politycznej*. Kraków and Warszawa: Narodowe Centrum Kultury.

Hardy, Jane. 2007. The New Competition and the New Economy: Poland in the International Division of Labour. *Europe-Asia Studies* 59 (5): 761–777.

Hartwell, Christopher A., and Mateusz Urban. 2021. Burning the Rechtsstaat: Legal Institutions and Protection of the Rule of Law. *Journal of Institutional Economics* 17: 105–131.

Hrycak, Jarosław. 2009. *Ukraina. Przewodnik Krytyki Politycznej. Z Jarosławem Hrycakiem rozmawia Iza Chruślińska.* Gdańsk, Warszawa: Wydawnictwo Krytyki Politycznej.

Hryniewicz, Janusz T. 2003. Polityczne i gospodarcze następstwa zaboru pruskiego. *Studia Regionalne i Lokalne* 3: 61–79.

Jedlicki, Jerzy. 2008. *Błędne koło 1832–1864*. Warszawa: Instytut Historii PAN, Wydawnictwo Neriton.

Kaczyński, Jarosław. 1976. *Rola ciał kolegialnych w kierowaniu szkołą wyższą: Praca doktorska napisana pod kierunkiem Stanisława Ehrlicha.* Warszawa: Uniwersytet Warszawski; Wydział Prawa & Administracji.

Kagarlitsky, Boris. 2008. *Empire of the Periphery: Russia and the World System.* London and Ann Arbor: Pluto Press.

Kamiński, Łukasz, and Grzegorz Waligóra. 2014. *Kryptonim „Pegaz": Służba bezpieczeństwa wobec Towarzystwa Kursów Naukowych 1978–1980.* Warszawa: Instytut Pamięci Narodowej.

Karpiński, Jakub. 2001. *Wykres gorączki. Polska pod rządami komunistycznymi.* Lublin: Wydawnictwo UMCS.

Kauffman, Jesse. 2013. Warsaw University Under German Occupation: State Building and Nation Bildung in Poland During the Great War. *First World War Studies* 4: 65–79.

Kennedy, Michael D. 1991. *Professionals, Power, and Solidarity in Poland: A Critical Sociology of Soviet-Type Society.* Cambridge: Cambridge University Press.

Kępińska, Ewa, and Oded Stark. 2013. The Evolution and Sustainability of Seasonal Migration from Poland to Germany: From the Dusk of the 19th Century to the Dawn of the 21st Century. *International Review of Economics and Finance* 28: 3–18.

Kilias, Jarosław. 2017. *Goście ze wschodu: socjologia polska lat sześdziesiątych XX wieku a nauka światowa.* Kraków: Nomos.

Kochanowicz, Jacek. 2014. The Curse of Discontinuity: Poland's Economy in a Global Context, 1820–2000. *Jahrbuch Für Wirtschaftsgeschichte / Economic History Yearbook* 55: 129–148.

Kola, Adam F. 2018. Marian Małowist's World History and Its Application to World Literature. In *The Routledge Companion to World Literature and World History*, ed. M. Hawas, 57–68. London: Routledge.
Koryś, Piotr. 2018. *Poland From Partitions to EU Accession: A Modern Economic History, 1772–2004*. Springer, Cham: Palgrave Macmillan.
Kosicki, Piotr H. 2016. *Personalizm po polsku: Francuskie korzenie polskiej inteligencji katolickiej*. Warszawa: Instytut Pamięci Narodowej.
Krajewski, Stanisław. 2019. My wszysy z niej: dezasymilacja polskich Żydów. In *Co się dzieje z polskim społeczeństwem? Księga jubileuszowa dedykowana Profesorowi Ireneuszowi Krzemińskiemu*, ed. U. Kurczewska, M. Głowania, W. Ogrodnik, and D. Wasilewski, 297–309. Warszawa: Wydawnictwa Uniwersytetu Warszawskiego.
Kryshtanovskaya, Olga, and Stephen White. 1996. From Soviet Nomenklatura to Russian Elite. *Europe-Asia Studies* 48 (5): 711–733.
Kuroń, Jacek. 1991. *Gwiezdny czas. „Wiary i winy" dalszy ciąg*. Londyn: Aneks.
Kwiek, Marek. 2021. Poland: A Decade of Reforms (2010–2020). *International Higher Education* 106: 36–38.
Landau, Zbigniew, and Jerzy Tomaszewski. 1995. *Bank Handlowy w Warszawie S.A. Zarys dziejów 1870–1995*. Warszawa: MUZA S.A.
Łazor, Jerzy. 2019. Kapitał francuski w Polsce międzywojennej. Stan badan i postulaty badawcze. *UR Journal of Humanities and Social Sciences* 1: 29–52.
Le Rider, Jacques. 2008. Mitteleuropa, Zentraleuropa, Mittelosteuropa: A Mental Map of Central Europe. *European Journal of Social Theory* 11: 155–169.
Leszczyński, Adam. 2017. *Leap into Modernity—Political Economy of Growth on the Periphery, 1943–1980*. Frankfurt am Main: Peter Lang.
Lewis, Daniel. 2020. Krzysztof Penderecki, Polish Composer with Cinematic Flair, Dies at 86. *New York Times*, March 29.
Łuczewski, Michał. 2012. *Odwieczny naród: Polak i katolik w Żmiącej*. Toruń: Wydawnictwo Naukowe Uniwersytetu Mikołaja Kopernika.
Mackiewicz, Józef. 2009. *The Triumph of Provocation*. New Haven: Yale University Press.
Marzec, Wiktor. 2020. *Rising Subjects: The 1905 Revolution and the Origins of Modern Polish Politics*. Pittsburgh: University of Pittsburgh Press.
Mażewski, Lech. 2004. *Powstańczy szantaż: Od konfederacji barskiej do stanu wojennego*. Warszawa: Książka & Wiedza.
McMeekin, Sean. 2011. Jihad-cum-Zionism-Leninism: Overthrowing the World, German-Style. *Historically Speaking* 12 (3): 2–5.

Mich, Włodzimierz. 2000. *Ideologia polskiego ziemiaństwa 1918–1939*. Lublin: Wydawnictwo UMSC.
Minakowski, Marek Jerzy. 2014. Leder się myli – rewolucja w Polsce była w 1990, nie 1946 r, April 14. http://minakowski.pl/leder-sie-myli-rewolucja-w-polsce-byla-w-1990-nie-1946-r/.
Miśkowiec, Małgorzata. 2013. Berlin: "polska metropolia kultury": niemiecka stolica jako ważny cel emigracji Polaków w XX i XXI wieku. In *Współczesne transformacje: Kultura, polityka, gospodarka*, ed. M. Banaś and J. Dziadowiec, 163–172. Kraków: Księgarnia Akademicka.
Morozov, Viatcheslav. 2015. *Russia's Postcolonial Identity: A Subaltern Empire in a Eurocentric World*. Houndmills, Basingstoke, and Hampshire: Palgrave Macmillan.
Naczyk, Marek. 2021. Taking Back Control: Comprador Bankers and Managerial Developmentalism in Poland. *Review of International Political Economy*. https://doi.org/10.1080/09692290.2021.1924831.
Nałęcz-Dobrowolski, Stanisław Marceli. 2015. Polityczny Zjazd Polaków w Moskwie. *Wspomnienia piórem i ołówkiem. Niepodległość i pamięć* 2: 273–293.
Naumann, Friedrich. 1915. *Mitteleuropa*. Berlin: Riemer.
Nelson, Stephen C. 2020. Banks Beyond Borders: Internationalization, Financialization, and the Behavior of Foreign-Owned Banks During the Global Financial Crisis. *Theory and Society* 49: 307–333.
Nowak, Andrzej. 2009. Intelligentsias in the Structure of the Empire: The Assimilative Function of the Central Counter-Elite. *Acta Poloniae Historica* 100: 31–55.
Nowak, Stefan, ed. 1991. *Studenci Warszawy: studium długofalowych przemian postaw & wartości*. Warszawa: Wydawnictwa Uniwersytetu Warszawskiego.
Osęka, Piotr. 2008. *Marzec '68*. Kraków: Instytut Studiów Politycznych PAN, Wydawnictwo Znak.
Osęka, Piotr. 2019. Dissidents and Police: The Polish March 1968 and a Tale of Two Generations. *East European Politics and Societies* 33 (4): 861–880.
Ossowski, Stanisław. 1963. *Class Structure in the Social Consciousness*. New York: Free Press.
Oswald, Ingrid, and Viktor Voronkov. 2004. The 'Public–Private' Sphere in Soviet and Post-Soviet Society. Perception and Dynamics of 'Public' and 'Private' in Contemporary Russia. *European Societies* 6 (1): 97–117.
Ozieranski, Piotr, and Larry King. 2016. The Persistence of Cliques in the Post-Communist State. The Case of Deniability in Drug Reimbursement Policy in Poland. *British Journal of Sociology* 67 (2): 216–241.

Pleskot, Patryk, and Tadeusz Paweł Rutkowski. 2009. *Spętana Akademia: Polska Akademia Nauk w dokumentach władz PRL. Materiały Służby Bezpieczeństwa (1967–1987)*. Warszawa: Instytut Pamięci Narodowej.

Ponarski, Zenowiusz. 2003. *Wokół sprawy polskiej na Wschodzie*. Toruń: Wydawnictwo Adam Marszałek.

Popiński, Krzysztof. 2018. System szkolnictwa wyższego w II Rzeczypospolitej i jego wpływ na funkcjonowanie uczelni polskich po 1945 roku. *Społeczeństwo i Ekonomia* 1 (9): 25–53.

Przeperski, Michał. 2021. *Mieczysław F.Rakowski: Biografia polityczna*. Warszawa: Instytut Pamięci Narodowej – Komisja Ścigania Zbrodni przeciwko Narodowi Polskiemu.

Rakowski, Mieczysław F. 1999. *Dzienniki polityczne 1958–1962*. Warszawa: Iskry.

Rolf, Malte. 2020. *Russian Rule in the Kingdom of Poland (1864–1915)*. Pittsburgh: University of Pittsburgh Press.

Roshwald, Aviel. 2001. *Ethnic Nationalism and the Fall of Empires*. London: Routledge.

Röskau-Rydel, Isabel. 2011. *Niemiecko-austriackie rodziny urzędnicze w Galicji 1772–1918. Kariery zawodowe - środowisko – akulturacja i asymilacja*. Kraków: Wydawnictwo Naukowe Uniwersytetu Pedagogicznego w Krakowie,

Róziewicz, Jerzy. 1984. *Polsko-rosyjskie powiązania naukowe (1725–1918)*. Wrocław: Zakład Narodowy im. Ossolińskich.

Schiller, Joanna. 2008. *Universitas Rossica: koncepcja rosyjskiego uniwersytetu 1863–1917*. Warszawa: Instytut Historii Nauki PAN; Uniwersytet Kardynała Stefana Wyszyńskiego.

Schoer, Karl. 1976. Origins of Junker Capitalism in Germany. *Social Scientist* 5: 33–46.

Secomski, Kazimierz. 1974. *Spatial Planning and Policy: Theoretical Foundations*. Warsaw: Polish Scientific Publishers.

Shields, Stuart. 2008. How the East Was Won: Transnational Social Forces and the Neoliberalisation of Poland's Post-Communist Transition. *Global Society* 22 (4): 445–468.

Sikorska-Kulesza, Jolanta. 1995. *Deklasacja drobnej szlachty na Litwie & Białorusi w XIX wieku*. Pruszków: Oficyna Wydawnicza "Ajaks".

Smoczyński, Rafał, and Tomasz Zarycki. 2017. *Totem inteligencki: Arystokracja, szlachta i ziemiaństwo w polskiej przestrzeni społecznej*. Warszawa: Wydawnictwo Naukowe "Scholar".

Smoczyński, Rafał, and Tomasz Zarycki. 2021. The Extended Family: Descendants of Nobility in Post-communist Poland. *Communist and Post-Communist Studies* 54: 157–175.

Sosnowska, Anna. 2018. *Explaining Economic Backwardness: Post-1945 Polish Historians on Eastern Europe*. Budapest: CEU Press.

Staniszkis, Jadwiga. 1999. *Post-communism: The Emerging Enigma*. Warsaw: Institute of Political Studies, Polish Academy of Sciences.

Stasiński, Maciej, and Andrzej Werblan. 2018. Natolińczycy, puławianie, partyzanci. Frakcje polityczne w PZPR: Wywiad z Andrzejem Werblanem. *Gazeta Wyborcza,* 26 lutego 2018. https://wyborcza.pl/alehistoria/7,121681,23056707,natolinczycy-pulawianie-partyzanci-frakcje-polityczne-w-pzpr.html.

Steinmetz, George. 2014. The Sociology of Empires, Colonialism, and Post-colonialism. *Annual Review of Sociology* 40 (1): 77–103.

Sułek, Antoni. 2010. "To America!": Polish Sociologists in the United States After 1956 and the Development of Empirical Sociology in Poland. *East European Politics and Societies* 24 (3): 327–352.

Sułek, Antoni. 2016. *Wspomnienie o Prof. Elżbiecie Tarkowskiej (1944–2016)*. Instytut Socjologii Uniwersytetu Warszawskiego. https://www.is.uw.edu.pl/pl/. Accessed 12 January 2020.

Surman, Jan. 2018a. *Universities in Imperial Austria 1848–1918: A Social History of a Multilingual Space*. West Lafayette: Purdue University Press.

Surman, Jan. 2018b. The Contexts of Early Polish Positivisms, 1840s–1900s. In *The Worlds of Positivism: A Global Intellectual History, 1770–1930*, ed. Johannes Feichtinger, Franz L. Fillafer, and Jan Surman, 239–272. Cham: Palgrave Macmillan.

Szwarc, Andrzej. 1991. Obóz ugody a inteligencja w zaborze rosyjskim w latach 1864–1905, In *Inteligencja polska XIX i XX wieku*, ed. R. Czepulis-Rastenis, 73–105. Warszawa: Wydawnictwo Naukowe PWN.

Turkowski, Andrzej. 2018. Polish Intelligentsia Totems in Elites' Struggles for Legitimization: The Case of Jerzy Giedroyc and Poland's Eastern Policy. *East European Politics and Societies: And Cultures* 33: 66–88.

Wajda, Kazimierz. 1990. Uwarunkowania polskiej myśli politycznej i społecznej na Ziemiach Polskich pod panowaniem pruskim 1864–1914. In *Studia z Dziejów Polskiej Myśli Politycznej. II - Polska Myśl Polityczna w Dzielnicy Pruskiej w XIX w*, ed. S. Kalembka, 55–66. Toruń: Wydawnictwa UMK.

Warczok, Tomasz. 2022. Polityka, system akademicki i antysemityzm: Dwa kryzysy w polskim polu władzy. In *Polskie nauki społeczne w kontekście relacji władzy i zależności międzynarodowych*, ed. T. Zarycki, 70–94. Warszawa: Wydawnictwa Uniwersytetu Warszawskiego.

Warczok, Tomasz, and Tomasz Zarycki. 2014. Bourdieu Recontextualized: Redefinitions of Western Critical Thought in the Periphery. *Current Sociology* 62 (3): 334–351.

Warczok, Tomasz, and Tomasz Zarycki. 2018. Polish Political Sciences in a Global Context. In *Oxford Research Encyclopedia of Politics*, ed. W. R. Thompson. Oxford University Press. http://politics.oxfordre.com/view/10.1093/acrefore/9780190228637.001.0001/acrefore-9780190228637-e-872.

Wasilewski, Jacek. 1995. The Forming of the New Elite: How Much Nomenklatura Is Left? *Polish Sociological Review* 2: 113–123.

Wedel, Janine R. 1986. *The Private Poland*. New York, NY: Facts on File Inc.

Wereszycki, Henryk. 1990. *Historia polityczna Polski 1864–1918*. Wrocław: Zakład Narodowy im. Ossolińskich.

Wesołowski, Włodzimierz. 1979, *Classes, Strata, and Power*. London and Boston: Routledge & K. Paul.

Wołek, Artur. 2014. Instytucjonalne Ramy Polskiej Polityki – Ćwierć Wieku Porządkowania Zamętu. *Horyzonty Polityki* 11: 51–71.

Zarycki, Tomasz. 2009. The Power of the Intelligentsia. The Rywin Affair and the Challenge of Applying the Concept of Cultural Capital to Analyze Poland's Elites. *Theory and Society* 38 (6): 613–648.

Zarycki, Tomasz. 2011. From Soviet to a Western-Dominated Political Scene: The Geopolitical Context of Politics in Poland. In *Cultural Transformations After Communism: Central and Eastern Europe in Focus*, ed. B. Törnquist-Plewa and K. Stala, 82–108. Lund: Nordic Academic Press.

Zarycki, Tomasz. 2012. The Embarrassing Russian Connection: Selective Memory of the Russian Heritage in Contemporary Poland. In *Russia's Identity in International Relations. Images, Perceptions, Misperceptions*, ed. R. Taras, 133–148. London: Routledge.

Zarycki, Tomasz. 2013. Aleksander Lednicki i los jego środowiska Polonii w Rosji na przełomie XIX i XX wieku jako zwierciadło przemian polskiego pola władzy. *Sprawy Narodowościowe* 42: 67–83.

Zarycki, Tomasz. 2014. *Ideologies of Eastness in Central and Eastern Europe*. London: Routledge.

Zarycki, Tomasz. 2015. Class Analysis in Conditions of a Dual-Stratification Order. *East European Politics and Societies* 29 (3): 711–718.

Zarycki, Tomasz. 2020a. Crises of the Communist and Neoliberal Orders 30 Years Later: Structural Comparison of Poland 1975 and 2019. *Social Science Information* 59 (3): 484–504.

Zarycki, Tomasz. 2020b. Why Geography in Poland Has Never Radicalized: Political and International Entanglements of Polish Geography Seen Through the Prism of Antoni Kukliński's Professional Trajectory. *Serendipities. Journal for the Sociology and History of the Social Sciences* 5 (1–2): 47–50.

Zarycki, Tomasz, and George Kolankiewicz, eds. 2003. *Regional Issues in Polish Politics*. London: School of Slavonic and East European Studies, University College London.

Zarycki, Tomasz, Rafał Smoczyński, and Tomasz Warczok. 2017. The Roots of Polish Culture-Centered Politics: Toward a Non-purely Cultural Model of Cultural Domination in Central and Eastern Europe. *East European Politics and Societies* 31 (2): 360–381.

Żor, Andrzej. 2011. *Kronenberg: Dzieje fortuny*. Warszawa: Wydawnictwo Naukowe PWN.

Żyłko, Bogusław. 2009. *Semiotyka kultury: Szkoła tartusko-moskiewska*. Gdańsk: słowo/obraz terytoria.

Zysiak, Agata. 2016. *Punkty za pochodzenie: Powojenna modernizacja & uniwersytet w robotniczym mieście*. Kraków: NOMOS.

3

The Field of Polish Linguistics and Literary Studies

3.1 Introduction

This part of the book attempts to reconstruct the formation of the Polish field of linguistics, and to some extent, literary studies. Linguistics and literary studies are here considered to be one broadly defined field. The internal divisions between linguistics and literary studies are extremely complex, and have important implications for the polemics between different factions within the field. Many of these tensions are part of broader struggles in the field of social sciences and have interesting institutional, ideological, and even political aspects. However, as space does not permit a discussion of all of them, a number of important and interesting aspects related to the field definition have been omitted. The primary focus is on selected cleavages and conflicts in the combined field of linguistics and literary studies. Reference is occasionally made to disputes in other areas of the social sciences, as well as divisions in the field of power, to illustrate how they broaden the context of the debates under discussion. The reconstruction presented here relies primarily on monographic works on the history of the two fields. I have made use of the collective volume, edited by Ernst Frideryk Konrad Koerner and

T. Zarycki, *The Polish Elite and Language Sciences*,
https://doi.org/10.1007/978-3-031-07345-8_3

Aleksander Szwedek, and published by John Benjamins (Koerner and Szwedek 2001), as well as several Polish studies, notably the volume edited by Andrzej Śródka (1999) on the history of the Polish philological sciences, Stanisław Urbańczyk's monograph (1993), and works by Artur Hutnikiewicz (1999), Irena Bajerowa (1999), Tomasz Karpowicz (2006), or Andrzej Karcz (2002). As for literary theory, an enormous amount of useful knowledge, as well as many interesting interpretations, can be found in the recent volume edited by Danuta Ulicka (2020a). I have also used a number of other sources. All of these studies provide precious factual knowledge and can be interpreted in several interesting ways. In addition, a number of other studies, including historical studies using Polish secret services (SB) documents, have been used and are referred to where appropriate. My main aim here was to extract relevant and illustrative elements from authentic historical accounts in order to present a general outline of the field structure and then draw more general conclusions pertaining to the dynamics of the formation of the field in a broader social and geopolitical context. My primary goal is therefore to reconstruct the main structural oppositions in the emerging field, rather than to provide a precise historical description of it. In this study, the key role is played by those figures who have become significant players in the field, especially those who set the course of its development, and who therefore constitute the points of reference for its crucial divisions. At the same time, attention is drawn to critical institutional frameworks and reference is made to the context of the political cleavages and international relations discussed in previous chapters. Here I show how they interact with, and manifest themselves in, the field of language studies.

As with the analyses of the field of power and the political field, an important hypothesis is that the dynamics of the Polish field of linguistics and literary studies has been strongly influenced by the international context, in particular, the context which can be considered in terms of peripherality, including the interface peripherality proposed by Stein Rokkan. As mentioned earlier, interface peripherality is eminently conducive to the emergence of unprecedented cultural phenomena; spheres of creative autonomy that produce cultural trends perceived as original and inspiring. These conditions in the development of Polish language and literature studies were especially propitious at the turn of the twentieth century, especially in the Russian Empire, and in the 1960s. The role of Russia in the development of the Polish cultural field has deep historical roots (Róziewicz 1984). For example, until the accession of Peter I, the Polish language was very popular among the upper classes of Moscow. It can be argued that this was largely due to the fact that Poland had been an intermediary between Russia and Western Europe. This configuration of mutual relations was repeated, albeit on a much smaller scale, in the 1960s and 1970s, when Poland's liberalization made it a window on the world for the Soviet intelligentsia. This resulted in an interest in the Polish language among the Soviet intelligentsia, especially in its main centers, viz. Moscow and Leningrad, but also in Belarusian and Ukrainian cities. This period, during which Polish books, records, and also scholarly works, including those on linguistics, were in great demand in the USSR, was abruptly cut short by the outbreak of the "Solidarity" revolution in Poland in 1980. However, as already mentioned, these conditions did not prevail at the University of Warsaw in the second half of the nineteenth century. It was, at least as far as the social sciences and humanities were concerned, a rather conservative institution that adhered to strict Russification policies. Consequently, it was unable to put scientific innovations into international circulation. This status, which was contingent on its political role, especially its mission to Russify Poland, can be interpreted in terms of external peripherality. Simply put, the University of Warsaw was a one-dimensional internal periphery of the Empire that was subject to strong political pressure from the center. However, in the depths of the Empire proper, relatively autonomous niches that can be considered in terms of interface peripherality were carved out—at least in some universities. This was a product of the clash between the centripetal forces of political pressure and other official Russian cultural influences, and the centrifugal forces of the influence of Western culture (in its broadest sense), including the trends of German, French, British (and later the United States) science, and the impact of intra-imperial cultural and political dynamics. Polish scholars, similarly to those belonging to other national minorities in the Empire, were carriers of the political forces of the national proto-fields of power and Western tendencies, to which they were often particularly well attuned. Their connection to Western science was often strengthened by having studied or worked in Western (primarily German) universities for extensive periods. Importantly, they often studied abroad because the University of Warsaw, where access was restricted for Poles, was not available to them or because they did not want to study at a provincial university. The Russian Empire helped others to go West, e.g. by funding scholarships. These overlapping influences, as well as opposing tendencies, emanating from different sectors of the Russian field of power, help explain the rise of the Kazan School of linguistics and the career of Jan Baudouin de Courtenay. De Courtenay was not only predominantly educated at German universities, but also worked in Kraków for a time.

The extent to which universities in the Austrian partition can be regarded as centers of unique creativity, which could be interpreted in terms of interface peripherality, is an open question. The universities of Lviv and Kraków might well have been. The Lviv academic center was much more dynamic than the conservative Jagiellonian University in Kraków at the turn of the twentieth century. Once Polish culture had been granted autonomy, it became a center for a rather traditional and normative academic culture with national identity as one of its central values. For all that, however, it was in Kraków that most of the infrastructure for the Polish social sciences and humanities was laid, even though Lviv boasted more original scientific personalities whose fame transcended the sphere of Polish culture. In fact, by all accounts, Lviv was an extremely creative Polish cultural and scientific center until 1939. In several scientific disciplines, it generated celebrities of international caliber. The most important figure in the literary sciences was arguably Roman Ingarden. His importance in overall world science, however, pales in comparison with that of Jan Baudouin de Courtenay—the only Polish linguist to warrant a mention in such concise and globally recognized accounts as *A Short History of Linguistics* by Robert Henry Robins (1997: 229). It is worth mentioning here that several Polish scholars had similar career trajectories, e.g. Leon Petrażycki springs to mind. The relatively limited international scale of Ingarden's career is discussed later in this chapter, but Lviv unquestionably exhibited a certain degree of interface peripherality. As with the Russian Empire, the influence of the capital (Vienna) was counterposed by international influences and strong internal centrifugal forces—especially Polish, but also Ukrainian.

Vienna provided a much more cosmopolitan stream of inspiration, but it is difficult to speak of a confluence of two distinct cultural forces (one conservative and the other left-liberal) similar to that which existed at many Russian universities.

3.2 The Nineteenth-Century Beginnings of the Field

I would like now to return to the initial phase of the field and the geopolitical context of the time. In 1807–1814, Samuel Linde (1771–1847) published his monumental, three-volume *Dictionary of the Polish Language*. It was an impressive and pioneering work by CE European (especially the "Slavophone" areas), and possibly even pan-European standards. Linde's dictionary not only played an important role in the systematization of the modern Polish language, but also became a reference point for other Slavic language dictionaries. The dictionary was financed primarily by private donations, but Tsar Alexander I and King Friedrich Wilhelm III both contributed financially to its publication. Russia and Prussia both supported Polish cultural autonomy at the time. Austria, by contrast, was engaged in cultural and (especially) linguistic unification, which obviously included the Germanization of its Polish territory. There were therefore no financial contributions from Emperor Francis I. However, the second edition (1854–1861) was published in Lviv, in the Austrian territory of Cisleithania, as Austria now had a much more liberal policy toward Polish culture than Prussia. For his efforts, Linde was made a hereditary noble, with the coat of arms "*Słownik*" (Dictionary), by Tsar Alexander I in 1826. He additionally received a number of other decorations from both Alexander I and his successor Nicholas I. Linde also lectured at the recently established University of Warsaw in 1816–1818. He was born in Toruń (then Thorn, Prussia) into a family of Swedish Lutheran immigrants. He studied in Leipzig, learned Polish, and soon established himself as a member of the Polish intellectual elite. According to Róziewicz (1984), Linde advocated the creation of a pan-Slavic language. However, the ideas of Slavic integration, especially Pan-Slavism, never took hold anywhere, least of all in

Poland, where after the uprisings of 1830 and 1861, they were viewed as tools of Russian expansion. This led to Linde mostly spending his final years with Russian officials and being snubbed by the Polish elite of Warsaw.

Linde's dictionary was a large scale, pioneering enterprise by the standards of the day. It is curious that its main sponsors were, on the one hand, Polish landowners, magnates, etc., i.e. the Polish political and financial elite, and on the other, two of the rulers of partitioned Poland. After all, the dictionary was an important tool in building modern Polish culture and yet the foreign rulers' contributions were substantial. This would seem to indicate the strength of the relationship between the dominant forces in the Polish proto-field of power and the imperial elites of at least two of the partitioning powers. Under this alliance, the Polish field of power had considerable autonomy. It was also the means by which an autonomous Polish science was born, especially in the Russian partition. Scholars such as the natural scientist Stanisław Staszic (1755–1826) and the historian Joachim Lelewel (1786–1861) belonged to this tradition.

Meanwhile, according to Urbańczyk (1993), the beginning of modern Polish linguistics was the generation of linguists educated at the Warsaw Main School (*Szkoła Główna Warszawska*), an important autonomous scientific institution established by the Russian authorities. The school was only in operation from 1862 to 1869, when it was replaced by the Russian-language Imperial University of Warsaw, but managed to educate a whole generation of Polish scientists and social activists. Urbańczyk contends that Lucjan Malinowski (1839–1898) played the most important role, followed by Adam Antoni Kryński (1844–1932) and Jan Niecisław Baudouin de Courtenay (1845–1929). Another important figure was Jan Karłowicz (1836–1903), although his professional trajectory followed a different path. Karłowicz began his studies in Moscow and continued them in Paris, Brussels, Heidelberg, and Berlin, and was more of a historian and ethnographer than a linguist. However, he also tried to defend his habilitation at the Warsaw Main School. Karłowicz was more active in scientific societies than universities. He was a member-correspondent of the Polish Academy of Arts and Sciences in Kraków, the Poznań Society of Friends of Science, *La Société des*

Traditions Populaires in Paris, the Lithuanian Literary Society, the Lviv Folklore Society, and several others. Karłowicz pioneered the study of Polish dialects, and compiled a reasonably comprehensive six-volume dictionary (*Dictionary of Polish Dialects*). The dictionary used data taken from collections of texts, tales, and folk songs from before 1890, and was published in 1900–1911. It suggested that the Polish language and Polish culture occupied a central position both geographically and in terms of differences between languages and dialects. This thesis was admittedly more ideological than empirical, but nevertheless found favor among the Polish intelligentsia, as it recognized the importance of the Polish language and justified the Polish, or at least Slavic, character of a good deal of Prussia. I would like now to return to Lucjan Malinowski, who headed the Department of Slavonic Philology at the Jagiellonian University from 1877 and was secretary of the Languages Committee and the Philological Faculty at the Polish Academy of Arts and Sciences. Malinowski proved to be a more systematic contributor to Polish dialectology than Karłowicz, although he and his students concentrated on monographs of localities, and it was only Kazimierz Nitsch who undertook a systematic effort to map all of Poland. Malinowski's monograph on Slavic dialectology *Beiträge zur slavischen dialektologie* (Contributions to Slavic Dialectology) was published in Leipzig in 1873. This was no accident, as he was closely associated with the *Junggrammatiker* (Neogrammarian) School (Gogolewski 2001). Malinowski was educated in Prague and Jena, and studied in Leipzig where he made contact with August Leskien (1840–1916), one of the chief founders of the Neogrammarian School. His seminary produced several future scholars, notably Kazimierz Nitsch (1874–1958).

3.3 Jan Niecisław Baudouin de Courtenay

3.3.1 Early Years

Jan Niecisław Baudouin de Courtenay (1845–1929) is Poland's most outstanding linguist and a key reference point for much of the discussion on the history of Polish linguistics. Baudouin de Courtenay was born into a Polish family of French aristocratic descent. His forebears had come to Poland to serve the king. He graduated from the Main School in Warsaw, where he began his studies in 1862. Similarly to many other students at the school, he did not support the January Uprising (Doroszewski 2016). He continued his studies in Prague, Jena, and Berlin on Russian government scholarships. These stays gave him the opportunity to meet many of Germany's leading scholars. Baudouin de Courtenay studied in St. Petersburg under Izmail Ivanovich Sreznevsky (1812–1880) in 1868–1870. It was Sreznevsky who encouraged his incipient interest in dialectology. In 1870, Baudouin de Courtenay spent some months in Leipzig where he worked with August Leskien, a former fellow student from Jena. He published a work on fourteenth-century Polish in Leipzig in 1870. For this the University of St. Petersburg awarded him a master's degree. In 1871, the historical-philological faculty of Kyiv University nominated him for the post of docent of comparative grammar, but the decision to employ him was vetoed by the university senate on account of his Polish ethnicity. This was related to the restrictions described earlier on the admission of Poles (lecturers and students) at Kyiv University. The Russian authorities were afraid that they would dominate this important university, in a city where they already constituted a considerable part of the elite. At the same time, Baudouin de Courtenay was unable to accept an offer of employment at the Jagiellonian University, as he had to work at a Russian state university to work off his earlier government scholarship. As the Russified University of Warsaw had no interest in hiring him, he was limited to more peripheral options. In 1874, the University of Kazan offered him the post of docent of comparative grammar, and as he was not exactly spoilt for choice, he accepted. He lectured in Kazan from 1875 to 1883, complaining about its provincialism, its isolation from European science, and the lack of opportunity to debate his ideas with front-rank scholars. The upside, however, was that he managed to gather a group of talented students whose creative seminars developed into a tradition that came to be known as the Kazan School. This circle became famous for experimental phonetics and quantitative linguistics. Mikołaj Habdank Kruszewski (1851–1887) was a key member of the group and Baudouin de Courtenay worked closely with him before their views diverged.

Kazan had an earlier tradition of linguistic research, and this was also conducted by a prominent Polish scholar, viz. Józef Kowalewski (1801–1878), who taught there from 1824 to 1862, i.e. before Baudouin de Courtenay arrived. Kowalewski, who had studied at Vilnius University before its closure, was exiled to Kazan in 1824. Once there, he studied Asian languages and traveled extensively through Asia, including China and Mongolia. He delivered his first lectures on Mongolian in Europe in 1833. He also served as dean and rector at Kazan, and in 1862, accepted the chair of universal history at Main School and moved to Warsaw. Róziewicz therefore argues in favor for a "Kazan school of linguistics" or "Kazan school of Polish linguistics". Baudouin de Courtenay's role in the development of linguistic research at the University of Kazan was also highly regarded, as evidenced by the fact that he was an honorary member in 1901 (Róziewicz 1984: 263).

3.3.2 The Tartu and Krakow Years

In 1883, Baudouin de Courtenay moved to Dorpat (now Tartu, Estonia) to take up the newly established Chair of Comparative Slavic Grammar. His social commitment was evident in his support for equal treatment of Estonians and Germans at the University, and for the institution of Estonian and Latvian electorates. He was made an honorary member of the Society of Estonian Writers in recognition. In 1893, after working in Russia for 25 years, he became eligible for early retirement, which allowed him to travel abroad. He accepted the position of professor of comparative linguistics at the Jagiellonian University in Kraków. He also became a very active member of the Polish Academy of Arts and Sciences. In Kraków he had a group of dedicated students (including Kazimierz Nitsch, Stanisław Szober, Henryk Ułaszyn, and Tytus Benni—all of whom went on to become distinguished scholars in their own right). His five-year contract, however, was not renewed. According to Baudouin de Courtenay's biographers (e.g. Doroszewski 2016; Adamska-Sałaciak 2001), the immediate cause was his public attack on widespread tax evasion, which seems consistent with his principled ethical stance. This attack on the conservative and hypocritical

Kraków bourgeoisie was not well received by the Stańczyk circles that dominated the city, both politically and intellectually. Their political leader, Stanisław Tarnowski (1837–1917), was the key ideologist of the Polish conservatives in Galicia (the Stańczyks), and a renowned literary historian. He was a lecturer at Jagiellonian University (serving as rector twice) and president of the Polish Academy of Arts and Sciences in Kraków. He sat in the Galician parliament and in the House of Lords in Vienna. He first studied in Kraków and Vienna, and received his doctorate at the Jagiellonian University in 1870. The conflict between Baudouin de Courtenay and Tarnowski carried over into the field of language studies, including its theoretical linguistic and literary components. As a professor at the Jagiellonian University for many years (and at one stage head of the Department of Polish Literature), Tarnowski educated a considerable number of outstanding students. In the view the adopted, the study of literary history was a form of the study of "national character" or even "national psychology". In this approach, influenced by Hippolyte Taine (1829–1893), the social environment of artists explains the nature of their work. Tarnowski dealt with the literature of the Renaissance and Romantic period, and had no time for the literature of Positivism and Young Poland. Their exponents returned his detestation in equal measure. Stanisław Brzozowski (1878–1911) is a case in point. This philosopher and political writer, inspired by Marxism, studied at the Faculty of Natural Sciences of the University of Warsaw. Brzozowski's critique of Tarnowski and the whole ideology of the Stańczyks is of a paradigmatic nature. He especially takes them to task for their parochialism and bourgeois lack of imagination. Another important Young Poland (*Młoda Polska*) writer was Stanisław Przybyszewski (1868–1927) who, as Karcz (2002) suggests, can also be seen as a precursor to formalist ideas, as he was an adherent of the radical autonomy of art and of the view of an artistic work as having value only for itself. The Baudouin de Courtenay-Tarnowski dispute continued into the interwar period. One of Tarnowski's successors on the conservative side was Ignacy Chrzanowski (1866–1940), whose *Historia literatury niepodległej Polski* (History of literature of independent Poland, Part I 1906, Part II & III 1916) categorically states that Polish literature is principally a manifestation of the struggle for a national existence and a means to achieve it. Baudouin de Courtenay's criticism might also have had something to do with the political confrontation between liberals and the dominant conservatives in the elite of Kraków. Tarnowski used all the political resources at his disposal in this dispute. In particular, he used his connections, especially Galicja's Lieutenant Governor, Leon Piniński (1857–1938), to prevent Baudouin de Courtenay's contract from being renewed. Another possible reason for Tarnowski's hostility toward Baudouin de Courtenay was his suspicion of his support of Pan-Slavism, given his interest in Slavic languages. This was anathema to Hungarian nationalists. The Hungarian Government considered Baudouin de Courtenay's research in Slovakia to be "pan-Slavic agitation" and called for his contract in Kraków to be terminated. Baudouin could have requested a pardon, but declined to do, as this would have required that he directly address the Stańczyk leaders, including aristocrats such as Stanisław Tarnowski and Leon Piniński (Czelakowska and Skarżyński 2011: 59).

3.3.3 Panslavism

It is worth discussing the role of pan-Slavism in the development of Polish linguistics in some detail. It was a bone of contention between the ideologies of the Austro-Hungarian and Russian Empires. Róziewicz discusses the interesting case of the *Encyclopedia of Slavonic Philology*, a transnational project that ended in failure, at the very end of the nineteenth century. The Imperial Academy of Sciences in St. Petersburg, especially its Department of Russian Language and Literature, addressed an invitation to the Polish Academy of Arts and Sciences in Kraków in 1903 to take part in the project. The Academy refused to participate out of fear that its Russian partners would follow directives of a pan-Slavic nature. In 1904, Włodzimierz Spasowicz (1829–1906), a well-known and well-connected Polish lawyer and literary historian from St. Petersburg, wrote to Baudouin de Courtenay, who was a professor in St. Petersburg at the time, to invite him to join the project. Polish Slavists were generally wary of Russian nationalism, but the Austrian authorities had especially strong fears, as evidenced by their refusal to extend Baudouin de Courtenay's professorial contract at the Jagiellonian University in 1899. The Academy of Arts and Sciences therefore decided to publish its own *Polish Encyclopedia*, which the Russians could use to create theirs. In the meantime, the initiative only succeeded in attracting the following Poles: Aleksander Brückner from Berlin, who wrote about the influence of German on Polish; Jan Baudouin de Courtenay from St. Petersburg, who was to write a study on the psychological basis of Slavic languages, a classification of Slavic languages and dialects, and a review of Slavic dialects in Northern Italy; Władysław Nehring (1830–1909) from Wrocław, who wrote on the historical grammar of Old Polish, the formation of contemporary Polish literary language, and Czech influences on Polish; Wiktor Porzeziński (1870–1929) from Moscow, who worked on a review of Polish dialects, the influence of Slavic languages on Lithuanian and Latvian; and Kazimierz Nitsch from Kraków, who wrote about the influence of Latin on Polish (Róziewicz 1984: 164) There were also attempts to involve Jan Rozwadowski (1867–1935), Karol Potkański (1861–1907), Marian Sokołowski (1839–1911), and Włodzimierz Demtrykiewicz (1859–1937). However, nothing came of this project, with Brückner withdrawing his submitted chapter. Vatroslav Jagić (1838–1923) from Vienna, then working in St. Petersburg, was hired for the Russian dictionary. This Croatian patriarch of Slavic studies complained about the Poles to Alexei Shakhmatov (1864–1920), an advocate of close cooperation among all Slavic peoples, as follows: "In Kraków, as it seems, they do not wish to enter into any relations with St. Petersburg, that is, with Russia". From a letter of Rozwadowski, who visited Jagić in Vienna, it followed that "In Kraków after the death of Lucjan Malinowski, there are no people with broader Slavic views. The Kraków Stańczyks are narrow separatists. I don't know how they relate to you, but they look at me with suspicion, thinking that I am a Russophile. Take, for example, their participation in the encyclopedia. Only Brückner and the late Nehring have kept their word, while Prof. Nitsch keeps postponing from year to year. Even Baudouin does not give the promised answer, so I have stopped asking him" (Róziewicz 1984: 165). For similar reasons, the Union of Slavonic Academies and Scientific Societies, established in 1912 on the initiative of the Department of Russian Language and Literature of the Imperial Academy of Sciences, was ignored by the Polish Academy of Arts and Sciences, which refused to join because of Russia's leading role in the project and its presumed pan-Slavism or neo-Slavism. The Czechs also declined, citing a lack of funds (Róziewicz 1984: 170).

3.3.4 Second Stage of Career in Russia

Baudouin de Courtenay therefore returned to Russia as a Russian citizen and in 1900 obtained a lecturing position at St. Petersburg University. He settled in St. Petersburg, where he lived in straitened circumstances, until he obtained a professorship. With his lectures, he exerted a great influence on young Russian linguists such as Lev V. Shcherba (1880–1944), Victor V. Vinogradov (1895–1969), and Nikolai S. Trubetzkoy (1890–1938). In addition to theoretical issues, he also dealt with specific linguistic areas. These often had a political dimension, e.g. the "Kashubian question". In the late nineteenth century, the status of the Kashubian language became the subject of academic disputes with a clear political dimension. Interestingly, discussions on this issue continue in Poland to this day. Some regard Kashubian as a separate language, while others contend that it is a dialect of Polish. The latter usually marshal nationalist arguments to counter what they see as a Kashubian attempt to use linguistic separatism to become a separate nation. But to return to the late nineteenth century, Stefan Ramułt (1859–1913) from Lviv University published his *Słownik języka pomorskiego czyli kaszubskiego* (Dictionary of the Pomeranian or Kashubian Language, 1893) in Kraków. His work assumed that Kashubian was a separate Slavic language. In the meantime Baudouin de Courtenay, although he did not conduct his own research in this area, published his text on the "Kashubian question", which was published in several language variants (Russian in 1897) in which he issued a kind of "Solomon's verdict". Avoiding the assessment of the current state of affairs, he stated that historically "the Kashubian language is more Polish than the Polish language". This phrase is still quoted by those who support the idea of Kashubian as a variant of Polish language. At the same time, however, it should be added that Baudouin de Courtenay had a peculiar approach to these kinds of issues. For one thing, he did not accept the objective existence of nations. He therefore rejected the existence of national languages and the notion of a nation's "soul". He was an ardent anti-nationalist. It could even be said that he was a proponent of the Benedict Anderson's thesis that a nation is merely an imagined community. For Baudouin de Courtenay, language only functioned on the level of individuals. This made him naturally suspicious of the notions of national and tribal languages. Society appeared to him to be a collection of individuals, so he rejected any personification of the nation. He viewed language as a property of humanity, while stressing the importance and value of every individual language and dialect. Thus he saw the "Kashubian question" not in terms of a national or regional language, but as a "linguistic area". This could be also seen as his emotional reaction to the works of nationalistically oriented linguists who glorified nations and their languages. While pointing out the geographical diversity of any language area, he recognized the importance of a sense of national belonging, but saw this as a separate issue. As for Kashubia (*Kaszuby*), he pointed out that most Kashubians were identified as Polish.

In 1907, Baudouin de Courtenay entered into debate with the famous Leipzig linguists Friedrich Karl Brugmann (1849–1919) and August Leskien, declaring himself a staunch supporter of artificial languages. This might well have been an expression of his strong anti-nationalism. In 1915, he was arrested by the Russian "*Okhrana*" (secret police) for supporting the autonomy of the smaller nations of the Russian Empire. This incident is worth examining. He was charged with inciting rebellion for publishing a brochure titled *National and Territorial Character in Autonomy* (St. Petersburg 1913), duly convicted, and sentenced to two years in prison. The Tsar reduced his sentence, which he spent in St. Petersburg's infamous Kresty Prison, to three months, and he was released in January 1915. The trial and conviction of such an eminent scholar and member of the Imperial Academy of Sciences obviously caused a sensation. Baudouin de Courtenay lost his university position until the February revolution in 1917 (Czelakowska and Skarżyński 2011). He returned to the university when the Provisional Government came to power, as the Constitutional Democratic Party (Cadets), of which he was a member, held a majority. Baudouin de Courtenay was sympathetic to the federalist-autonomists, who called for the federalization of Russia in order to give Russia's many national and ethnic groups maximum autonomy. It is worth noting that, in this dimension, Baudouin de Courtenay belonged to a larger group of Polish intellectuals at the intersection of the Polish and Russian fields of power. His biography is strikingly similar to that of another prominent Polish scholar, Leon Petrażycki (1867–1931). Petrażycki, although a graduate of Kyiv University, had also served internships at Germany's most prestigious universities and even took part in important German public debates. Although Petrażycki was a lawyer, the two scholars adopted similar psychologizing approaches. Most interestingly, however, they had similar political views. Petrażycki was even a member of the Duma, where he fought for the rights of minorities and women. He also spent three months in a Tsarist prison in 1908. Like many other "Cadet" Party deputies, he had signed the Election Manifesto of Viborg in 1906. Both Baudouin de Courtenay and Petrażycki became professors at St. Petersburg University at roughly the same time. They also left Russia during the Bolshevik Revolution to take up professorships at the University of Warsaw. Neither could find his place in interwar Poland, not least because they professed extremely liberal views that were out of step with the politics of the day. In an earlier text (Zarycki 2013), I argued that they were both members of a larger sector of the former Polish elite who, as a result of the Bolshevik Revolution, lost their previous social positions and also the reference points for their political positions, which were defined in terms of the Russian field of power. In the former set of coordinates, they were members of an intermediate faction between the bourgeoisie and the intelligentsia that stood in sharp opposition to the conservative elites of the Russian Empire. They criticized these elites from liberal positions and demanded democratization in all dimensions, but de facto they also represented the interests of a large part of the bourgeoisie, especially its dynamic and non-entirely Russian circles.

This earlier text (Zarycki 2013) discusses the Polish faction of Russian liberals to which Baudouin de Courtenay and Petrażycki belonged through the prism of its political leader, Aleksander Lednicki (1866–1934). Lednicki was also a member of the Duma representing the

Constitutional Democratic Party. He was a wealthy lawyer who maintained direct relations with all the liberals in the Russian field of power, as the lawyer and literary historian, Włodzimierz Spasowicz, had done before him. After moving to Poland from Moscow in 1918, Lednicki lost his privileged status, partly as a result of having a large part of his estate confiscated, but mostly due to the Polish field of power being structured in such a way that there was no place for liberals of his kind. It should be added, however, that Aleksander Lednicki's son, Wacław Lednicki (1891–1967), had a successful academic career. Wacław Lednicki was a literary historian who studied in Moscow, where he was also born, and in Kraków, while he received his habilitation in Vilnius in 1926. During the war, he emigrated to the United States and was a professor of Slavic literature at the UC Berkeley in 1944–1962. It is remarkable that Aleksander Lednicki is seldom mentioned in mainstream accounts of Polish history. There is also very little discussion in contemporary Poland about the circle of Polish liberals in the Russian field of power.

3.3.5 Late years and Legacy

However, to return to Baudouin de Courtenay. During the Bolshevik Revolution he lost his rich linguistic archive. After the revolution, he worked briefly in St. Petersburg, but moved to Poland in 1918, aged 73, to take up the chair of Indo-European linguistics at the University of Warsaw. In 1922, he ran for President of Poland. His candidature was nominated by the country's national minorities, especially the Jewish community. It is worth mentioning that he advocated that Yiddish be taught in Polish schools and that Polish be taught in Jewish schools where Yiddish was the language of instruction (Róziewicz 1984: 184). Unlike most of the mainstream intelligentsia he was an atheist. As a radical liberal, he was detested by conservative, clerical and nationalist circles, but found it hard to identify with any political party. He also provoked numerous controversies in the academic field. The best evidence of this is the fact that both the University of Warsaw and the University of Lviv refused to award him an honorary doctorate.

In the 1920s, Baudouin de Courtenay spent some time in Prague, lecturing on the classification of languages, and was invited to lecture in Copenhagen in 1923. His international reputation was considerable, as attested by the fact that the organizers of the 1927 Hague International Linguistic Congress asked him to provide the names of notable Polish linguists worthy of being invited. The organizational committee told him that the following names were on their list: Jan Rozwadowski, Tadeusz Jan Kowalski (1889–1848), Jan Łoś (1860–1928), Kazimierz Nitsch (1874–1958), Wiktor Porzeziński (1870–1929), Stanisław Szober (1879–1938), Aleksander Brückner (1856–1939), and probably Jerzy Kuryłowicz (1895–1978). Baudouin de Courtenay sent back the following additions: Karol Appel (1857–1930), Mojżesz Schorr (1874–1941), Stanisław Słoński (1879–1959), Witold Doroszewski (1899–1976), Stanisław Wędkiewicz (1888–1963), Władysław Kotwicz (1872–1944), Tytus Benni (1877–1935), Jan Otrębski (1889–1971), Stefan Gixelli (1888–1938), Olgierd Chomiński (1884–1943), Tadeusz Lehr-Spławiński (1891–1965), Adam Kleczkowski (1883–1949), Stanisław Łempicki (1886–1947), Edward Klich (1878–1939), Henryk Oesterreicher (1892–1943), Zenon Klemensiewicz (1891–1969), Mikołaj Rudnicki (1881–1978), Witold Taszycki (1898–1979), Jan Janów (1888–1952), Henryk Ułaszyn (1874–1956), Ignacy Stein (1875–1964), Roman Zawiliński (1855–1932), Antoni Śmieszek (1881–1943), and Stefan Saski (1888–1974) (Czelakowska and Skarżyński 2011: 73).

A characteristic feature of his approach was a strong psychologism, as evidenced by, e.g. his belief that people had certain feelings about their languages and that these feelings were objectively verifiable and therefore testable. He saw the importance of society for the functioning of language, but did not study it in detail. This led him to become involved in what is now known as psycholinguistics rather than what was to become sociolinguistics. This probably detracted from his later influence as a structuralist. However, his approach was strongly positivist, with a clear cult of "scientific fact", an interest in mathematics, and a tendency to view linguistics as a natural science. He thus had a lot in common with de Saussure. Polish scholars in particular point to the convergence of their thinking. For instance, in his inaugural lecture at St. Petersburg in 1870, Baudouin de Courtenay insisted that the "static" aspects of sounds in a language were every bit as important as the "dynamic" aspects. This made him one of the first to speak in favor of this dichotomy, which de Saussure later termed synchrony and diachrony. In the same lecture, he mentioned the distinction between the two forms of language, viz. abstract and concrete ("language" vs. "speech"). This corresponds with the opposition between langue and parole, introduced in Saussure's lectures of 1907–1911. This led Wąsik to argue for "Baudouin's contributions to structuralist views of language, some fifty years before its first theses were formulated" (Wąsik 2001: 17).

Baudouin de Courtenay's contribution to the development of the structural theory of language cannot be considered in isolation from the achievements of his younger colleague. Mikołaj Kruszewski (1851–1887). Kruszewski graduated from the Imperial University of Warsaw, and was a colleague of Baudouin de Courtenay at the University of Kazan (Berezin 2001). While there, he developed what can be called positivistic phonology. Its basis was the idea of describing "linguistic facts" as perceived solely in terms of the physiology of sounds, i.e. "anthropophonics". Kruszewski was introduced to positivist thinking at the University of Warsaw by the eminent Russian psychologist and logician Matvey Mikhailovich Troitsky (1835–1899), who was close to materialism. Through Troitsky, he learned John Stuart Mill's system of logic. Kruszewski argued that linguistic phenomena are governed by laws of nature that have no exceptions, and so he favored the hypothesis that sound laws were universal. His enthusiasm for the positivist inductive approach was in clear opposition to the deductive orientation of the Neogrammarians and was not understood by them. Kruszewski's depiction of language as a constant clash between progressive association through similarity and conservative assimilation through contiguity was later popularized by Roman Jakobson (1896–1982) and became one of the foundations of twentieth-century structuralism. His major work, according to Fedor Berezin (2001), was prevented from becoming a classic by his premature death. Jakobson wrote about Baudouin de Courtenay and Kruszewski in this context as follows: "These two Polish researchers were the only linguists in the world who approached a genuine theoretical conception of language, according to Saussure's acknowledgement of 1908, and his Geneva courses in general linguistics evince a deep and fruitful influence of Kruszewski's thought. Yet in such crucial questions as, for instance, the creative aspect of language and the relationship between verbal signs and concepts, Kruszewski stands closer than Baudouin and Saussure to the scientific vistas of today" (Jakobson 1971: 449–450). One of the most obvious traces of their influence on Saussure is the fact that in 1881 and 1882 Baudouin de Courtenay and Kruszewski presented their work at four consecutive meetings of the *Société de Linguistique de Paris* and de Saussure attended three of them (Wąsik 2001: 7).

However, although Baudouin de Courtenay is Poland's best known linguist internationally, his influence is not as great as it might have been, especially if he is to be considered a theorist on par with de Saussure. One reason may be that he never produced a synthesis, a work summarizing his approach, or even some of his views. At the same time, he dealt with an extraordinarily large number of topics, from highly theoretical to detailed empirical studies. In the Polish context, however, he remains the greatest figure in the entire history of the field and his influence on its history has proven to be very strong. Baudouin de Courtenay seems to have infected a whole generation of students with his psychologism. His students included Henryk Ułaszyn, Witold Doroszewski, Wiktor Porzeziński, and Jan Rozwadowski, of whom are more below. His most famous student was probably Kazimierz Nitsch, who could hardly be called a theoretician, as he mostly developed and applied dialectological methods.

3.4 The Institutional Infrastructure of the Field at the Turn of Centuries

If we take a longer-term perspective, one that began in the nineteenth century, the building of the modern Polish nation-state proceeded in a partially reversed order compared to the classical Western model. This is because the building of a national culture largely preceded the building of the state. As mentioned in the second part of this book, Polish national culture was not only built by the universities in Austrian Galicia, but also by independent cultural and academic institutions (concentrated in the same region). It is worth recalling the role that these proto-state institutions played in the development of scientific Polish linguistics, and subsequently, the role of this linguistics in the development of a standard, modern Polish language.

As discussed earlier, of particular relevance in this context was Kraków and the *Akademia Umiejętności* (Academy of Arts and Sciences, since 1919, the Polish Academy of Arts and Sciences, PAU). This gave birth to the *Komisja Językowa Akademii Umiejętności* (Commission of Language) in 1874. Its first secretary was Lucjan Malinowski. The Commission (with Rozwadowski, who replaced Malinowski in 1898) increased in importance during the presidency of Baudouin de Courtenay (1894–1900). After Baudouin left Kraków in 1900, Rozwadowski acted as president between 1908 and 1935. The next secretary (1908–1935), and its last president (1935–1952), was Kazimierz Nitsch. The Commission included not only scholars from Kraków, but also from other cities (and even from abroad). It therefore played a central role in Polish culture by integrating all the streams of Polish linguistics. The PAU took charge of the *Słownik Staropolski* (Old Polish Dictionary), *Atlas językowy polskiego Podkarpacia* (the Linguistic Atlas of the Polish sub-Carpathians by M. Małecki and K. Nitsch), and published *Prace Komisji Językowej* (Works of the Commission on Language). It should be borne in mind that the Warsaw Scientific Society, the Poznań Scientific Society, the Lviv Scientific Society, the Vilnius Scientific Society, and others also had linguistic (or philological) committees and publishing houses. However, the Kraków center was clearly stronger than the Warsaw one, especially because the PAU and the Jagiellonian University were both located in the city. In Warsaw, *the* Mianowski Fund was an important institution. The Warsaw Scientific Society was active from 1907. Jerzy Karłowicz also conducted private seminars in Warsaw as part of the Society for Educational Courses (*Towarzystwo Kursów Naukowych*, TKN), but their scale was much smaller than that of the Jagiellonian University and the PAU in Kraków.

This institutional infrastructure was taken over by the Polish state that emerged in 1918. At the same time, independent institutions continued to develop. The Society of Polish Language Lovers (*Towarzystwo Miłośników Języka Polskiego*, TMJP) was founded in 1920, and the Polish

Linguistic Society (*Polskie Towarzystwo Językoznawcze*, PTJ) in 1923. The latter grouped professional linguists, the former—as its name implies—was open to everyone. It had numerous branches around the country. The TMJP published booklets from its library and a monthly magazine *Język Polski* (Polish Language), which occasionally featured serious scientific treatises in addition to its regular popular science articles. In 1929, a separate Society for the Correctness of the Polish Language was also founded in Warsaw. Urbańczyk (1993) calls this the "Warsaw schism". Its original name was the Society for the Promotion of the Culture of the Polish Language (*Towarzystwa Krzewienia Kultury Języka Polskiego*, TKKJP). It was reestablished as the Society for the Culture of Language (*Towarzystwo Kultury Języka*) in 1966. It continues to publish *Poradnik Językowy* (Language Guide). It approaches the level of a scientific magazine, but has been written in a more accessible style since the editorial office was moved to Warsaw in 1932.

This rivalry between Kraków and Warsaw had much deeper historical roots and manifested itself most clearly in the orthographic reforms of the early twentieth century. The PAU refused to approve it, declaring itself to be "a research institution in the field of scientific theory, not a legislative body in matters of a practical nature" (Bajerowa 2002: 10). According to Irena Bajerowa, this reaction already showed a contrast between the Warsaw center, oriented more toward solving practical issues, and the more theoretical Kraków center. In 1892, a reform of Polish spelling was finally introduced, but there was no consensus between the two centers. Kraków opted for more historical language forms, while Warsaw adopted a reformed spelling adapted to modern pronunciation. The Warsaw option was supported by Baudouin de Courtenay and Aleksander Brückner, who disagreed with the Kraków proposals and even signed a protest against them. As a result of this conflict, two separate systems of Polish orthography (Austrian and Russian) were used in parallel until 1918 (Bajerowa 2002: 10).

Another important body organized by the PAU was the Orthographic Committee which implemented orthographic reform in 1935–1936. Work on the unification of Polish was also carried out intensively throughout the interwar period and after the war. New rules of spelling and pronunciation were issued soon after the establishment of the new

Polish state. In the late 1970s, Zdzisław Stieber gave the following evaluation of the state of affairs in this field in Poland in his paper on "Spelling and correct pronunciation in independent Poland" (Stieber 1982: 247–248): "The work of Polish linguists in determining correct pronunciation produced better results than could have been expected. Even linguists now find it difficult to determine, on the basis of pronunciation, which part of Poland his or her interlocutor comes from. This is especially the case with people with tertiary education, although it is also true of those with secondary education, or even only elementary education. A fairly uniform pronunciation is used throughout the country" (quoted after Grucza 2001: 98). The unification of the Polish language, which proceeded over approx. 100 years, can therefore be rated a success. This was largely due to the resources that the communist authorities devoted to it, as well as the work of at least the three previous generations of linguists and the institutions they created. Their commitment to the standardization of national culture, in particular the Polish language, gave them additional social prestige and earned official recognition.

It may also be noted that the field of Polish linguistics included several influential scholars working outside of Poland. An excellent and worth mentioning example is Aleksander Brückner (1856–1939), a contemporary of Baudouin de Courtenay. Born into a Polonized family of German origin, he completed his studies in Lviv, and then gained experience in such famous academic centers as Leipzig, Berlin, and Vienna under the supervision of such eminent Slavists as Leskien, Jagić, and Franc Miklošič (1813–1891). He obtained both his doctorate (1876) and habilitation (1878) at the University of Vienna. In 1881, he was offered the chair of Slavic studies in Berlin as the successor to Vatroslav Jagić, where he taught until his retirement in 1924. He died in Berlin in 1939. Brückner was very active in the Polish academic field and remained loyal to it throughout his career. He tried to raise the prestige of old Slavic culture in the eyes of the Germans among whom he worked and among his fellow Poles. He was a member of the Polish Academy of Arts and Sciences and the Lviv Scientific Society. In 1929, the University of Warsaw awarded him an honorary doctorate. Brückner specialized in older periods of Polish and Slavic culture. He discovered, interpreted, and published the oldest known manuscript in Polish, the *Kazania*

Świętokrzyskie (Holy Cross Sermons). Although Brückner remained in Berlin after Poland regained its independence, he remained de facto a full member of the Polish linguistic field. This somewhat "remote" form of participation seems to have remained after the partitions, when the dispersion of Polish scholars between different countries was natural.

3.5 Baudouin de Courtenay's Legacy and Successors

3.5.1 Andrzej Gawroński and Jan Michał Rozwadowski

Baudouin de Courtenay had his detractors as well as well as his followers. The most famous critic of Baudouin de Courtenay was Andrzej Gawroński (1885–1927). Gawroński graduated from the University of Lviv, and also studied in Leipzig, where he received his doctorate in 1906. He then became an assistant professor at the Jagiellonian University. In 1917, he assumed the chair of comparative linguistics at the University of Lviv. He was a member of the Polish Academy of Arts and Sciences and he co-founded the Polish Linguistic Society. Gawroński took issue with Baudouin's view that the language of an individual was the only reality (by implication, the language of a nation was not). What attracted his criticism was the definition of language as a system of signs, the division of linguistics into synchronic and diachronic, the concept of language as a group property, and the division of speech into langue and parole, among others. However, like Baudouin de Courtenay, Gawroński subscribed to psychologism. He believed, for example, that language is a phonic and articulated form of the psychic depth of the individual, a means of exteriorization of the human spirit, and as such it can be compared to the plastic arts, which make themselves visible through expression. He also argued that there is no language of community just as there is no collective spirit or collective intelligence. He considered this a fiction based on the content of communication.

Jan Michał Rozwadowski (1867–1935) was an influential student of Baudouin de Courtenay and an older colleague of Gawroński. Rozwadowski also criticized his teacher for doctrinairism and radicalism. Most of Baudouin de Courtenay's alumni, however, more or less shared his views on language. Milewski (1972) considers Rozwadowski to have been Poland's most important young grammarian, along with Jan Łoś (1860–1928). Rozwadowski graduated from the Jagiellonian University, where he received his doctorate in 1891 and became a professor in 1899. He was the President of PAU (1929–1929) and chairman of its Orthographic Committee of the PAU, a co-founder of the Polish Linguistic Society (1925), and the first president of the East European Institute in Vilnius. The Institute was active between 1930–1939 and is considered a pioneering center of what later became known as Sovietology (Kornat 2000). Rozwadowski studied in Leipzig, among other places, and took over the chair in Kraków from Lucjan Malinowski. He was a consummate language theorist best known for his theory of the duality of the structural divisions of words. Rozwadowski proposed the law of bipartition, which asserts that every linguistic entity consists of two elements that are semantically distinct but formally identical. This approach was later developed by Doroszewski in his theory of word formation. In opposition to Wilhelm Wundt (1832–1920), who claimed that objects are usually named after one unique feature, Rozwadowski contended that the basis of every language product consists of two parts: a distinguishing component pertaining to its semantic aspect; and an identifying component pertaining to its formal aspect. This can also be considered a psychological approach. His "law of automation" states that the reason for evolution in linguistic creation, and in the whole of cultural realm, is the constantly felt need to have human creations refreshed by inserting emotional elements into them, and that this need is diminished in the process of habitual automation. Rozwadowski was at the peak of his scientific accomplishments when he delivered a lecture before the audience of the *Société de Linguistique de Paris* in 1925. He presented linguistics as a discipline in the highest degree predestined, among the other humanistic disciplines, to formulate generalizations about the main mechanisms of human thinking. He proposed great undertakings for linguistics, including the construction of a general grammar that was to have included all the linguistic categories and structures of every language in the world, and a general lexicon containing all possible lexical entries, with etymological explanations and semantic values. As for his views, they were significantly influenced by the WWI. The bloody war, and then the Bolshevik Revolution, not only frightened him, but made him wary of any revolutionary trends, pushed him toward conservatism, and made him a catastrophist. This attitude was not all that unusual at the time. Marian Zdziechowski (1861–1938), a literary scholar who taught in Vilnius during the interwar period, expressed similar views. This former Jagiellonian University professor was an expert on Slavic literatures with a particular interest in Russian literature, and he knew many authors personally. Although more conservative than Lednicki, he also belonged to the Polish elite of the Russian Empire and was well-connected in both the Russian and Polish fields of power. He regarded the fall of the Russian Republic in 1917 as a tragedy of metaphysical proportions, although his views can also be interpreted in sociological terms as having resulted from the collapse of the social system of the empire to which he belonged. In this context, it is worth mentioning that Aleksander Lednicki and Leon Petrażycki, who lost their status with the revolution, both committed suicide in the 1930s. This was at least partly due to their inability to integrate into the field of power of the newly created Republic of Poland. Zdziechowski in turn complained about the lack of recognition for his deeply critical views on communism and Soviet Russia. His final years were also full of pessimism, as he expected an imminent Bolshevik invasion of Poland.

3.5.2 Kazimierz Nitsch

Another major figure in the history of Polish linguistics, and another student of Baudouin de Courtenay, was Kazimierz Nitsch (1874–1958). Nitsch graduated from the Jagiellonian University, where he had been Malinowski's youngest pupil, and earned his doctorate in 1898. He became an associate professor at the Jagiellonian University in 1911 and a full professor in 1920. He maintained constant contact with this university, except in 1917–1920, when he was a professor at the

University of Lviv. Nitsch was above all a dialectologist, arguably the greatest that Poland has ever produced. He was directed toward this field of study by Rozwadowski. Nitsch's first dialectological work was devoted to the Kashubian region and published in 1904. In 1898, he became an associate of the Language Commission of PAU. In 1907, he published *Dialekty polskie Prus Wschodnich* (Polish dialects of East Prussia), and formed working relationships with "progressive" German scholars. In 1908 he founded the periodical *Rocznik Slawistyczny* (Slavic Yearbook) with Rozwadowski. In 1915 his first attempt to systematize Polish dialects was published in PAU's *Polish Encyclopedia*, mentioned above. This research qualified him to become an expert adviser for the Paris Peace Conference in 1919, which provided empirical data to legitimize Polish territorial claims on ethnic borderlands (Romer 2010). Maciej Górny (2017) notes that when the borders of the nascent Polish state were being negotiated, most of the Polish expert advisers were anthropologists and geographers. Steven Seegel (2018) describes the key role of the latter. Of the Polish geographers, the great Polish cartographer, Eugeniusz Romer (1871–1954), played a key role. As Morgane Labbé (2019) shows, Romer provided key statistics and maps to Roman Dmowski, the chief Polish negotiator at Versailles. Nitsch devoted a great deal of time and effort into systematically mapping all the dialects of Polish. He began with Prussia, where the question of Polishness was particularly sensitive, as the government was implementing a Germanization policy. Nitsch dealt primarily with the classical dialects of Polish, i.e. the ethnographic core of the Polish language area. At the same time, however, he dealt with the extremely delicate issue of the contact areas of the Polish language with neighboring languages and cultures. To this end, he studied border regions, which separated Polish language areas from German, Czech, Slovak, Ukrainian, Belarussian, and Lithuanian language areas. A particularly politically fraught subject in this context was the status of Kashubian. His position was somewhere between that of those who considered it a dialect of Polish and those who considered it a separate language. Nitsch made a clear distinction between what he considered the two major branches of Polish. He saw one (continental) as "Polish" in the strict sense of the word, and the other (coastal), he saw as "Pomeranian-Polish". In 1934, Nitsch published the above mentioned

A linguistic atlas of the Polish sub-Carpathians with Mieczysław Małecki (1903–1946). The Atlas of the Sub-Carpathians was conceived as a general attempt before work on an atlas of Polish dialects began. In the spring of 1939, the two authors presented a paper entitled "A plan of a nationwide language atlas to the Linguistic Commission of the Polish Academy of Arts and Sciences". The atlas had an impressive size of 500 maps, some of which extended into Czechoslovakia.

Nitsch adapted quite well to the new communist order, although he did not join the PZPR. He can justifiably be described as a "statehooder" (*państwowiec*), i.e. he felt loyalty to the state and had an instinct for leaning toward "central", i.e. dominant and strongly naturalized, variants of Polish culture. According to Joanna Okoniowa, he could have been described as a socialist even before WWI, although he once said: "Although I was strongly opposed to Endecja politically and socially, I've been a 'Pan-Pole', in the geographical sense, from a young age" (Okoniowa 2011: 21). The term "Pan-Pole" (*wszechpolak*) was a nationalist term that denoted an expansionary and universalist Polish identity. In 1952–1957, Nitsch was even nominated vice-president of the Polish Academy of Sciences (PAN), which indirectly demonstrated and elevated the status of linguistics in the hierarchy of academic disciplines. He certainly benefited from the fact that, after the war, dialectological research gained a new relevance and a broader front. Another expert in dialectological syntheses was Zdzisław Stieber (1903–1980) from the University of Łódź. According to Karpowicz (2006), Stieber's *Rozwój fonologiczny języka polskiego* (The Phonological Development of Polish), published in 1952, was the first Polish linguistic study to implement the structuralist research method. In 1973, Karol Dejna (1911–2004), also from the University of Łódź, presented a clearly structuralist synthesis. Earlier syntheses, especially those of Nitsch, were strongly influenced by the historical school of the neogrammarians, even though he distanced himself from it. However, it was primarily historical dialectology that Nitsch developed most intensively in the 1930s, with Witold Taszycki (1898–1979) as his main collaborator, at the Jagiellonian University. Historical research was divided between approaches focused on historical texts and on vanishing dialects. There was also an important dispute over the geographical location of the roots of literary Polish. One camp claimed they were in Małopolska (Lesser Poland), which is largely coextensive with the Western part of former the Austrian province of Galicja. This conviction was shared by Aleksander Brückner, Tadeusz Milewski (1906–1966), and Stanislaw Szober. The other camp claimed they were in Wielkopolska (Greater Poland). This group included Kazimierz Nitsch, Tadeusz Lehr-Spławiński (1891–1965), Jan Łoś (1860–1928), Stanislaw Urbańczyk, and Mikołaj Rudnicki (1881–1978).

That cartographic synthesis of the knowledge on Polish dialects amassed by Polish scholars, which comprised twelve large-format volumes, was published in Wrocław in 1957–1971. Nitsch edited the first two volumes. In 1959, Mieczysław Karaś (1924–1977) was appointed head of the Dialectological Department of the PAN Polish Language Institute (*Instytut Języka Polskiego*, IJP) and edited the remaining volumes. In 1962–1968, a six-volume Dialectal atlas of Kielce province by Karol Dejna was published. Kazimierz Nitsch was also the initiator of *Słownik gwar polskich* (the Dictionary of Polish Dialects) which began to be published in 1964, although work to replace Karołowicz's dictionary had begun in 1954. In 1953, the Department of Atlas and Dictionary of Polish Dialects of the Polish Academy of Sciences was set up in Kraków, headed by Nitsch, and operated until 1958. It is worth noting that after the Polish border changes in 1945, dialectological research in the so-called Recovered Territories, as those Polish territories that were in Germany before the war were known during the communist period, also gained importance, especially in those regions where part of the native population remained in Poland after the forced population resettlements, e.g. Upper Silesia, Warmia, and Masuria (*Mazury*). Thus, it can be said that the achievements of Nitsch and his alumni were also used by the new Polish state to legitimize its presence in the new territories. The school of dialect research of which Kazimierz Nitsch was a leader is probably the most distinct manifestation of the important role of linguistics in what can be called "state building".

Before concluding the discussion of Kazimierz Nitsch's significance, one more dimension of the disputes in which he participated is worth noting, viz. his clash with the literary historian and critic, Stefan Kołaczkowski (1887–1940), in the 1930s. Kołaczkowski studied in Leipzig, Freiburg, and Berlin under Aleksander Brückner. He earned his doctorate (1914) and habilitation (1929) at the Jagiellonian University, where he became a professor of Polish literature in 1933. Kołaczkowski fought to have the Polish studies program reformed. In particular, he demanded that less time be devoted to linguistics and more to other subjects, specifically sociology, psychology, and cultural history. He accused linguistics of not moving with the times, and decried their lack of interest in the school of Karl Vossler (1872–1949) and his followers. Nitsch was supported by Rozwadowski and Ułaszyn in this dispute, which was based on the longstanding resentment and rivalry between the sub-fields of history of literature and linguistics, and had previously confined to collegial disputes. Kołaczkowski's action ultimately came to nothing. He did not receive enough support from his fellow literary historians, although a committee appointed by the Ministry of Religion and Public Enlightenment of the Second Republic made minor changes to the study regulations just before the outbreak of WWII. This dispute nevertheless reveals an interesting dimension of the structure of the wider field of language studies. Linguists still present Kołaczkowski's position as resulting from his "ignorance" in modern linguistics and its broader relevance. For Kołaczkowski, his campaign against "linguistic imperialism" was part of a wider struggle against what he saw as lowering the level of "general culture" of the younger generation of the intelligentsia. He regarded the "mechanical" teaching of linguistics, especially of complex grammatical and dialectical theories, as well as structuralism, as part of the technicization and massification of higher education, which was losing its humanistic and erudite qualities, which he also identified with the ethos of the intelligentsia. Behind what was ostensibly a dispute between a group of literary scholars led by Kołaczkowski and a majority of Polish linguists led by Nitsch, was a broader controversy over what constituted an "ideal member of the intelligentsia". Another manifestation of this tension was the dispute between Nitsch and Kołaczkowski over spelling reform in the 1930s. Kołaczkowski argued this time "against the orthographic dictatorship" of linguists (Szymański 2000).

3.5.3 Wiktor Porzeziński and Stanisław Szober—Toward a Systematic Standardization of Language

Baudouin de Courtenay's students included Wiktor Porzeziński (1870–1929), who was a professor at Moscow University until he moved to the University of Warsaw in 1921. Porzeziński had previously been a student of Philip Fortunatov (1848–1914), the founder of the Moscow school of linguistics. According to Róziewicz, Porzeziński, together with Witold Doroszewski, were among the most active organizers of Polish academic life in Moscow during WWI. They set up the Polish Scientific Circle in Moscow, whose president in 1915–1917 was Doroszewski, and then Porzeziński (Róziewicz 1984: 222). In Warsaw, Porzeziński took over the chair of Indo-European linguistics. He was somewhat less psychologizing than Baudouin de Courtenay. In contradistinction to his predecessors, Porzezinski held that only grammatical (i.e. morphological and syntactical) criteria should be taken into account in defining parts of speech.

Porzeziński's most illustrious student, Stanisław Szober (1878–1939), subscribed to a similar vein of psychologism as Gawroński, becoming yet another language theorist who considered it an expression of the human psyche. This drew criticism from Porzeziński, who found his psychologizing in matters of language excessive. Szober was educated at the Imperial University of Warsaw, but received his post-graduate training in Moscow as a pupil of Porzeziński. During his studies in Warsaw, he simultaneously attended private classes with Jan Karłowicz, who ran informal courses in linguistics and "folklore studies" for Polish youth (Wieczorkiewicz 1959). It bears repeating that was a dual system of education for the Polish intelligentsia in Warsaw, with some being taught in Russian at the university, while being educated in Polish, and in a patriotic spirit, privately. This was typical of Polish science in the Russian partition in the early twentieth century. Szober turned down an offer from the rector of the Imperial University of Warsaw, Efim Karski (1860–1931), because of the Polish boycott of Russian schools at the time. Instead of lecturing at the University of Warsaw, he became involved in the independent Polish Teachers' Association. It was not until

1915 that he accepted an offer to lecture in Indo-European linguistics at the University of Warsaw, which was reactivated by von Beseler. In 1919, Szober accepted the Chair of Polish Language at the now fully Polish University of Warsaw in 1919, and in 1930, after Porzeziński's death, the Chair of Indo-European Linguistics. Significantly, his *Zarys językoznawstwa ogólnego* (Outline of General Linguistics, 1924) consisted of a first part, *The Psychological Foundations of Language*, and a second, much smaller part, *Language and Society*, devoted to the conventional nature of language and to the participation of the individual in linguistic creativity in terms of style. Szober was also a key figure in standardizing and teaching the Polish language. He authored classical textbooks on descriptive grammar and wrote the famous *Słownik poprawnej polszczyny* (Dictionary of Correct Polish Speech) which had tens of editions starting in 1937 as *Słownik ortoepiczny* (Orthoepic Dictionary). Descriptive grammar and a normative approach to language, which Szober incarnated, played a key role in the field throughout the interwar and communist periods. Standardization involved a constant struggle against "contamination" by foreign influences, especially in the registers of officials and craftsmen. Moreover, colloquial language had many Germanisms and Russianisms. It was therefore deemed necessary to constantly remind Poles of their patriotic duty to remove unnecessary borrowings, to integrate society by introducing a uniform language for the public service, the armed forces, etc., and to standardize the teaching of the Polish language in schools. Polish linguists were very active in all these fields. Szober was, therefore, a prototypical figure for a whole range of linguists "fighting" for the "purity" and "correctness" of the Polish language. His like can still be found in the public sphere. Zenon Klemensiewicz (1891–1969), who studied linguistics and literature under Jan Michał Rozwadowski, Józef Kallenbach (1861–1929) and Igancy Chrzanowski (1866–1940) at the Jagiellonian University in 1910–1914, can also be mentioned in this context. He wrote a historical grammar and a history of the Polish language, as a notable member of the orthography committee, and took part in work on diverse types of dictionaries. In the 1960s, he held a position at the Jagiellonian University and at the PAN. Klemensiewicz's daughter, Irena Bajerowa (1921–2010), was also a noted linguist. The Polish linguistics mainstream was relatively normative throughout the entire communist period. According to Karol Janicki "Polish language socio-linguistics was largely normative between 1970 and 1989. One of the reasons for this was that the longstanding tradition of extolling and cultivating the standard variety of the language was maintained during the communist period. This was curious and surprising as the prestige of the various working-class non-standard varieties might have been expected to be raised" (Janicki 1995: 166–167). At the same time, there was a moderate but growing interest in current language in the 1970s. The *Słownik gwary studenckiej* (Student Dialect Dictionary), published in 1974, was an example of this tendency. However, for political reasons, its circulation was minimal.

3.5.4 Linguistics in the Service of the State: Polonization of Place Names

Another important "state-building front" on which the role of linguists was crucial was the Polonization of place names. The renowned linguist, Mikołaj Rudnicki (1881–1978), whose particular area of activity was Wielkopolska, which had been in the Prussian partition, was an extremely eminent scholar active in this area. Rudnicki graduated from the Jagiellonian University, and from 1911, was an assistant professor of Indo-European linguistics there. He was one of the founders of the University of Poznań, and was appointed to the Chair of Linguistics there in 1919. In 1921, he was instrumental in establishing the Western Slavonic Institute at the University of Poznań. The institute published the periodical *Slavia Occidentalis*. The aim of the institute was to study the history and culture of the Western Slavonic tribes, particularly on the right bank of the Elbe, i.e. in areas which had long since been completely Germanized. The institute's programmatic approach led to arguments with German scholars, who usually disputed the importance of the Slavic legacy. In the 1930s, the institute's activities were viewed with suspicion and disdain in Germany. After the German invasion of 1939, several of the institute's researchers were persecuted for having worked there. Some even paid with their lives.

Rudnicki primarily studied the area of contact and competition of Polish and German cultures, i.e. between the Vistula and Oder rivers. Interestingly, he was to declare that, after taking up the chair in Poznań, he considered the work on empirical studies important for the reborn Poland to be much more important than his theoretical work, which he therefore marginalized. It was with this intention that he focused on linguistic research into Western Slavic studies. Their clear goal was to prove the Polishness of the Western lands of the country, which had long been under German (Prussian) (Bańczerowski 2001b).

Many of the areas that Rudnicki studied have been part of Poland since 1945, when Poland's Western border was moved westward to the Oder-Neisse line and its northern border moved to include a continuous stretch of Baltic coastline and the southern half of East Prussia. Linguists were charged with the politically important task of Polonizing toponyms, landscape features, etc. They appealed to the government, through the PAU, to revive the pre-war Main Commission for the Establishment of Place Names. The Commission for the Definition of Place Names and Physiographic Names was duly established. In addition to Rudnicki, it included Kazimierz Nitsch, Witold Taszycki (1898–1979), Jan Safarewicz (1904–1992), Stanisław Rospond (1906–1982), Ludwik Zabrocki (1907–1977), and Stanisław Urbańczyk. The Commission published its decisions on name changes in an official "Monitor", which listed the old German names and the new Polish names. It obviously employed public servants, and people trained in other scientific disciplines, including geographers. These changes were discussed in scientific papers. Among them a series entitled *Ziemie Staropolskie* (Old Polish Lands) was published by the newly founded *Instytut Zachodni* (Western Institute) in Poznań. Bajerowa (2002: 222) notes that contacts were established with Czechoslovakia, and in 1948, Polish and Czechoslovakian toponymists worked together on Slavicization and came to an agreement of naming border areas, such as the Karkonosze Mountains, the Liberec area, the Wałbrzych area, and Kłodzko County. This made an interesting contrast with the Soviet Union, where a completely different naming policy was adopted, particularly in the northern part of East Prussia (now Kaliningrad Oblast). No attempt was made to refer to any local traditions. The names were entirely ideological, referring primarily to historical, and even living heroes of the Communist Party, the War with Germany, Soviet symbols and ideals, and to classical Russian natural and cultural motifs. This contrast between the Soviet model, based on the undisguised ideologization and arbitrariness of naming practice, which was particularly pronounced in Kaliningrad Oblast, but evident everywhere in the USSR, also seems to be linked to differences in the role of the intelligentsia in the two countries. Poland, although communist, had only recently reemerged as a nation-state. The humanist intelligentsia, by highlighting the country's 1,000 years of statehood and culture, played an important role in legitimizing the state. These name changes helped legitimize the post-war borders, and having them based on detailed and reliable research by internationally recognized specialists played no small part in serving this purpose. It scientifically established the Slavic, and more specifically Polish, nature of these lands. In the USSR, historical legitimacy was secondary to political and military legitimacy. Here, the references were to historical justice, and the Party and the Red Army predominated. There was therefore no point in having specialists prepare new names. Vitaly Maslov (2016) reports that the Soviet government occasionally consulted linguists, geographers, and ethnographers from the relevant institutes of the Soviet Academy of Sciences, and the retention of the historical roots of regional toponyms was sometimes humbly suggested. However, such suggestions were invariably ignored. For example, a 1947 proposal from a Lithuanian scholar, Povilas Pakarklis (1902–1955), to use historical Lithuanian toponyms in a part of the Kaliningrad Oblast was rejected on the grounds that it could have justified the inclusion of the region (or part of it) into the Lithuanian SSR. Most importantly, the role of scholars was not made public. By contrast, the Polish renaming effort considerably revived onomastic research and the results were openly discussed and published. Moreover, the issue become a central research topic for some Polish scholars, e.g. Witold Taszycki, who published a monograph entitled *Słowiańske nazwy miejscowe* (Slavic local names, 1946).

Another notable student of Baudouin de Courtenay was Henryk Ułaszyn (1874–1956). Ułaszyn also found employment in Poznań in independent Poland. Ułaszyn studied in Kyiv (1895–1897), Kraków (1897–1901), Vienna (1898–1899), and Leipzig (1901–1903). It was in Kraków that he came into contact with Baudouin de Courtenay. In 1919, he was employed at the University of Lviv, and from 1921, at the University of Poznań. After WWII, he worked at the University of Łódź. Possibly owing to Baudouin de Courtenay's influence, his convictions were liberal. He spent a lot of time in Poznań, where he remonstrated with Endecja and the conservative clergy over their racial and religious intolerance. Meanwhile, Mikołaj Rudnicki was a National League sympathizer, so Poznań boasted two prominent linguists whose political and ideological orientations were diametrically opposed.

3.5.5 Literary Studies at the Turn of the Century

I would now like to briefly examine the literary part of the field. Hutnikiewicz (1999) believes that, at least until the emergence of the Second Republic, the University of Lviv was the leader in the field. This was where Antoni Małecki (1856–1874), author of the first Polish literary monograph based on well-defined methodological principles, dedicated to the poet Juliusz Słowacki (1866), worked. Another important Lviv scholar was Roman Piłat (1874–1905), who organized the first literary seminar in the Polish field of culture. Hutnikiewicz points out that the students of Piłat and Małecki, which included Wilhelm Bruchnalski (1907–1931), also attended the lectures and philosophical seminars of Kazimierz Twardowski (1866–1938). Twardowski greatly influenced most of his students, who came from various departments, by instilling the positivist rules of "logical thinking" and "clear" scientific language into them. The circle of his students was very wide, and mainly comprised of philosophers and psychologists, but there were also many literary scholars and linguists (e.g. Jerzy Kuryłowicz). Lviv also produced such notable figures as Julisz Kleiner, Eugeniusz Kucharski (1880–1952), Manfred Kriedl, Zygmunt Łempicki (1886–1943), Roman Ingarden, and Stefania Skwarczyńska (1902–1988). Juliusz Kleiner (1886–1957) was one of Poland's leading literary historians. He graduated from the University of Lviv, where he received his habilitation, in 1912. After WWII, Kleiner settled in Lublin, where was a professor at the Catholic University of Lublin (KUL) in 1944–1947. He then moved to Kraków, where he was a professor at the Jagiellonian University. He was particularly interested in the poets Juliusz Słowacki (1809–1849) and Zygmunt Krasiński (1812–1859). As a literary theorist, he combined his studies with philosophy and took some steps toward structuralist research. In particular, his work *Charakter i przedmiot badan literackich* (The Nature and Object of Literary Research, 1913) argued that the study of literature is a study of texts, and that the psychological qualities of literature were not merely part of the genesis of literary works, but a component of their structure (Karcz 2002: 44). The institutional infrastructure of Lviv played a significant role. This included the Adam Mickiewcz Literary Association, created by Piłat at the end of nineteenth century, the Ossoliński Scientific Institute (*Zakład Narodowy Ossolińskich*) Library, and *Pamiętnik Literacki* journal, which was first published in Lviv.

The interwar period was marked by a final break with nineteenth-century historicism, which treated works of fiction as documents of culture and social custom. The final liberation of the history of literature as an autonomous discipline was thus completed. Literary studies became the history of art and writing skills. Literary works were increasingly seen as autonomous creations, independent of their authors and having lives of their own. Comparative literature studies went far beyond the old-fashioned, contemptuously termed "influenceology" (*wpływologia*), based mechanical analyses and juxtapositions of presumed or imaginary connections and relationships between authors. A miscellany of classic works of Polish literature was regarded as the basis for the development of the field. It was assumed that there could be no history of literature without a precise determination of the object of study, and that the object of study were the texts in their original versions. However, none of these publishing projects ever came to fruition, as none of the planned editions of the works of prominent Polish writers were ever completed. Some were interrupted by the outbreak of the WWII, others faced insurmountable obstacles of various kinds.

Now to discuss the other key figures in the interwar literary field, starting with Julian Krzyżanowski (1892–1976). He graduated from the Jagiellonian University, where he received his habilitation in 1926. In 1925, he became a professor at KUL. Between 1927 and 1930 he taught Polish literature at the School of Slavonic Studies at the University of

London, and from 1930 to 1934, he was a lecturer of Slavonic literature at the University of Latvia in Riga. In 1934, he was appointed head of the Chair of Polish Literature at the Faculty of Polish Studies at the University of Warsaw. He held that post until he retired in 1957, but continued to work until his death in 1976. In the 1957/1958 academic year, he taught Polish literature at the Mickiewicz Chair at Columbia University in New York. He was a member of the Scientific Council of the Institute of Literary Research, chairman of the Committee on Literature and the Committee on Slavonic Studies at the PAN, a full member of the Philology Department of the PAU, and from 1952, a full member of the PAU. In 1964 he was a signatory of Letter of 34, but under duress from the Polish authorities, Krzyżanowski, along with nine other signatories of Letter of 34, wrote to *The Times* daily to protest the alleged campaign against the People's Republic of Poland and to deny that the authors of Letter 34 had suffered any reprisals. He is considered one of the most versatile scholars of Polish literature, an impressive polyhistoric lecturer and a dominant figure in Warsaw after in the post-war period. Vilnius had such figures as Kazimierz Kolbuszewski (1922–1931), newer literature by Józef Kallenbach (1919–1920) and Stanisław Pigoń (1921–1930), and in the second decade, Manfred Kridl (1932–1939), a former professor of Slavonic literatures in Brussels, and Konrad Górski (1934–1940). The only Chair of Comparative Literature in Poland was headed by the pre-war Jagiellonian University professor Marian Zdziechowski (1919–1931).

3.5.6 Witold Doroszewski—Consolidation of the Warsaw Center of Language Studies

Witold Doroszewski (1899–1976) was an institutionally important Polish linguist who played a key role in the development of the Polish language education system. Doroszewski is considered the founder of the Warsaw school of linguistics, which emerged in the 1930s and almost lasted until the end of the Polish People's Republic. Doroszewski studied in Moscow until 1918, then in Warsaw under Baudouin de Courtenay, where he obtained his doctorate in 1923. From 1930, he was a professor at the University of Warsaw, and in 1935- 1939, he was the director of the Phonetic Institute of the University of Warsaw. In 1939, he succeeded Szober as the Chair of the Polish Language Department at the University of Warsaw, and was Head of the Department from 1945 until 1968. He also managed the PAN Dialectological Laboratory from 1954 until his retirement in 1969. Doroszewski's main achievements were in dialectology and lexicography. He is best known for introducing the quantifiable isogloss method internationally. This has proven to be an extremely effective dialectological research tool. Doroszewski's first paper on this method was published in 1935, but has only been fully appreciated since the development of sociolinguistics in the 1960s. His method could be applied in registering the instability of language norms under the influence of a standard variety of a language. Word formation was another area where Doroszewski's achievements were widely recognized. He published on this topic between 1928 and 1931. In particular, he built on Rozwadowski's bipartition concept regarding the division of word-stems into word-formatives and word-bases according to their historical formation by theorizing the distinction between structural and semantic specialization of derivational formatives, and between logical and syntactic derivations. He also made important contributions to geolinguistics. In addition to the isogloss method, he pioneered the use of statistics in dialectological studies.

Doroszewski's Warsaw school assumed that a dialect consisted of concrete, real-life utterances. For this reason, the presence and intensity of various linguistic tendencies (including secondary ones) that belied the supposedly uniform character of a given dialect, were carefully observed. This school pioneered the use of dot distribution maps and statistical analyses. It did not so much study the dominant aspects of a dialect, as its internal differentiation and constant change (its "linguistic becoming"), and avoided abstract generalizations. By contrast, the Kraków school, with Nitsch as its leader, aimed to establish isoglosses based on sounds and the type of language prevalent at a given location, without considering individual fluctuations and deviations. The aim was to distinguish certain dialectal groups. At the same time, dialect data were used to explain and reconstruct general language phenomena. Maps were drawn using the isoglossic (less frequently—plane) and dot method. Statistical methods, however, were not used in Kraków. These two dialectological schools clashed at the 1937 and 1938 Congresses of the Polish Linguistic Society (*Polskie Towarzystwo Językoznawcze*, PTJ). In Kraków, according to Irena Bajerowa, the focus was on history, tradition, and pure science, whereas in Warsaw, it was on modernity, topicality, and applied linguistics (Bajerowa 2002). Bajerowa called this approach "linguistic realism", as it was based on behaviorism and was the outgrowth of the positivist attitude of the Warsaw milieu. Doroszewski rejected an "idealistic" treatment of linguistic phenomena.

On the theoretical level, Doroszewski advocated monism, i.e. he did not separate the historical and current descriptions of language. He clearly distanced himself from psychologisms, but subscribed to pragmatism and utilitarianism, according to which language was to be perceived as a "social activity". This supposedly stood in opposition to the Durkheimian-Saussurian approach of treating language as a "social fact". Interestingly, Doroszewski argued that de Saussure's insights into the social nature of language were derived from the work of Durkheimian. Although he never adduced any textual evidence for these claims, his suggestion gained wide currency (Wąsik 2001). His attitude to structuralism was ambivalent if not decidedly critical. He declared himself to be a continuator of Baudouin de Courtenay, whom he admired for his contribution to the development of structuralism. Doroszewski's veneration for Baudouin de Courtenay and his repudiation of the major part of his structuralist legacy suggests that Baudouin de Courtenay had become a classic "intelligentsia totem", i.e. an idealized historical figure to whose heritage the dominant part of a given intelligentsia group refers, even if they are unaware of, or disagree with, much of what that figure said (Turkowski 2018).

Doroszewski also frequently referred to de Saussure's *Cours de Linguistique Générale*, while distancing himself from it, just as he maintained a tactical distance from the Prague School, which was not spared his criticism. Interestingly, Doroszewski wrote the introduction to the first Polish translation of Saussure's *Cours* (published in 1961) and he left no doubt as to his critical disposition toward Saussurean structuralism. The revised second edition of *Cours*, which appeared 30 years later (in 1991), was supplemented with a more distanced and informative introduction by Kazimierz Polański (1929–2009). Doroszewski also promoted the normative standards of the Polish language through, e.g. his contributions to *Poradnik Językowy*, of which he became editor-in-chief in 1932. He was widely regarded as an exemplary teacher of "correct Polish", and was familiar to most educated Poles through his media appearances, including his own radio broadcasts on the Polish language. He also became known to the public as the editor-in-chief of the 11-volume *Słownik Języka Polskiego* (Dictionary of the Polish Language). Work on the dictionary began in the 1950s, with subsequent volumes coming out between 1958 and 1969. By this time, the dictionary had become a standard item in the apartments of the intelligentsia elite. It can be seen in the background of many photographs of intellectuals in 1970s and 1980s, especially famous opposition figures.

Władysław Kupiszewski (2016) and Renata Grzegorczykowa (2016) highlight the importance of Doroszewski in Warsaw linguistics and Polish studies. Kupiszewski writes that Doroszewski "monopolized many of the classes at the Polish studies department of the University of Warsaw, thereby consolidating and disseminating his views" (Kupiszewski 2016: 189). Grzegorczykowa points to both the institutional and informal dimensions of his central position. Regarding the latter, he created a broad social circle that extended far beyond the institutions with which he was affiliated. Members of this network recall the integrating role of his name days, which were celebrated in style on his yacht, which was moored in the Vistula River in Warsaw. In the institutional dimension, he combined the management of the organizational units at the University of Warsaw and the Polish Academy of Sciences. Additionally, he was able to adapt to the political situation and coexist with the communist authorities. Although he was not a Party member, he was seen as loyal to the authorities, who also appreciated his critical attitude toward the Catholic Church (Grzegorczykowa 2016). It would therefore be fair to describe him as a "man of the center", i.e. as someone who tried to retain a neutral position toward the main political camps or as someone located (in social fields) in positions homological to the center of the field of power. Holding important positions at the University of Warsaw and the PAN was common among the academic intelligentsia elite whose careers peaked in the 1960s. The political revolution of 1968 brought an end to such dual roles. Those not denied the possibility to work as academics were forced to limit themselves to a single job, and often to resign their managerial positions. This was broadly linked to the weakening of the power of the liberal faction in the field after 1968.

3.5.7 Jerzy Kuryłowicz—World-Famous Polish Linguist with no Successors

Jerzy Kuryłowicz (1895–1978) one of Poland's most internationally renowned linguists. Kuryłowicz studied Romance and Germanic Philology at the Faculty of Philosophy of the University of Lviv. He completed his doctorate in three years (1923) and then traveled to Paris on a French Government scholarship to continue his Romance language studies. After listening to Antoine Meillet (1866–1936) at the *Collège de France*, he abandoned Romance studies for Indo-European linguistics (Smoczyński 2001). Kuryłowicz concluded his Paris studies with a diploma from the *École Pratique des Hautes Études*. He received his Habilitation in Indo-European linguistics at the University of Lviv in 1926, where he was appointed Chair of Indo-European Linguistics in 1927 following the death of Andrzej Gawroński. Kuryłowicz spent a lot of time abroad in the 1930s. A fellowship from the Rockefeller Foundation took him first to the United States (Yale, 1931–1932), where he met Edward Sapir (1884–1939) and Bronisław Malinowski (1884–1942), then to Vienna (1932), and London (1935–1936). He presented papers at every Linguistic Congress held before World War II, including in The Hague (1928), Geneva (1931), Rome (1933), and Copenhagen (1936). During WWII, he taught general linguistics at the Soviet-controlled University of Lviv (1939–1941), and again at the now renamed Ivan Franko University of Lviv (1944–1946). While there, he gave a lecture on the "stadial theory" of language propounded by Nikolai Marr (1865–1934), and the famous doctrine known as Marrism, which was then in vogue among Soviet linguists. Marrism held that the Indo-European languages were at a more advanced stage of development than the Caucasian languages. Marr's pupils built up a theory of the stages of development through which all languages supposedly passed as a result of changes in social structure. Language, in turn, was supposed to be an element of an ideological superstructure. In his lecture, Kuryłowicz subjected Marr's doctrine to a detailed analysis and devastating criticism. According to Kuryłowicz's biographers (Smoczyński 2001: 257; Urbańczyk 1993), this lecture influenced Stalin's decision to repudiate Marr's doctrine in a 1950 *Pravda* article, in which he claimed that language was a tool of communication without a class character. Only a short time earlier, Marrism was gaining adherents and being propagated in Poland as well. If this really happened, it would have been a rare case of a Polish scholar having a spectacular influence on the Soviet scientific field. However, the very fact that Polish scholars worked at the Sovietized University of Lviv in 1939–1941 and then from 1944 until being forcibly evicted to Poland, is a rare example of the Polish academic community in the institutional system of the USSR. This makes it interesting from a geopolitical perspective. The Lviv group, which also included the renowned Polish mathematician Stefan Banach (1892–1945), obviously constituted the most distinct Polish presence in the Soviet system. However, there were many more individual cases of Poles lecturing or studying at Soviet universities. In 1946, Kuryłowicz joined the newly founded University in Wrocław. In 1947, he was elected a foreign member of the Royal Academy of Sciences in Copenhagen. This might have been a gesture of solidarity from Louis Hjelmslev (1899–1965) with whom he was well acquainted. In 1948, Kuryłowicz was appointed Professor of General Linguistics at the Jagiellonian University in Kraków, where he introduced a fresh wave of structuralist language theory. Kraków was the strongest center of linguistic research in Poland at the time, but it was largely focused on historical studies, mostly drew on factual material, and was close to the *Junggrammatiker* in spirit.

Comparative Indo-European linguistic departments were closed down during the Stalinist period. Kuryłowicz took up lecturing on historical linguistics and related subjects while working intensively in his own field privately. Smoczyński (2001) even suggests that he was in a kind of intellectual exile during the Stalinist period, communicating with the world through his own extensive library. His important monographs were written during this period, particularly those on accent and apophony. The fact that Kuryłowicz won a Lviv City Council seat in 1939, running on the Christian Nationalist list, indicates right-leaning political views. His behavior during the post-war period was therefore close to that described in the Polish historical field. As mentioned in previous chapters, Valentin Behr (2017, 2021) shows how a large group of historians who had been educated before the war, and whose views were seldom consonant with those expected by the authorities, quietly survived the Stalinist period. This was partly because they worked on non-political topics and focused on improving their scholarship. This is how Jerzy Axer (2013) described the strategy of the eminent classical philologist, Kazimierz Kumaniecki (1905–1977), during this period. Axer wrote about what he called "resistance through academic excellence", and compared Kumaniecki to Aleksander Gieysztor (1916–1999), who "became an unofficial ambassador of Polish scholarship after 1956 and was treated as such by academic communities in Western Europe" (Axer 2013: 200). Axer argued that the circle of relatively conservative scholars, to which Kumaniecki and Gieysztor belonged, did not join the armed anti-Soviet resistance, but busied themselves in grassroots academic work. Axer considered Home Army (AK) colonel, Jan Rzepecki (1899–1983), the best exponent of this strategy. Rzepecki made the following declaration "We are not going into the forest, we are building a university" (Axer 2013: 200). However, the field of literary studies was configured differently than the field of history. There was no clean-cut division analogous to that between "party" and "classical" (pre-war) historians. Instead, during the first post-war decade, scholars who had graduated in the years immediately preceding the outbreak of WWII, and who were consequently well prepared for what they had to do, dominated. They deftly managed to meet political requirements while upholding and building on the traditions of Polish linguistics. Immediately after the thaw of 1956, Kuryłowicz threw himself into teaching. He reinstated the English studies program in Kraków and traveled abroad. His list of international contacts was long and impressive. His many foreign trips included lecture invitations to Ann Arbor, MI (1957), Hamburg (1960), Stanford, CA (1961), Cambridge, MA (MIT,

1962, Harvard, 1964, 1965), and Innsbruck (1967). He also received several honorary doctorates. If he had been oriented toward Paris before WWII, he was focused on the United States in the 1960s. There, he became associated with such renowned linguists as Roman Jakobson and Morris Halle (1923–2018). He contributed substantially to the popularization of American linguistics in Poland. Kuryłowicz retired in 1965. His Festschrift boasted essays by such luminaries as Émile Benveniste (1902–1976) and Roman Jakobson. He continued to work for many more years in retirement. His final years are eloquently summarized in his *Problèmes de linguistique indo-européenne*, published in 1977 by Ossolineum in Wrocław.

Kuryłowicz is regarded as one of the greatest promoters of classical structuralism, which referred intensively to both the Prague and Copenhagen schools. The flourishing of this study in Poland, which began in the 1960s, is largely due to Kuryłowicz's inspiration and support. Smoczyński describes his approach to historical linguistic studies as "diachronic structuralism". This combination enabled Kuryłowicz to develop structural tools for analyzing the evolution of language within the traditional field of comparative history. The distinction between lexical derivation and syntactic derivation was another of his innovations. He used the concept of hierarchical organization of language, and drew attention to isomorphism in text structures (between syllables, words, and sentences, which may be seen as compositional parallelisms). These were all trailblazing ideas that have inspired subsequent generations of scholars. Kuryłowicz, however, did not create a school of his own. One of the reasons that historians give for this is that, similarly to Baudouin de Courtenay, he did not synthesize his general linguistic views. The problem might have been his individualistic style or his limited cooperation with literary theorists.

Leon Zawadowski (1914–2018) is regarded as Kuryłowicz's most outstanding student. Zawadowski's underwhelming career trajectory offers some clues as to Kurłowicz's failure to found a Polish structuralist school. He graduated from the University of Warsaw before WWII and worked at the University of Wrocław, where he met Kuryłowicz, after the war. His doctoral dissertation entitled *Zagadnienia teorii zdań względnych*

(Issues of the theory of relative sentences, 1948) applied recent structuralist approaches in comparative research. His later frequently cited works, e.g. *Rzeczywisty i pozorny wpływ kontekstu na znaczenie* (The Real and Apparent Influence of Context on Meaning, 1949) and *On the Elements of Semantic Systems* (1950), also date from this period (Wąsik 1999). He also published *Lingwistyczna teoria języka* (Linguistic Theory of Language, 1966), which was clearly influenced by the Copenhagen school. According to Andrzej Bogusławski (2014), this book "constituted a kind of citadel of the maturing and already mature Polish linguistic structuralism" (Bogusławski 2014: 16). Zawadowski began to be regularly invited to Warsaw in the 1960s, especially by the group formed around Maria Renata Mayenowa (1908–1988) at the IBL. He worked closely with Anna Wierzbicka (b. 1938), whose work he clearly influenced. They kept in touch for a long time when they were both in exile, although intellectually their paths had diverged earlier. Anna Wierzbicka's international success may also be indirectly attributable to Kuryłowicz's achievements. In any case, according to Bogusławski, Doroszewski was very reluctant to accept Zawadowski's work. This is probably why Zawadowski could not find a position in Warsaw, although he already had a wide circle of followers there. In 1968, Zawadowski went to the United States on a scholarship from which he never returned. Initially, he lectured at Indiana University in Bloomington, but later, as Bogusławski writes, he found a job (albeit with difficulty) at Lakehead University in Thunder Bay, ON, Canada. Jakobson assisted him in finding employment, despite their theoretical disagreements toward him. Bogusławski also mentions a meeting between Zawadowski and Jakobson at a semiotic symposium in Kazimierz Dolny, during which it transpired that their views were completely incompatible. Bogusławski also describes an episode from a 1967 seminar in Warsaw, where Zawadowski strongly emphasized the importance of truth as the key value in science. This position was not well received by many in the audience, although it was perfectly consistent with the positivist belief in truth and logical thinking that typified the Warsaw-Lviv School. The online portal of the Polish community of Canada (tp 2018), in its obituary, describes Zawadowski as a socially committed Catholic, and the founder of the Society for Catholic Life & Culture in Canada, which still exists. This religious commitment would correspond with the closeness to the Catholic Church exhibited by many of those associated with the Warsaw-Lviv School during the second half of the twentieth century. Bogusławski hails Zawadowski as "the greatest linguistic and theoretical talent among early twentieth-century Poles after Baudouin de Courtenay and Kuryłowicz" (Bogusławski 2014: 27). This is intriguing as Zawadowski did not have an important career in Canada and is largely forgotten in Poland. Another student of Kuryłowicz was Adam Heinz (1914–1984). Heinz was based in Kraków, but his work did not receive wide publicity. His most important work is arguably his meticulous *Dzieje językoznawstwa w zarysie* (Outline of History of Linguistics, 1978).

3.5.8 Later Baudouin de Courtenay's Successors

Mention should also be made of another follower of the Baudouin de Courtenay tradition, viz. Ludwik Zabrocki (1907–1977), whose mentor was Henryk Ułaszyn (1874–1956). Zabrocki was a Germanist, although he graduated from the University of Poznań with a degree in Polish Studies in 1931 and received his doctorate there in 1945 (Bańczerowski 2001a). According to Franciszek Grucza, Zabrocki created the first "Polish school of applied linguistics" in Poznań (Grucza 2001). The school was formed in the late 1950s within the Department of Germanic Languages (which had been established in 1952 with Zabrocki as the Chair) at the University (renamed Adam Mickiewicz University in 1955). In 1968, traditional chairs were abolished and institutes were created in their stead. Grucza sees this as a restriction of academic autonomy. This reform is discussed in a previous chapter. Thus, Zabrocki was head of the Institute of English Philology in Poznań in 1970–1990. This was where the "Polish-English Contrastive Project" was implemented in collaboration with the Center for Applied Linguistics in Washington in 1971–1988. Jacek Fisiak (1936–2019), who graduated from the University of Warsaw with a major in English, was its leader and Zabrocki's heir. Fisiak was a Fulbright scholar in 1963–1964 and moved to Poznań in 1965. He and his circle transformed the Institute of English

Philology into an important study center in the generative grammar paradigm. Its final achievement was a syntactic-generative dictionary of Polish verbs, compiled between 1980 and 1992 (Karpowicz 2006). Fisiak served as rector of the Adam Mickiewicz University in Poznań in 1985–1988 and was Minister of Education in the last communist government (1988–1989). He also took part in the Round Table talks as a representative of the PZPR. He probably identified with a liberal and technocratic faction of the PZPR, which had several younger scholars with Fulbright scholarship experience in the United States.

Tadeusz Milewski (1906–1966) was also interested in structuralism and had theoretical ambitions. He found inspiration in phonology and stylistics, especially as taught by the Prague School. Milewski earned his doctorate at Lviv University in 1929, then moved to the Jagiellonian University, where he habilitated in Slavonic philology in 1933. In 1937 he also defended a second degree in Indo-European linguistics, after serving an internship in Paris. After WWII, he worked at the Jagiellonian University until his retirement in 1960. Milewski published several ambitious synthetic works, in particular, *Wstęp do językoznawstwa* (Introduction to Linguistics, 1954) and *Językoznawstwo* (Linguistics, 1965, 1971).

According to Zdzisław Wąsik (2001), Polish structuralism came to an end in the 1970s, following the death of Milewski in 1966 and the emigration of Zawadowski in 1969. Wąsik writes about the end of an era in Polish general linguistic theory, as both Milewski and Zawadowski "were able to encompass the whole scope of the linguistic discipline, both in terms of 'data orientation' and 'theory-orientation'. They familiarized themselves with the tenets of the various competing structuralist schools, but went on to develop their own postulates as to how to delimit the domain of linguistic study and to go about their self-imposed tasks. Starting from the 1970s, more energy was devoted in Poland to conduct empirical investigations on the basis of more recent Western theoretical trends embracing more specific - and restricted - areas of linguistic inquiry" (Wąsik 2001: 39).

Wąsik (2001: 14) also discusses the role of the generation gap that emerged after World War II, claiming that groups of relatively young scholars formed around a few of the older professors who had survived the war. There being few middle-aged scholars, the younger generation had no option but to accept the dominant conservative approaches of their masters. This conservatism, Wąsik argues, was most evident in Doroszewski's approach to structuralism, which tended to adhere to the interpretations of Rozwadowski and Szober. For this reason, the latter's classical grammars of Polish remained in force until the end of the 1980s. Methodological conservatism, however, protected Polish linguists from any politically motivated tendency to decry Saussurean and Hjelmslevian linguistics as not conforming to Marxist-materialistic philosophy by virtue of standing in opposition to historicism. Be that as it may, the view of the field held by Wąsik, and several other linguists, e.g. Irena Bajerowa (2002), evidences a rift between the legacy of linguistics proper and that of literary studies. This probably explains why the traditions of the Polish formalist school of the 1930s which was strongly associated with the Prague Linguistic Circle and the Russian Formalists, are often ignored or marginalized by historians of the field. That tradition is discussed below, as it seems to be an inspiring legacy which led to the development of what some call the Polish school of structuralism in the Institute of Literary Research, which was most active in the 1960s.

3.6 The Configuration of the Field in the First Half of the Twentieth Century

3.6.1 Developments Prior to WWII in a Wider Perspective

Before proceeding, I would like to summarize the discussion so far, especially the part concerning the development of Polish linguistic and literary studies in the nineteenth century. Polish linguistics primarily developed under German influence, especially that of the Young Grammarians. Most of the important Polish linguists of the nineteenth century studied or trained in Leipzig or Jena, i.e. the main centers of the *Junggrammatiker* School, although some studied at other German universities, which led the world in the development of the social sciences. As mentioned above, Lucjan Malinowski pioneered this trend. Most Polish linguists built their careers, at least partly, at Austrian universities, mostly in Galicia, but also in some other provinces of Cisleithania, including (obviously) Vienna. However, many Polish scholars built their careers in the Russian Empire, where it can be argued that the Polish language sciences developed in a somewhat different direction. Zdzisław Wąsik (2001) argues that there were two camps of linguists, which can be divided geographically into the following networks: Warsaw-Vilnius-Lublin and Lviv-Kraków-Poznań. Vilnius and Lublin were mostly developed after 1918, and they attracted staff from Warsaw and elsewhere in former Russian Empire. Poznań was also a new academic center created in 1918 that mainly drew its staff from Kraków and Lviv. Most of the civil servants in Greater Poland in the Second Republic were "imported" from the former Galicia, as there were no universities in the Prussian partition.

Galicia, often referred to as the Polish Piedmont, generally developed within the framework of a powerful institutional infrastructure which served to build the cultural basis of the future Polish nation-state from the end of the nineteenth century. Linguistics and the literary sciences were primarily directed toward establishing the canons of Polish literature and documenting its history, and formalizing the Polish language, defining its grammar, and interpreting its history. To this end, numerous dialectological and dictionary works were compiled. The Young-Grammatist tradition continued to hold sway in Kraków, but it was modernized with structuralist and psychologist elements, or more accurately, structuralist elements were harnessed within a peculiar synthesis to systematically work on the modern Polish language and its history. At the same time, literary histories were being written to help define the "Polish national spirit". Linguists also worked on justifying the geopolitical interests of the future Polish state, especially the demarcation of its borders. These efforts proved extremely useful at the Paris Peace Conference (1919–1920), where they were presented by Kazimierz Nitsch (Romer 2010: 25). This can probably be interpreted in the context of the Polish proto-field of power, i.e. as an infrastructure created in, and partly by, the Austro-Hungarian Empire to generate the key elements legitimizing the creation of a Polish nation-state.

In the Russian part of the Polish field, structuralism can be seen as having developed in opposition to, or at least in competition with, the Young Grammatical school, and as evolving concurrently with, and partly independently of, the West. The Kazan School (Baudouin de Courtenay and Mikołaj Kruszewski) not only developed simultaneously with, and in opposition to, the traditional historical Young-Grammatist approach, but also in political opposition to nationalism, especially the Russification espoused by the conservative faction of the Russian Empire's field of power. It was also opposed to the analogous ambitions of the nationalist factions of the Polish proto-field of power. Thus, Galician linguistics and literary studies were generally conservative and eminently suited to the needs of the newly established Polish state. The opposite was true of liberals from the Russian Empire, e.g. Baudouin de Courtenay or Petrażycki (legal philosophy). Their orientation was inconsistent with building a classic nation-state. Doroszewski reports that, in his inaugural lecture at the University of Warsaw in 1918, Baudouin de Courtenay said that "Poland was not reborn as a state in order to become an imperialist hyena" (Doroszewski 2016: 85). This statement provoked opposition of some and won support of others. In any case, Baudouin de Courtenay was to play no serious political or intellectual role in the reborn Polish state, becoming instead a naturalized, and politically neutralized, totem. Thus, he became a symbol of the greatness of Polish science, whose achievements have since won universal praise. References to his work, however, do little more than pay lip service and is intended to support different agendas. As mentioned above, this is probably how Doroszewski "made use" of Baudouin de Courtenay. On the one hand, he built a cult around this great scholar, but on the other, he distanced himself from structuralism, developing a school of descriptive grammar and dialectology in competition with the Kraków school, although it had many similarities. His was perhaps slightly less prescriptive and more oriented toward contemporary rather than historical language. Nevertheless, it was primarily conceived as an ideological apparatus for the modern nation-state, and was later employed in this capacity by Communist Poland as well. Baudouin de Courtenay, although marginalized in terms of his position in the field of power, clearly influenced the development of Russian formalism through his Russian students. Although he was opposed to the dominant forces of the Russian, Soviet, and Polish fields of power, he fits much better into the system of liberal Russia, in particular the short-lived Russian Republic established by the Provisional Government in 1917, where he occupied a dominating position. Had the Russian Republic survived, he would almost certainly have been not only a linguist of greater renown, but also an important ideologist of Russian liberalism, which was the ideology of the bourgeois-intelligentsia elite. In any case, Russian liberals (esp. the Cadets) can be seen as homologous allies of liberals of the Baudouin de Courtenay type in the scientific field. It should be noted that their relative strength stimulated the emergence of areas of academic autonomy in the Russian Empire from which Baudouin de Courtenay and a number of other Polish scholars sprang. Similar tensions also existed in the Austro-Hungarian Empire, but Vienna's policy of alliance with the Polish conservative elite narrowed the scope of possible sectors of the academic field that would develop in opposition to the dominant bloc of the "old" landed-aristocratic elite.

Now I would like to reconstruct synthetically the development of two broader theoretical currents, which can be regarded as the most distinct forces forming the two main poles of the field in the first half of the twentieth century. These are, respectively, the school of Polish formalism and the so-called Warsaw-Lviv school. As it seems, their opposition is to a large extent homologous to an important division in the Polish field, particularly one between the progressive and conservative factions of the Polish intelligentsia.

3.6.2 Polish School of Formalism of the Interwar Period

I now wish to take a synthetic look at the history of the Polish structuralist school which emerged in the 1930s, primarily among the young literary scholars of Warsaw and Vilnius. Kazimierz Wójcicki (1876–1938) stands out as the key figure in this school. Wójcicki studied Polish philology at the Jagiellonian University and law at the University of Warsaw. He did not take up professional academic work, but devoted himself to high school teaching. He was an official at the Ministry of Education in 1917–1938, i.e. he began his bureaucratic career under German administration, and mostly worked in the Warsaw secondary school inspectorate. He also taught commissioned classes at various institutions of higher education, including the University of Warsaw. Andrzej Karcz (2002) calls him the "John the Baptist" of Polish formalism. Wójciki's most important formalist works were written in 1911–1914, so they were even slightly ahead of the development of the Russian formalist school, whose heyday was between 1915 and 1916. This was when the first Russian Formalist texts, e.g. *The Resurrection of the Word* by Victor Shklovsky (1893–1984) appeared. It seems paradoxical that Wójciki inspired the development of a scholarly current that was largely defined in opposition to the dominant tendencies in the field of literary studies, and that his position corresponded homologically in the field of power to orientations that were critical of the ruling Sanacja, and which even sympathized with communism. As a public servant, he was considered a loyal official and decorated with state orders, but perhaps his attitude to his creative work as an additional occupation, carried out largely on a free footing, was the source of his autonomy. Wójcicki's view that one of the major problems with the Polish science of literature and Polish literary criticism was that, throughout "partitions period", they were subordinated to the "national cause" is telling. He opposed this dominant "patriotic" approach to the concept of unifying all the components (substantial and formal) of a literary work. At the same time, the conception of a literary work as a separate, self-contained whole, with its own inner life, its own laws, and special relations between its constitutive parts, was crucial for him. He called the dominant approaches, which he criticized, an "external theory of literature", whereas he proposed to develop an "internal theory of literature". In this view, literature appeared as a system, a world closed in on itself, which only existed by and for itself. Andrzej Karcz (2002: 65) points out the similarity of this understanding of literature to the approach of the Russian formalists, in particular– Yury Tynyanov (1894–1943), and the notion of literary fact, or Jakobson, with his notion of "literariness" (*literaturnost*) and the central role of the notion of a "device". There are several other parallels. In his precursory *Forma dźwiękowa prozy polskiej i wiersza polskiego* (The sound form of Polish prose and poetry), published in 1912, Wójcicki distinguished between the language of imaginative literature and demotic language, as well as subject and content, and claimed that the evolution of style was a crucial task for literary theory.

Manfred Kridl (1882–1957) was as important to this circle as Wójcicki. He was born in Lviv and was a product of the multiculturalism of the Austro-Hungarian Empire. His father was an Austrian army officer of Czech nationality, who, similarly to many of the city's elite, was gradually Polonized. Kridl graduated in Lviv, and pursued further studies in Freiburg and Paris. In 1921, he habilitated at the University of Warsaw, where he began working. In 1929, Kridl became a professor of Slavic studies at the University of Brussels, and from 1931, after the publication of his *Główne prądy literatury europejskie. Klasycyzm, romantyzm, epoka poromantyczna* (The Main Currents in European literature: Classicism, Romanticism, and the Post-Romantic Era) in 1932, he began working at Vilnius University. During his time in Vilnius, he created the publishing series *Z zagadnień poetyki* (From Poetics Issues), which he opened with his own book *Wstęp do badań nad dziełem literackim* (An Introduction to Studies on literary works) (Vilnius, 1936), which became a leading manifesto of Polish structuralism. He was politically active during his time in Vilnius, serving as chairman of the Vilnius branch of the "Democratic Club" party (1937–1939), a bloc critical of Sanacja and nationalist tendencies. He protested the introduction of separate benches for Jewish students at universities in the 1937–1938 academic year (the "bench ghetto"). In 1940, he left for Brussels via Sweden and then went to the United States. In 1945, he published *Literatura polska na tle rozwoju kultury* (Polish Literature Against the Background of Cultural Development), a book intended to introduce Polish culture to Americans. A second (modified) English edition of this work, titled *A Survey of Polish Literature and Culture*, followed in 1956. In the United States, he first worked at Smith College in Massachusetts, and in 1948, he took over the Adam Mickiewicz Chair at Columbia University in New York. The funding of the chair by the Polish government caused some Poles in the

United States to dislike Kridl, who was suspected of being a communist sympathizer. Works published abroad to which Kridl contributed include the collective work *A. Mickiewicz—Poet of Poland* and a volume on Juliusz Słowacki. He retired in 1956, but continued his teaching and research work until the end of his life.

Kridl's approach emphasized anti-historicism and anti-psychologism, which may be seen as an influence of Benedetto Croce (1866–1952). He suggested that identical research methods should be applied to analyzing both Polish and world literature, and that, contrary to what more conservative scholars argued, this would not deprive Polish literature of anything, but would make it easier to emphasize its universal elements. In essays he wrote in 1934 and 1935, Kridl referred to most of the key Russian Formalists, including Viktor Zhirmunsky (1891–1971), Yury Tynyanov, Lev Yakubinsky (1892–1945), Boris Eikhenbaum (1886–1959), Osip Brik (1888–1945), Roman Jakobson and Boris Tomashevsky (1890–1957), and gave an extensive presentation of Russian formalism that constituted its first systematic introduction to Polish readers. Interestingly, this occurs when formalism has already been under attack in the USSR since 1926 by government Marxists. In 1935, Kridl presented his Vilnius group of formalist studies at a conference in Lviv. His *Wstęp do badań nad dziełem literackim* (*Introduction to the Study of a Literary Work*, 1936) provoked a passionate polemic. The methodological proposals formulated in the book, which the author described as an integral method, were an attempt to adapt the principles and ideas of Russian formalism, mixed with elements of Ingarden's philosophical views and the linguistic and literary theories of the Prague structuralist school. Kridl not only cut himself off from his positivist heritage, as others did, but also eliminated the concept of personality of the author, along with the entire historical and cultural context, from the scope of his interest. His Lviv speech was received positively, albeit with reservations, by Roman Ingarden. The strongest negative reaction came from the adherents of the traditional philological method, e.g. Wacław Borowy and Julian Kołaczkowski. The latter, in his *Bilans estetyzmu* (1937), even went so far as to accuse Kridl of preparing the ground for Bolshevism, seeing in formalism a primitive Marxist scheme designed to "automate" the interpretation of culture and deprive it of deeper humanistic meanings. Significantly, Kołaczkowski' was indignant that Kridl had undermined the importance of "truth" as a value. His criticism was therefore largely a classic intelligentsia critique of relativism, seen as an important threat by the mainstream of the Polish cultural elite posed by the formalists, as bent on sapping the strength of national culture and undermining spiritual and humanistic values. For his part, Ingarden saw Kridl's approach as built on insufficiently uniform philosophical and epistemological foundations, which he considered indispensable to any deliberations on methodology. Kridl's detachment could be said to have reflected Ingarden's position, which can be described as being fascinated with formalism while remaining firmly rooted in traditional humanist ontology and phenomenology.

Now for a brief look at Roman Ingarden (1893–1970). Ingarden studied at the University of Lviv under Kazimierz Twardowski, then moved to the University of Göttingen to study philosophy under Edmund Husserl (1859–1938). In 1933, he was promoted to professor of philosophy at Lviv University. From 1945, he served at Nicolaus Copernicus University in Toruń, then moved to the Jagiellonian University in Kraków, but was soon banned from teaching in 1949 due to his alleged "idealism", which supposedly made him an "enemy of materialism". In 1957, he was reappointed to the Jagiellonian University after the ban was lifted. Ingarden specialized in ontology, cognition, and phenomenology. However, although he was a disciple of Husserl, he did not subscribe to Husserl's transcendental idealism. His books, especially *Das literarische Kunstwerk* (1931), and its Polish version, *O poznawaniu dzieła literackiego* (1937), offered a theoretical reflection on the essence of literature. Both were philosophical treatises on ontology and the theory of cognition, but the analyses they offered, both of the internal structure of literary works and of the processes involved in their reception, although they could not be applied directly to the research work of literary historians, contributed a great deal to the theory of literature, and were taken up by several scholars. Ingarden understood a literary work to be an intersubjective intentional object that owed its qualitative features to the creativity of the author. Of special interest to literary theorists was his analysis of the structure of literary works, in which he distinguished four superimposed layers (verbal sounds, meanings, schematic appearances, and depicted objects), which combined to form a peculiar structure that contained "places of indeterminacy", which were only filled in by the individual concretizations of the reader.

In the late 1930s, two groups of young scholars crystallized and began to develop their academic interests in structuralism, primarily under the influence of Russian formalism and the Prague Circle. The first was the Vilnius group, which included Maria Renata Mayenowa, Eugenia Krassowska (1910–1986), Jerzy Putrament, Maria Rzeuska (Rzewuska) (1908–1982), Irena Sławińska (1913–2004), and Czesław Zgorzelski (1908–1996). The second was the Warsaw group, which included Kazimierz Budzyk (1911–1964), Dawid Hopensztand (1904–1943), Franciszek Siedlecki (1867–1934), and Stefan Żółkiewski. Manfred Kridl became the intellectual leader of both groups in 1935, although they were formally self-organized. The two groups began to work together, resulting in a collection of essays dedicated to Kazimierz Wójcicki, regarded by both as the founder of the Polish school of formalism (*Prace ofiarowane Kazimierzowi Wóycickiemu*, Wilno-Vilnius 1937). Along with members of both groups (Kridl, Budzyk, Mayenowa, Maria Rzeuska, Siedlecki, Żółkiewski), the volume also included texts by members of the Prague Linguistic Circle, notably Roman Jakobson, Nikolai S. Trubetzkoy, and Josef Hrabák (1912–1987). This clearly indicated foreign recognition of the Polish formalists. The volume also contained some texts which distanced themselves from formalism or were critical of it. These were written by Wacław Borowy, Henryk Elzenberg (1887–1967), Jerzy Kuryłowicz, Zygmunt Łempicki (1886–1943), and Karol Zawodziński (1890–1949). The volume is thus a confirmation of the cooperation of the two Polish circles with the Prague Circle. As in most such cases, it was based on close personal and professional contacts, especially between Budzyk and Siedlecki and members of the Prague school (Baluch 1998). One of the results of these relations was the publication of Budzyk's article in the Czech journal *Slovo a slovesnost* in 1937. Kazimierz Budzyk (1911–1964), who can also be regarded as one of the precursors of structuralism in Poland, graduated from the Polish Studies Department of the University of Warsaw in 1934 and already had a decidedly leftist worldview. He was interested in new approaches to basic stylistic problems. In his *Gwara a utwor literacki* (The Use of Dialect in a Literary Work, 1936), he dealt with the function of dialect and the motives for using it, rather than with the traditional question of the correctness of its usage. During WWII, Budzyk defended his doctorate at an underground university in 1944. In 1948, he co-founded the Institute of Literary Research (IBL), the history of which is discussed below. He was also the founding head of the Department of Literary Theory at the University of Warsaw (est. 1955).

As for the rest of the Warsaw-Vilnius formalist network, Andrzej Karcz (2002) points to the pioneering research of Franciszek Siedlecki (1906–1942) and Dawid Hopensztand (1904–1943). Siedlecki graduated from the University of Warsaw in 1936 with a degree in Polish, and after a very intense and productive life, died of exhaustion during WWII. His major innovation was formalist research methodology. His most outstanding achievement was *Studia z metryki polskiej* (Studies in Polish Metrics, 1937). This presented a precise description of Polish syllabic verse in its historical development, and was full of revealing observations and hypotheses. Siedlecki pointed to the role of context in comprehending utterances. He also analyzed the cyclical nature of changes in versification conventions. He saw the entire history of Polish versification as a chain of "canonizing" and "de-canonizing" cycles that succeeded one another. He referred to the Jakobson model of stress and rhythm deformation as the basis of poetic language and discussed the role of defamiliarization as the basis of art. Siedlecki also discussed the technique of inducting an expectation and its nonfulfillment (breach of metrics). This may be related to the mechanisms of homology discussed in the first chapter. The fact that homology is always partial, which seems parallel to Siedlecki's suggestion that art can only partially fulfill expectations to be meaningful. His claim that functionalists were "laying bare the device" is applicable to the modern notion of deconstruction.

No less innovative were the works of Dawid Hopensztand, who ran the section of sociology of literature in the Warsaw circle. Hopensztand was especially interested in "sociological stylistics", in which he was inspired by the Russian formalists, whom he went on to surpass. He combined a formalist analysis of literary style and composition with a Marxist or near-Marxist interpretation of social and economic phenomena as the context of literature. He also considered the author's linguistic consciousness under the influence of Vinogradov, in particular his notion of the device of "defamiliarization" employed by means of the stylistic structure of the "crypto-quotation" (Karcz 2002: 174). In the 1937 volume dedicated to the memory of Kazimierz Wójcicki (discussed above), Hopensztand published an analysis of the novel by Juliusz Kaden-Bandrowski (1855–1944) that would now be classified as a piece of discourse analysis. Hopensztand pointed out the key features of the political language of Marshal Józef Piłsudski's circle (the "Legionaries"). Hopensztand died in the Warsaw ghetto in 1943.

The political contexts of the formation of the Polish formalist school are extremely interesting. As Andrzej Karcz (2002) points out, whereas the Bolsheviks suppressed formalism in the USSR soon after consolidating their power, in Poland it was paradoxically attacked as a Bolshevik idea. These attacks mostly came from the extreme right. However, the Warsaw-Vilnius formalist circle was not directly persecuted, although it was denied state subsidies. This can best be explained by the fact that most of the members of this circle held leftist or even communist views (Hopensztand). However, Jerzy Putrament (1910–1986), a writer (and later secretary of the Polish Writers' Union) and a member of the Warsaw circle, was the only one who explicitly declared that he was interested in the Russian formalists because they came from the USSR; the others declared their interest in the content of the ideas expressed. Some were Jewish, so they suffered from rising anti-Semitism. On the other hand, Karcz claims that the less politically active Vilnius circle also had members associated with Catholicism (Sławińska, Zgorzelski). Markiewicz (1979) also points that the Vilnius circle was more receptive to Ingarden's phenomenological approach, while the Warsaw circle leaned toward neopositivism and was receptive to Marxism. The picture is further complicated by Budzyk's obituary of Franciszek Siedlecki. As Budzyk testifies, the Warsaw circle was supported by Zygmunt Łempicki (1886–1943), a professor of German literature at the University of Warsaw. Budzyk maintains that Łempicki did so despite his diametrically opposed theoretical scientific views and even though he supported the right and was a columnist for the conservative *Kurier Warszawski* (Budzyk 1946). Nevertheless, he supported the publishing activities of the Polish Studies Circle at the University of Warsaw. Kazimierz Nitsch from Kraków was also said to be friendly toward the Circle, although formally he did not participate in its activities. This is all the more intriguing because, according to Budzyk, the Circle attacked many of the authorities of the contemporary science of literature, e.g. it fought against German idealism, aestheticism, and most historical and literary studies canons, and attacked many recognized professors (Budzyk 1946). This combination of factors makes it possible to speak of the emergence of a relatively autonomous sector in the field of language and literature studies. This sector, however, was relatively weak, marginalized, and along with the entire Polish state, was broken up in 1939 before it managed to rise to the status of an internationally recognized "school" or milieu.

It can also be noted that the weakness of the Polish formalist milieu in comparison to the corresponding Soviet milieu was due to the lack of clear homology between the intellectual field and the field of power. By this I mean that, in contrast to Poland, disputes within the Soviet power elite translated into creative tension in the field of social theory. On the other hand, the same connection between the two fields meant that these inspiring theoretical debates were soon terminated and consigned to oblivion. Galin Tihanov writes about the clear competition between Marxists and Formalists that flourished in the intellectual field of the Soviet Union in the 1920s (Tihanov 2019: 34–35). Each of the two circles attempted to formulate the dominant cultural ideology of the fledgling communist state. According to Tihanov, the apogee of this dispute was the debate that took place in 1927. This turned out to be a manifestation of the clear dominance of the Marxists, behind whom stood the dominant bloc in the Soviet field of power. The Formalists, like other representatives of the avant-garde of the time, e.g. the Futurists and the Constructivists, were accused by Marxists of pointless experimentation with form. However, the accusations leveled at the Marxists by the Futurists, e.g. their primitive sociologism, went unanswered. And the Marxism of the time did indeed refer to a rather superficial sociologism—at least in the field of literary studies. By emphasizing the vision of a work of literature as an instrument of class struggle, and of social change by means of the dictatorship of the proletariat, it became a normative legitimization of the new communist realism promoted by the dominant faction of the field of power. In this context, Formalism, with its concept of the autonomy of the literary work, which in turn implied the autonomy of the literary field, was an oppositional orientation. Tihanov describes the two regimes of relevance that Marxism and formalism advocated: one that directly represented reality vs. one that valued the analysis of its hidden structures. That is not to say, however, that there were no attempts to link the two. Tihanov writes about the dialogue between these directions which took place in Russian émigré circles in Western Europe at the turn of the 1930s. This dialogue pointed to the possibilities of mutual inspiration between the two currents. In particular, Marxism's recognition of the importance of the autonomy of literature (both as a postulate and as a social fact) was considered, while the Formalists took greater account of "external", and therefore social, factors in the development of literature. It is symbolic, however, that in 1939, Émilie Litauer (1902–1941), whom Tihanov identifies as one of the main voices in this dialogue, emigrated from France to Moscow, where she was executed for espionage in 1941. This degree of intellectual tension in the theoretical dispute between the purely formalist school and the Marxists was unknown in Poland. This can be attributed primarily to the small scale of the Polish Formalist milieu and to the weakness of the homology of its disputes with its opponents in the field of power.

3.6.3 The Warsaw-Lviv School and Its Legacy

Understanding the origins of some of the important players in the development of the Polish structuralist school, particularly Kridl and Ingarden, requires some knowledge of the traditions of the Warsaw-Lviv School, whose roots go back to the late nineteenth century and are primarily connected with the University of Lviv. Through the efforts of Kazimierz Twardowski (1866–1938), a school of philosophy was established at the end of the nineteenth century. It is now known as the Warsaw-Lviv school, and became the main center of Polish philosophical thought, playing a significant role in philosophical education and quickly gaining dominance. The school's pupils and successors were not only philosophers, logicians, and methodologists of sciences, but also psychologists, linguists, classical philologists, historians of literature and culture. According to Jan Woleński (b. 1940), while there was a strong affinity between the Vienna Circle and the Warsaw-Lviv school, there were also significant differences (Woleński 2014). The Vienna circle rejected introspective psychology, the *vestehende* theory of humanities, absolutism, and objectivism in metaethics, which was present in the Warsaw-Lviv school, which valued logical positivism and focused on the study of language. Moreover, the Vienna Circle embraced the idea of reducing philosophy to a logical theory of language, without any references to metaphysics, which was present in the Warsaw-Lviv school.

What most typified the Warsaw-Lviv school was rationalism, or rather anti-irrationalism, the centrality of classic logic, a strong emphasis on methodology, and no a priori assumption of demarcation between science and metaphysics. The Warsaw-Lviv school was rather minimalist and analytical but, importantly, it was not anti-metaphysical. It was interested in semantics and in language in general, but also in language as a tool for analyzing reality. The Vienna Circle and the Warsaw-Lviv school shared a belief that science was the highest cognitive authority—a stance that was tantamount to classic scientism. They were both part of the analytical movement, and interested in the philosophy of science and semantics. Typical Warsaw-Lviv school approaches included subjecting factual issues to semantic analysis and studying the relationship between logic and language and between logic and reality. Semantics was at the center of the school, as was the notion of truth, which was Alfred Tarski's (1901–1983) central topic. The influence of Kazimierz Twardowski on the Warsaw-Lviv school was clearly dominant, and it included the centrality of the ideas he championed, e.g. the "precision of notions" and the "rigor of thinking". His supporters and students proposed tools that were deemed certain and purely objective, thus coinciding with the requirements of legal positivism, and the normative approaches to many other disciplines, including linguistics. The following literary scholars can be considered as belonging to the school: Juliusz Kleiner, Eugeniusz Kucharski, Manfred Kridl, Zygmunt Łempicki, Roman Ingarden, and Stefania Skwarczyńska, with Ingarden being the most prominent.

The legacy of the Warsaw-Lviv school is significant and multidimensional. It is inscribed in the dominant canon of the Polish intelligentsia mythology, both ideologically and as a kind of totem. It was primarily embodied by the members of the school, then their students, and finally by symbolic objects, e.g. the books and places associated with it. The city of Lviv assumes pride of place in this last category. Lviv is particularly important for contemporary Polish culture, especially its elite intelligentsia variant. An important aspect of the mythical role of Lviv is its seizure from Poland in 1945 and the virtual loss of contact with it for the entire communist period. In Lviv, apart from the historical buildings of the university and Lychakovsky cemetery, there is the former Scottish Café (*Kawiarnia Szkocka*), which was closed during the war. Interestingly, another café was opened in the same building under its historical name in 2015. This was probably done with Polish tourists in mind. If so, it eloquently illustrates the power of mythology. Although the café did not exist until recently, it has remained present in narratives on the history of Polish science, in which it features as a meeting place of Lviv mathematicians, who wrote down fascinating problems and their solutions in notebooks, which have survived (compiled in *The Scottish Book*). The legend of the Scottish Café revolves around meetings of the city's refined scientific elite, outstanding minds, who were able to achieve outstanding scientific results, using only "pure", but extremely intensive, thinking. Many of the patrons have become internationally famous. This is a positivist myth of the power of "precise" thinking and intellectual willpower manifested in consistent application of thought. Stefan Banach, Kazimierz Kuratowski (1896–1980), Hugo Steinhaus (1887–1972), and Stanisław Ulam (1909–1984) were among the frequent participants of the mathematical meetings in the Scottish café. The Lviv school of mathematics, to which they belonged, was a branch of the broader Warsaw-Lviv school. Their common patron was Kazimierz Twardowski (Stefan Banach's teacher). The connections between the Warsaw-Lviv and the Vienna schools allowed the Polish group to shine with the additional reflected glow of the globally respected Vienna, the successes of its philosophical school, and the "Westness" of Austria. The totem of the Warsaw-Lviv school also revolved around its positivist identity, which pragmatically implied that competing orientations were "less logical" and therefore less rational. In fine, the Warsaw-Lviv school was considered a symbol of moral and methodological reliability, and placed

Polish culture firmly in the realm of "Western civilization". This ethos permeated many Polish social science and humanities disciplines.

The Warsaw-Lviv school has bequeathed Polish intellectual culture an extremely powerful legacy. It would not be going too far to say that it continues to shape the identity of a whole spectrum of theoretical and ideological orientations. Polish Marxism inherited analytic philosophy. As Bogusław Wójcik (b. 1967) put it, analytic-linguistic Marxism was a uniquely Polish creation that emerged from the clash between and subsequent combination of the achievements of the Warsaw-Lviv school and Marxist ideology. Its foremost exponent was Adam Schaff (Wójcik 1993: 32). In 1968, his party loyalist critics charged him with calling the scientific nature of the Marxist dialectic into question. In particular, he was accused of trying to synchronize it with rules of the formal logic (Gawrak 1968). Wójcik and Stanisław Borzym (Borzym 2006) also illustrate the ways in which the influence of the Warsaw-Lviv school manifests itself on the religious, conservative side of the field of philosophy. In particular, at the Catholic University of Lublin (KUL) a clear opposition developed between its two professors, Stefan Świerzawski (1907–2004), who had been Ajdukiewicz's assistant in Lviv before WWII, and Mieczysław Albert Krąpiec (1921–2008), a Dominican priest. Borzym shows that this was a dispute between two projects intended to update Thomism, in which Krąpiec represented existential Thomism. Świeżawski represented personalist Thomism, which was rooted in his connection with Jacques Maritain, a Catholic philosopher he had met before the war. Both later participated in the Second Vatican Council. Another dispute over the interpretation of Thomism took place between Świeżawski and Józef Tischner (1931–2000), another priest and philosopher. The debate reached its apogee in the pages of *Tygodnik Powszechny* weekly under the rubric "The Intellectual Form of Christian Culture and the Role of Philosophy in Culture". There was also a dispute between Tischner and Krąpiec. The former propagated a dialogical philosophy, although both recognized the authority of John Paul II. Alfred Gawroński (1929–2021) took part in those discussions, appealing to Polish Thomists to rethink the basic concepts in reference to linguistic studies, by analyzing the philosophical formation of concepts taken from everyday language. Gawroński saw an opportunity for classical philosophy in linguistic philosophy (e.g. John Langshaw Austin 1911–1960). He cited Jan Maria Bocheński (1902–1995) and Thomists analysts to argue for a new approach to synthesizing St. Thomas and consolidating his achievements by invoking the personalistic and anti-neopositivistic current of linguistic philosophy. This put him at odds with Tischner, who strongly relied on existentialism, phenomenology, and hermeneutics (Husserl and Heidegger). Gawroński considered these schools irreconcilable with Thomism. For his part, Krąpiec eschewed the philosophy of language altogether. He accused linguists of trespassing on the field of philosophy. In his opinion, linguistic philosophy and linguistics did not allow for any judgments about reality. Krąpiec also wrote that, with semantics and pragmatics, linguistics "carelessly enters the field of philosophy and utters the usual 'foolishness'" (Krąpiec 1987: 134). This was a clear repudiation of the legacy of the Warsaw-Lviv school from a radically conservative perspective, but it was also an offer to negotiate the boundary between the two disciplines.

Hanna Dębska argues that the influence of the analytical philosophy of the Warsaw-Lviv school on Polish legal theory is indisputable and that its dominance may have contributed to the highly normative nature of Polish legal thought (Dębska 2019). To a large extent, the thought of Petrażycki stood in opposition to the school, so that when he returned to Poland in 1919 and was offered a chair at the University of Warsaw, a clear opposition was formed in the Polish field of philosophy of law between the normativists influenced by the Warsaw-Lviv school and Petrażycki's disciples and supporters, who promoted a sociological and psychological theory of law and called for the sociology of law to be recognized as an empirical science. This tension seems to be homologous to the tension that was evident in the development of Polish structuralism, i.e. between the heirs of Baudouin de Courtenay, who were committed to empiricism and a de facto relational approach to language, and the heirs of the Warsaw-Lviv School headed by Roman Ingarden. In both cases, the result was a certain synthesis that weakened the potential radicalism of structuralist thinking. These hybrid paradigms are useful tools for legitimizing the dominant position of the intelligentsia in Poland, particularly as a normative teaching tool. At the same time, however, this tendency to moderate the potential tools for deconstructing the social system limits the creative and critical potential of specific intellectual currents. The development of Jerzy Pelc's (1924–2017) thought can be interpreted this way. At one time, much hope for the development of a Polish school of semiotics was pinned on him. Pelc was also brought up in the tradition of the Warsaw-Lviv school. He initially combined his interests in philosophy and Polish studies (Borzym 2006). In 1961, he defended his habilitation at the Institute of Philosophy and Sociology (IFiS PAN) with a book titled *On the Notion of the Subject* (*O pojęciu tematu*). He then switched to logic and semiotics, but remained strongly influenced by Janina Kotarbińska (1901–1997) and Tadeusz Kotarbinski and Kazimierz Ajdukiewicz. In 1968, he helped found the Polish Semiotic Society. He served as president of the Institut International de Philosophie in 1984–1986 and as president of the International Association for Semiotic Studies in 1984–1994. Quite characteristically, the volume he edited, viz. *Semiotics in Poland 1894–1969* (Pelc 1979), which presented the achievements of the discipline in Poland, contained mostly texts by classics associated with the Warsaw-Lviv school, especially by such authors as Kazimierz Twardowski, Tadeusz Kotarbiński, Izydora Dąmbska (1904–1983), and Maria Ossowska (1896–1974).

Another example of the enduring legacy of the Warsaw-Lviv school is a recent paper entitled "The Polish School of Argumentation: A Manifesto" (Budzyńska et al. 2014) signed by over 50 Polish scholars. The paper attempts to identify the dominant characteristics of the Polish school of argumentation and identify those strengths that show promise. Significantly, the authors see the main historical basis of their school of argumentation in the legacy of the Warsaw-Lviv School with its logico-methodological precision and rigorous thinking. They also refer to the works of Jerzy Pelc, a Warsaw-Lviv alumnus. The three basic elements of the proposed paradigm of studies on "the power of argument" are reason, trust, and cognition. They also recommend that this school specializes in logic, linguistics, rhetoric, psychology, cognitive science, AI, and law. Sociology is conspicuously absent. The lack of any mention of the problems associated with power relations and the social contexts of rhetoric is also telling. Be that as it may, this manifesto, which was published in the international journal *Argumentation* has received little citation.

3.6.4 Polish Emigrants During WWII

I would like to conclude this discussion of the interwar period by examining the role of those scholars who emigrated to the West during WWII. The most important Polish linguists and literary scholars to emigrate were Kridl and Wacław Lednicki, who became professors at US universities—Kridl at Columbia University (NY) and Lednicki at UC Berkeley. Lednicki continued his work on Russian literature and on Polish-Russian cultural contacts, which was important for the visibility of Polish literature in the United States, but was not particularly innovative from a methodological or theoretical standpoint, although to be fair, he was never a theoretician. However, the situation was different with Kridl, who had played an extremely important role in the development of the Polish formalist school in the interwar period. Kridl tried to propagate this tradition at Columbia. In 1951, he published an article titled "The integral method of literary scholarship" based on his 1936 manifesto. This was not his first attempt to describe Russian formalism. However, according to (and others) the tenets of Russian Formalism mostly became known in the United States through the work of such scholars as Rene Wellek (1903–1995) and Austin Warren (1899–1986) (Theory of Literature) and Victor Erlich (1914–2007) (Russian Formalism) (Ulicka 2018). According to Ulicka, Kriedl was not well received by the US Slavists and literary scholars. His position was weakened by his poor command of English and by the fact that the chair he held at Columbia was funded by the Communist Polish government. He also disappointed many of the Polish intelligentsia living in exile in the United States, who expected him to promote more traditional variants of Polish literature, with a particular emphasis on "martyrological" themes. However, he did make some concessions. It was their money that financed Kridl's popular *Literatura polska na tle rozwoju kultury* (Polish Literature against the Background of Culture), which was published in Polish and English in 1945. However, because of these concessions, his book *Adam Mickiewicz—Poet of Poland* (Kridl, 1951) failed to live up to the hopes of people like Czesław Miłosz (1911–2004), who wanted Poland's greatest poet portrayed as a modern thinker. Ulicka therefore concludes that Kridl failed in his role as an émigré scholar, in contrast to

Jakobson, who was able to establish contacts with the American scientific elite, as was Rene Wellek, who together with Austin Warren, translated a textbook on literary theory into more than a dozen languages (Wellek and Warren 1949). As a result, it was mainly through these authors that knowledge of the achievements of Eastern and Central European linguistics and literary studies reached the global stage. Jakobson, who is discussed later in this book, played a special role here. While he gave great impetus to the development of structuralism in Poland in the 1960s, he popularized its achievements in a rather limited way, giving much more publicity to the Russian formalists and the Prague school of linguistics in his capacity as a leading exponent of both. It is with great disappointment that Ulicka details the absence of classic Polish scholars from the classical works of Western literary studies. None of them is referred to by e.g. Umberto Eco (1932–2016), Jacques Derrida (1930–2004), or Tzvetan Todorov (1939–2017). Ulicka finds it especially curious that Polish literary studies are only represented in Joseph Shipley's (1893–1988) *Dictionary of World Literature* by Maria Gołaszewska (1926–2016) and Jan Zygmunt Jakubowski (Ulicka 2007: 191). Jakubowski (1909–1975), as is discussed in more detail below, is regarded by most of the elite of the Polish scientific field as a mediocre scientist whose servile attitude toward the communist authorities called his integrity into question. At the same time, Ulicka notes with obvious disappointment the virtual absence of Roman Ingarden, hailed as one of the greatest Polish scholars of the twentieth century, in classical Western textbooks. This absence of Ingarden in global cannons is especially painful to the more conservative exponents of the Polish social science and humanities fields, but his greatness is universally recognized in the Polish scientific field. This is sadly emblematic of the failure of Polish language and literature studies to gain international recognition, and may help explain why their exponents have largely turned their backs on the global scientific field and gone their own way.

3.7 The Post-WWII Reconstruction of the Field

3.7.1 Construction of the New Institutional System

Let me now turn to post-WWII developments. As already mentioned, the wartime losses of Polish science in 1939–1945 were enormous. Many scientists perished in concertation camps or were directly or indirectly killed, and infrastructure, buildings, book collections, and archives were destroyed on a massive scale. However, the post-war reconstruction and reactivation of the field of science and higher education proceeded quickly. The first Chair of Polish Literature was established as early as November 1944 at KUL with Juliusz Kleiner as the incumbent (1944–1948). The Jagiellonian, Warsaw, and Poznań Universities resumed their activities in 1945. In Kraków, the chairs of literature were taken up by Stanisław Pigoń, and from 1948, Kazimierz Wyka, and professors and assistant professors from the Lviv milieu—Kleiner (from 1948), Stanisław Łempicki (1945–1947), Karol Badecki (1886–1953) and Mieczysław Piszczkowski (1901–1981). In Warsaw, Julian Krzyżanowski (1945–1960), and Wacław Borowy (1945–1950) returned to their pre-war chairs, and Kazimierz Budzyk (professor from 1948), was appointed. In Poznań the chairs of literary studies were held by Roman Pollak (1945–1960), Zygmunt Szweykowski (1946–1965), and Wacław Kubacki (1949–1953). While the old universities were being rebuilt, new universities were being established in Wrocław, Łódź, and Toruń. This was necessitated by the redrawing of national borders and the emergence of new intellectual life. These new universities mostly employed professors from Lviv and Vilnius, but they also drew staff and students from Warsaw and Kraków, which were overcrowded as a result of forced migration from the east. In Wrocław, the chairs of literature were taken up by from 1946 Władysław Floryan (1907–1991), Kucharski's pupil from Lviv, Stanisław Kolbuszewski (1945–1965), an alumnus of Poznań University who had been a professor in Riga before the war, and Tadeusz Mikulski (1946–1958) from Warsaw. In Łódź, the largest intellectual center in the first years after the war as a result of the destruction and depopulation of Warsaw, a university started to be organized along the lines of the Free Polish University. The Theory of Literature chair was held by Stefania Skwarczyńska (1945–1972), and the chairs of literary history by Stanisław Adamczewski (1945–1950) and Andrzej Boleski (1945–1948). In 1945, the Nicolaus Copernicus University (*Uniwersytet Mikołaja Kopernika*) was founded in Toruń, and mainly employed academics from Vilnius, Lviv and Warsaw. The chair of the Theory of Literature and Comparative Literature was held by Eugeniusz Kucharski (1946–1952), and the chairs of literature by Konrad Górski (1945–1965) and Tadeusz Makowiecki (1946–1952). University libraries were established in Toruń, Łódź, and Wrocław, and a major part of the collection of the Ossolineum in Lviv was transferred to Wrocław. At the same time, other institutional forms of Polish science began to be rebuilt. The PAU and pre-war scientific societies resumed their activities.

The Institute of Literary Research was established in Warsaw in 1948 pursuant to a decision made at the highest level of the Polish Workers Party (PPR), the governing party at the time. As discussed in more detail below, it was predominantly staffed by young literary historians and theoreticians. It was generously funded by the state, and had resources that were not available to philological circles operating within university structures. The Institute hosted not only individual, but also team-based research, which was something of a novelty. Its first director was Stefan Żółkiewski and his significance would be hard to overestimate. Żółkiewski perfectly illustrates the key role played by the pre-war intelligentsia in shaping the new system. Born into an old intelligentsia family of noble (*szlachta*) origin in Warsaw in 1911, Żółkiewski graduated from the University of Warsaw in 1934. Żółkiewski was an important member of the formalist "Polish Studies Circle" (*Koło Polonistyczne*) at the University of Warsaw, serving as president in 1932–1934. Similarly to Siedlecki, his chief interest was methodology. In 1937, Żółkiewski published *O podstawach metodologii badań literackich* (On the Foundations of the Methodology of Literary Research, 1937), in which he postulated that literary studies should be connected with linguistics, especially as developed by the Prague Linguistic Circle. At the same time, he saw phenomenological propositions in methodology, especially those of Ingarden, as rather weak, as they mostly concerned modes of existence. Even then, Żółkiewski wanted to build a science of literature as an autonomous, rigorous discipline, conscious of its tolls. He became involved in the communist movement during WWII, joining the Polish Workers' Party (PPR) in 1943. In 1945, he became a member of the PPR Central Committee, and from 1948, of the PZPR Central Committee, on which he sat until 1968 (as a "deputy member" until 1954). This was a key political role that gave Żółkiewski immense power. He headed the Cultural Department of the Central Committee in 1948–1949 and the Science Department in 1955–1956. He was the founding director of the IBL in 1948 and headed it until 1952. From the late 1940s, Żółkiewski was the main animator and codifier of the new methodology in literary studies (discussed in more detail below). He was the minister of higher education in 1956–1959. He had previously improved his formal academic qualifications. In 1951, he defended his doctorate in Wrocław under Tadeusz Mikulski (1909–1958). Shortly thereafter, he took over the chair at the University of Warsaw after Julian Krzyżanowski and Wacław Borowy. In 1952, he defended his habilitation, and was awarded the title of professor in 1954. However, he had already been a "deputy professor" at the University of Warsaw since 1948 (Lam 2016). Somewhat typical for this period, he combined full-time positions at the University of Warsaw and the Polish Academy of Sciences (PAN), a multi-positional configuration that was called a "personal union". In 1952–1961, he was a correspondent member of the PAN, then became a full member. He served as scientific secretary of PAN in 1952–1954. No less important were his roles as an editor of the militant journal *Kuźnica* (Hammer Mill) in 1945–1948, and then as the first editor-in-chief of *Polityka* in 1957–1958. This weekly, which still exists, was the organ of the Central Committee of the PZPR and was intended to plot a "middle course" for the left-liberal intelligentsia, i.e. it sought a compromise between pro-Soviet hardliners and extreme liberals whose deviation from the party line was deemed a threat (these radical tendencies led to the closure of the extremely popular weekly *Po prostu* in 1957). Żółkiewski's successor as editor-in-chief of *Polityka* for many years was Mieczysław F. Rakowski under whose stewardship the magazine became widely read by intellectuals of various orientations. Żółkiewski also edited such magazines as *Nowa Kultura* (1958–1961, created at the beginning of the 1950s from a merger of the closed *Kuźnica* and

Odrodzenie magazines) and the well-known scientific journal *Kultura i Społeczeństwo* (1958–1968).

I would like now to return to the Stalinist period. The PAN was established pursuant to a Sejm resolution dated October 30, 1951, and incorporated the IBL. The assets and agencies of the PAU and the Warsaw Scientific Society were transferred to this new leading Polish scientific institution. The group of scholars gathered around Żółkiewski had been developing new canons since 1945. At the Second Congress of the Polish Writers' Union (ZZLP), held on October 26–27, 1947 (Gosk 1985: 93), intellectual leftists clashed with Catholic intellectuals. Żółkiewski wrote at the time: "The morality of the writer today requires a clear political declaration - not only in life - but also in literature [...] As the structure of society is political, so too is the function of literature and its moral judgment about man and society, both in its theses and in its silences" (Gosk 1985: 96). He added that "the dividing line runs through all areas of life, thought, culture [...] one cannot be neutral" (Gosk 1985: 100) The value of work was to be determined by the ideological criterion of whether it conveyed the "truth" about life and the world. The "truth" on which Żółkiewski laid so much importance was understood, after Lukács, to be a reflection of reality with an emphasis on cognitive values, and generally known as the "realist method". The term "sociological construction of human fate", which was contrasted with the condemned "psychologism", also appeared in his writing. Żółkiewski repudiated what he considered the false humanism that typified Catholic literature (Żabicki 1966). The 4th Congress of the Polish Writers' Union, held in Szczecin in 1949, proclaimed socrealism as the official style. An even more significant event for the field of linguistics and literary studies, however, was the all-Polish Congress of Polish Studies in Warsaw in the spring of 1950. The Congress papers were published in a book entitled *O sytuacji w historii literatury polskiej* (The Situation Regarding the History of Polish Literature, 1950). The achievements of the Polish science of literature were severely criticized, and a complete volte face was announced. And so opened a new chapter in the history of Polish literary historiography.

Initially, the main forum for the programmatic activities of Żółkiewski's circle was the magazine *Kuźnica*. It was here that a new paradigm was forged. Their work largely consisted in reinterpreting the existing literary canon. Although many of the proposed interpretations were quite radical, and were soon to be imposed, their elaboration, especially in the initial stage, often involved numerous polemical exchanges. This could therefore be described as a dialogical process (at least to some extent) that did not quite fit the classical anti-communist stereotype of interpretations being imposed by Soviet fiat during the Stalinist period. It obviously has to keep in mind that hardline supporters of the new regime and its opponents were not evenly matched. Scientists considered skeptical of communism were marginalized in various ways, and sometimes even removed. Some could not withstand the psychological pressure exerted by the new authorities, e.g. Wacław Borowy (1890–1950) died immediately after an aggressive political attack on his person (Lam 2016). Borowy received his doctorate in philosophy at the Jagiellonian University in 1914. From 1930 to 1935 he was a lecturer in Polish literature at the London School of Slavonic and East European Studies (SSEES). In 1936, he was appointed director of the University of Warsaw Library. From 1938, he was a professor at the University's Department of Polish Literature. He mostly published classical texts on Polish literature in the United Kingdom. After the war, he headed the Historical Seminar of Polish Literature.

3.7.2 Establishment of New Canons

The question of reinterpreting historical canons and cultural patterns was often the subject of genuine debates in the late 1940s. Tomkowski cites the dispute over the interpretation of Prus's *Lalka* (The Doll) in *Kuźnica* (Tomkowski 2006). It began when Jan Kott, who identified Warsaw positivism with the ideology of capitalism and explained artistic phenomena by referring to economic and social facts. This triggered a wave of polemics. Tomkowski believes that this brilliant but provocative piece was probably deliberately extreme in its conclusions (Tomkowski 2006: 159). Now let me take a brief look at the weekly *Kuźnica*, which was first published in Łódź on June 1, 1945 (Gosk 1985). Żółkiewski wrote the main program articles. His papers, especially from 1948 onward, stated the official party line. It was Kott who created the hierarchy of new literary values. Adam Ważyk, Mieczysław Jastrun and Kazimierz Brandys (1916–2000) wrote dissertations, and Paweł Hertz (1918–2001) wrote press reviews. Reviews were published by young writers and critics such as Wiktor Woroszylski (1927–1996), Andrzej Braun (1923–2008), Maria Janion (1926–2020), Maria Żmigrodzka (1922–2000), and Henryk Markiewicz (1922–2013). *Kuźnica* fought fiercely for the new culture. The common opponent of *Kuźnica* and other party-sponsored magazines, e.g. *Odrodzenie*, was *Tygodnik Powszechny*, where several prominent conservative authors, e.g. Stefan Kisielewski (1911–1991), Jerzy Zawieyski (1902–1969), Paweł Jasienica (1909–1970), Antoni Gołubiew (1907–1979), and Zofia Kossak-Szczucka (1889–1968) published. *Dziś i Jutro* was the organ of Piasecki's PAX. It published nationalists loyal to the authorities, as well as Jerzy Dobraczyński (1910–1994) and Aleksander Bocheński (1904–2001). *Kuźnica* also published numerous texts by Adam Schaff, which bore such titles as *Konsekwencje Rewolucji Październikowej* (Consequences of the October Revolution, 1946), *Uspołecznienie przemysłu* (The Socialization of Industry, 1946), and *Humanizm socjalistyczny* (Socialist humanism, 1947) (Żabicki 1966). Among its numerous campaigns was a battle with the culture of the nobility in the Polish national canon (*kultura szlachecka*). Józef Chałasiński was one of those who published on the topic. Soon afterward, Chałasiński's paper was published in an extended form as a separate book, which is still frequently quoted (Chałasińki 1946). *Kuźnica*'s debates were often quite sophisticated, and not all the postulates mentioned in them could be written into the oppressive cultural model of the day. However, its ideological offensive was commonly associated with the dogmatic tendencies of the prevailing cultural policy. For this reason, it is now believed today that *Kuźnica* contributed to limiting the possibilities of genuine literary debates for some time (Żabicki 1966: 261). According to Żabicki, *Kuźnica* waged a battle to "rationalize" the worldview of the Polish intelligentsia by saturating it with historical thinking (Żabicki 1966: 262).

It may be mentioned, by way of digression, that similar motifs in the Polish intellectual debate return like a kind of boomerang, serving entirely new disputes between selected factions of the intelligentsia. In particular, it should be noted that currently, in the early 2020s, there is another installment of the debate about the culture of the nobility, or more precisely, its counterpoint, viz. the culture of the peasantry. A number of books have appeared criticizing the centrality of noble culture, the backgrounding of peasant culture, and the plight of the peasantry, in the canons of Polish culture. Books on this subject have been published by Adam Leszczyński (2020), Michał Rauszer (2020), Kacper Pobłocki (2021), and others. Obviously, the context, and thus the implications, of the current debate are quite different from that which resulted from the peculiar ideological offensive that accompanied the introduction of the Stalinist interpretative framework. The current debate is taking place in a democratic environment, and neither criticism of nobility culture's domination of the canons nor attempts to remind the peasantry of the injustices meted out to them are coming from scholars supported by the authorities, but rather from those close to the left-wing faction of the opposition. However, there are certain analogies, particularly in the tension between the dominant intelligentsia circles, especially between those with right-wing or traditional views and leftists. The latter are young researchers associated with the new left, who see the change of canons they postulate as part of a broader struggle for social justice. These tensions exhibit a certain cyclicality. One recent wave involved many of those involved in mainstream Polish social sciences and humanities who began as leftists shifting to more traditionalist positions with the re-traditionalization of the 1970s, as mentioned earlier. On the other hand, one theme evident in the programmatic manifestos of the Stalinist period that is now used on the opposite side of contemporary ideological struggles, i.e. at the right end of the spectrum, is that of the struggle against cosmopolitanism.

Now to return to the Stalinist period. A number of programmatic articles of a very orthodox nature also appeared in academic journals. Examples include Jan Otrębski's (1889–1971) dissertation, from the University of Poznań, *Józef Stalin o marksizmie w językoznawstwie* (Joseph Stalin on Marxism in Linguistics, 1950), and his article "Towards the Realization of Materialistic Linguistics" (Otrębski 1951), printed as first positions in the journal *Lingua Posnaniensis*. Obrębski's interpretations, which were rejected as being incompatible with Marxism, referenced Saussurianism and its structuralist offshoot in the Copenhagen School, because they opposed historicism in the treatment of linguistic phenomena. Otrębski expressed his conviction that "Polish linguists will undoubtedly join the defense of historicism in linguistics. They even have a tradition of their own in this respect: Jan Baudouin de Courtenay and Jan Rozwadowski were Poles" (Wąsik 1999: 49). Both de Courtenay and Rozwadowski, now classics, were thus used for every conceivable political purpose. This amply illustrates how strongly the canon of Polish linguistics (which was recognized even in interpretations such as the above) was naturalized during the Stalinist period.

The basic dispute regarding the theory of literature was the debate about realism. This was opposed to psychologism, creationism, avant-garde art, and bourgeois culture. Possible meanings ascribed to realism included the study of sociological and economic changes, and critical reflections on reality. Marxist scholars put considerable effort into shaping a new image of Adam Mickiewicz, Poland's national poet. He acquired a new image of a revolutionary poet, calling for the overthrow of the old order, drawing inspiration from folk art, and being a friend of Pushkin. For Mickiewicz's anniversary, Żółkiewski prepared a book presenting new interpretative proposals, *Spór o Mickiewicza* (The Dispute Over Mickiewicz, 1952). This was also the basis of his habilitation (Witkowska 2002: 16). The "Young Poland" period, i.e. Polish Modernism, was generally rejected, as were most literary works of the Interwar Period, with minor exceptions, such as revolutionary poetry and social reportage novels. After the thaw, this brand of Marxism was dubbed "vulgar sociologism" (Tomkowski 2006). It has to be emphasized, however, that there was no wholesale replacement of canons, i.e. traditional Polish national culture was not supplanted by some new proletarian or Soviet culture. The main effort of the new elite of the field was focused on reinterpreting the classical works of Polish culture and rearranging the hierarchies of its canons. It can be said, then, that the communists, as represented by the new generation of scholars did not so much overthrow as revise the old canons and proclaim themselves to be their new custodians. The legitimizing strategy that the new authorities pursued after the war was not to forcibly overthrow the old culture, but to better protect its legacy from its predecessors.

This can be linked to the structural nature of the new system in Poland, about which I wrote in the second of my previous books. The communist regime in Poland proved to be largely dependent on the traditional intelligentsia elite. The situation in the social sciences perfectly illustrates this. Their reorganization was overseen by, e.g. Stefan Żółkiewski and Adam Schaff, who came from multi-generational intelligentsia families. Symbolically, the former came from a family with noble traditions, while the latter came from a Polonized bourgeois Jewish family. They can therefore be regarded as symbols of the two main historical branches of the Polish intelligentsia. They and their colleagues and associates imposed a strongly intelligentsia-centered sub-culture on the new institutions that marked them for generations to come. John Connelly (2000) and Agata Zysiak (2016) have both analyzed this phenomenon. As they showed, even politically vetted youngsters coming from upward social mobility quickly adopted an intelligentsia identity after joining the ranks of academia. This co-optation was usually followed by a gradual transformation of views and the adoption of a sense of intelligentsia mission. As already mentioned, this intelligentsia ethos in Poland always included responsibility for national culture.

These tendencies can also be linked to the broader political concept on which the recreated post-war Poland was based. It was to be a socialist (or communist) state, but at the same time it was to be ethnically homogeneous. This was achieved by forced population exchanges, especially with the former Soviet Union and Germany. The ruling Communists believed that the ability to build Poland as a nation-state demonstrated the strength of Communism and the importance of Soviet support for this project. A new cultural canon was planned to meet the needs of this ethnically pure Poland. This was to be drawn up by the new intelligentsia, which, as already mentioned, was only partly new, and mostly traditional in its basic structures and inclinations. In this new Polish state, which was socialist with communist aspirations, while simultaneously being ethnically heterogeneous, linguists and literary scholars played an important role in strengthening its culture, and especially its language, from the outset. In this setting, it should not be surprising that the "struggle for the Polish language" in various epochs was a popular research topic, e.g. Tadeusz Milewski studied Enlightenment-era

Polish, while Renata Mayenowa went back even further in her *Walka o język polski w życiu literaturze staropolskiej* (The Struggle for the Polish Language in Life and Old Polish Literature, 1955). Another characteristic trend in this respect was the dynamic development of dialectology in 1945–1956 (esp. by Nitsch and Doroszewski) and lexicography (e.g. Nitsch's *Studia z historii polskiego słownictwa* [Studies in the History of Polish Vocabulary, 1948]). The contributions of people like Nitsch illustrate that the research effort of the entire field resumed from where it had left off before the war. Research topics that resumed after the war included work on monumental dictionaries and Polish language atlases. The magnitude of the forces of Polish linguistics concentrated on these studies and works, which were aimed at strengthening the scientific base of the Polish language, shows how strongly the new state sought legitimacy through activities focused on erecting the edifice of Polish national culture. However, as mentioned, this tendency could be interpreted as giving the intelligentsia, as the key Polish elite, the tools that enabled them to fulfill their fundamental role as gatekeepers of Polish culture more effectively than ever.

3.7.3 Institute of Literary Research (IBL)

Let me now focus on the central institution for the field in question, viz. the Institute of Literary Research. Its history also provides a particularly interesting insight into the main transformations which took place in Polish science during the communist period. Janusz Sławiński's (1934–2014) chronological division of the IBL's history is both meaningful and useful (Sławiński 1994: 146). He distinguished three periods. The first, which he called "the IBL in the trenches of the ideological front", lasted from 1948 until the end of the 1950s. The second, which was highly significant, he called "the golden years of the IBL". He dated this period from the early 1960s until the late 1970s. The third, from the late 1970s until the early 1990s, he called "the period of the IBL's spiritual drift". He also predicted that this period, which he considered barren, would last a long time (Sławiński 1994). I would now like to take a closer look at the IBL during the first period from a personnel perspective. The

Institute was created mostly by young researchers, especially the formalist groups who had studied in Vilnius and Warsaw in the 1930s. Their political orientation was generally left and they were clearly influenced by Russian formalism and Czech structuralism, with Manfred Kridl as their leader in Vilnius. Stefan Żółkiewski, Kazimierz Budzyk, Dawid Hopensztand, and Franciszek Siedlecki were among their leaders in Warsaw. An institute along the lines of the IBL was being mooted even before the war, primarily in the deliberations of Manfred Kridl and Franciszek Siedlecki. Due to Siedlecki's death and Kridl's emigration to the United States, it was designed and implemented after the war—mainly by Jan Kott and Stefan Żółkiewski respectively. Witkowska describes this as the "realization of a certain utopia" (Witkowska 2002: 12). Budzyk helped found the IBL as a committed Marxist. He focused on the Old Polish language (Renaissance) and in 1955 set up the Chair of the Literary Theory at the University of Warsaw. Budzyk was a devoted protagonist of Stalin's reforms. In his programmatic works, Żółkiewski wrote of Budzyk's "victorious" polemic with Krzyżanowski, where he supposedly "proved that Mikołaj Rej (1505–1569) is a Renaissance writer and not a representative of the late, dead sounds of the Middle Ages in the sixteenth century" (Żółkiewski and Stradecki 1955: 29). It is worth recalling that Krzyżanowski was Budzyk's pre-war lecturer.

Żółkiewski relied on a circle of collaborators that included Jan Kott, Kazimierz Wyka, and Henryk Markiewicz. Many of their personal histories were not politically unambiguous, but in 1945, they unequivocally declared their loyalty to the new authorities. Jan Kott was born in 1914 and graduated in law studies at the University of Warsaw in 1936. Before the war he was involved in various ideological initiatives, from communism to Catholicism. In Paris, he met the circle of Jacques Maritain. Kazimierz Wyka (1910–1975) graduated from the Jagiellonian University in 1932 under the guidance of Ignacy Chrzanowski, Stefan Kołaczkowski, Kazimierz Nitsch, and others. In the 1930s, he was one of Poland's most important up-and-coming literary critics. In 1934, he became an assistant at the Jagiellonian University's Seminar of the History of Polish Literature. In 1937, he received his doctorate on the basis of his dissertation *Studia nad programem Młodej Polski* (Studies on the Young Poland program). After graduation, he worked at the university as an assistant; he was habilitated after the war (1946) and became a professor in 1948. In 1963–1965, he served as vice-rector of the Jagiellonian University. He headed the Institute of Literary Research in 1953–1970, and was a member of the Polish parliament in its 1952–1956 term. He became a full member of the Polish Academy of Sciences in 1954. In 1964, he was a signatory to the Letter of 34, which criticized censorship in the People's Republic of Poland. He later withdrew his signature. The literary seminar led by Kazimierz Wyka at the Jagiellonian University's Faculty of Polish Studies in the 1950s, known as the Kraków School of Criticism, was one of the first to adopt the "scientific worldview" derived from the Marxist theory of literature. This focused on the campaign against "idealism", "formalism", and "psychologism" in literary studies. Wyka and his students, were inspired by Marx in their literary studies, and leaned toward historical rather than formal approaches, and toward sociologism rather than psychologism. Kazimierz Wyka's approach was syncretic, and he often warned against the doctrinaire and simplistic nature of the new methodology. Henryk Markiewicz (1922–2013) graduated from the Jagiellonian University in 1950. In 1955, he received his Ph.D. from the Institute of Literary Research (IBL), and in 1956, he was made an associate professor of the history of Polish literature at the Jagiellonian University. He was appointed to a full professorship in 1964. His work mostly focused on the history of Polish literature between 1864 and 1939, the theory of literature, the methodology of literary research, the history of science of literature, and literary criticism. In 1977–1984, he was the director of the Institute of Polish Philology at the Jagiellonian University, and in 1986, he was elected to the Senate of the Jagiellonian University. Markiewicz was additionally the editor of the Polish Biographical Dictionary (*Polski Słownik Biograficzny*). From 1946 to 1948, he was a member of the Polish Workers' Party (PPR), and then the PZPR until its dissolution in 1990.

Żółkiewski also chose a circle of young collaborators whose careers soon developed very rapidly. One of them was Maria Janion (1926–2020), who began studying Polish philology at the University of Łódź in 1945. From 1946, she attended a critical-literary course run by Stefan Żółkiewski at *Kuźnica* weekly. In 1948, she joined the Polish Institute of Literary Research, initially as an assistant, and worked there until her retirement in 1996. In 1949, she joined the PZPR, from which she was expelled in 1979. In 1951, she took a master's degree at the University of Warsaw; in 1955, at the IBL, she was awarded the degree of candidate of philological sciences, the equivalent of a doctorate, written under Żółkiewski's supervision. She was awarded the title of professor of humanities in 1963.

Żółkiewski had a number of pre-war scholars whose approaches were considered incompatible with the new regime dismissed. These included Wacław Borowy (1890–1950), Juliusz Kleiner, Stefania Skwarczyńska (1902–1988), Roman Ingarden, Konrad Górski, Stanisław Pigoń, and Zygmunt Szweykowski (1894–1978). The degree of their degradation varied, but few were expelled from the academic sphere altogether. Stalinism in Poland, especially when compared with other countries under Soviet control, was therefore relatively mild. There were also significant curricula changes. For instance, university English and German departments were closed, although Romance departments remained. However, Russian studies were privileged. Although the level of scholarship was poor, they were need to train teachers, as Russian had been made obligatory in Polish schools (Smoczyński 2001).

When he set up the IBL, Żółkiewski prepared a set of program documents that became the basis for reforming the vast area of language and literature studies in Poland. He published several programmatic papers, including *Stare i nowe literaturoznawstwo* (Old and New Literary Studies, 1950) and *Badania nad literaturą polską. Dorobek, stan i potrzeby* (Studies on Polish Literature: Its Achievements, Condition and Requirements, 1951). The latter was a general treatise on "bourgeois" literature studies and a blueprint for a Marxist takeover. These were fairly concise studies that set out new interpretative directions. Żółkiewski laid down general guidelines, stressing the necessity to fight for "systemic Marxism", and proposed reinterpretations of specific areas, such as Polish Romanticism. Żółkiewski coordinated work on this issue in his capacity as head of the "Laboratory of Romantic Literature of the IBL" in 1950–1953. His proposals covered many areas of linguistics and also touched on methodology. In *Stare i nowe językoznawstwo* Żółkiewski quite seamlessly combined Marxism with formalism, and "hard" ideologization with subtle analyses of artistic structures (Witkowska 2002: 14). As Witkowska notes, he drew his inspiration from Soviet sources, including the formalism of Shklovsky, Eichenbaum, and Tymianov, and from the trivial Marxism of the Soviet scholars of the time. Żółkiewski argued that versification could be seen as a fight for innovation (poetic, ideological, and artistic), and that, as with any ideology, it was determined by social factors. While his sophisticated forms of combining Marxist ideology with officially criticized currents (e.g. formalism), and his attempts to legitimize a relatively diverse range of theoretical approaches evoke admiration—however begrudging—Żółkiewski is also viewed critically as a brutal apparatchik. For example, at the 1950 Congress of Polish Studies Scholars (*Zjazd Polonistów*), he responded sharply to Kleiner's demand that the classics of Polish literature not be censored, arguing that this would be contrary to the interests of the working class. Sławiński argued that Żółkiewski was driven "not so much by fidelity to his youthful predilection for the methodology of the Viennese neo-positivists or the linguistics of the Prague structuralists [...] as fidelity to the strategic goals and tactics of the PPR in its drive to control the artistic, literary, and scientific communities. This influential party activist, intellectual apparatchik [...] editor, and leading columnist of *Kuźnica* - a magazine whose primary mission was to convert the Polish intelligentsia to the new faith - undertook the ambitious task of building a bastion to strengthen the 'ideological front' in the field of humanistic studies" (Błażejowska 2018: 14). Sławiński was also critical of Żółkiewski's successor, Kazimierz Wyka, describing as a talented and ambitious "typical intelligentsia-member fellow-traveler (*poputchik*)" (Sławiński 1994: 595).

There was no sign of the impending rejection of structuralism by the new regime immediately after the war. In a text published in *Kuźnica* in 1946, Żółkiewski argued that structuralism helped determine the social background of a literary work. He also maintained that structuralism allowed for the creation of a uniform system of morphological notions covering both language and literary works. In 1946, Budzyk attempted to revitalize the formalist series *Z zagadnień poetyki* (On the issues of Poetics) in which some formalist works were published in 1948–1949. Gradually, however, structuralism was marginalized, although it was never completely abolished. The memory of structuralist traditions was preserved, and structuralist assumptions, although seldom explicitly referred to as such, remained the methodological basis for much of the work carried out at IBL. Mayenowa especially was strongly committed to continuing the pre-war formalist tradition. Thus, despite structuralism not being considered fully legitimate, it was not entirely eliminated, but rather relegated to the background. This relatively mild treatment of this legacy can again be considered a Polish peculiarity and yet another manifestation of the relatively mild course of Stalinism in Poland. This stands in stark contrast with the Soviet Union, or even Czechoslovakia, where e.g. Jan Mukařovský was forced to gradually make his methodological assumptions reconcilable with Marxism (Gierowski 2018: 44). The pressure was so great that he declared official self-criticism in 1951 and categorically repudiated his previous work. The situation regarding structuralism was complicated by the fact that it had been close to Marxist theory in some respects before the war. Czech Marxists spoke highly of structuralist research methods. Thus, in contrast to Poland, where structuralism was co-opted, albeit in very marginalized positions, there was something of a "chief revolutionary" competition between Marxism and structuralism in Czechoslovakia (Gierowski 2018: 48). Official Czechoslovakian Marxism saw an existential threat to its own axiological system in structuralism, and so it was depreciated as a pseudo revolution, as a dangerous and insidious enemy, and as a progressive façade hiding bourgeois ideologies. The greatest challenge facing the Czech communists was the worldwide fame of the Prague Linguistic Circle. They attempted to deal with this by extricating structuralism from the tradition of the Circle and to separate what was theirs (and therefore good) from what was foreign (and therefore bad), including structuralism, even though it had once been theirs. This negative structuralist element was embodied by Roman Jakobson, who was portrayed as an anti-Soviet émigré, disparaged as a cosmopolitan, and reviled as an undercover Trotskyist, as Petr Sgall wrote in 1951 (Gierowski 2018).

3.7.4 The First Congress of Polish Science (1951)

The First Congress of Polish Science (*I Kongres Nauki Polskiej*) took place at the turn of July 1951. This was a highly symbolic event during which the official programmatic visions for all the social sciences were presented. I will now discuss the documents of the congress in more detail as they provide a fascinating case study of a communist project to redesign and adapt the traditions of an entire academic field with a view to integrating it into its state-building project. The linguistics paper was published anonymously. The main speaker was Witold Doroszewski. The linguistics sub-section was chaired by Jerzy Kuryłowicz and Zdzisław Stieber, which indicates the importance of continuity in the field (Karpowicz 2006). The state of the discipline was given a reasonably positive assessment. Baudoin de Courtenay was criticized for his psychologism, but was nevertheless recognized as a leading figure in Polish linguistics. This shows a more general tendency (already mentioned) to recognize most of the classics of Polish literature, literary studies, and linguistics. While the proposed changes might strike many as revolutionary, they were mostly changes to the interpretation of the canon, rather than a complete overhaul of that canon. Baudoin de Courtenay was declared to have been ahead of his time in many respects, e.g. in his attitude to the theses of the young grammarians on the uniqueness of vocal rules, and as the creator of "psycho-phonetic" concepts, which only became the subject of lively discussion in Europe after his death. He was also considered to be right in his moral stance, especially that of a passionate seeker of truth not only in science but also in life (striving to apply the same criteria of conduct in both domains) and to maintain integrity based on "logical thinking and a sense of justice". According to the authors, this striving for a uniform attitude and maintaining the links between theory and practice made Baudoin de Courtenay close to his contemporaries.

Rozwadowski's concepts were presented as idealistic and dualistic, which was considered a negative trait, but his scholarly greatness, which was on par with that of Baudoin de Courtenay, was not questioned. His international stature, and his role in the history of the Polish linguistics were acknowledged. Wide horizons, supposedly protected Rozwadowski against dogmatism, according to the authors. Andrzej Gawroński was presented as a talented linguist. His primary thesis was presented as the primacy of the emotional factor. This was seen as one-sided psychologism and thus condemned as irrationalism. In summary, the paper attacked dualism (ideal language vs social-communicative language in reference to Plato), psychologism, as linguistics should be concerned with social facts, not internal feelings and idealism. What was called for was dialectic materialism, which assumed that the world was knowable and that language was a social and communicative activity, a tool of thought, and a reflection of reality that was dependent on physical phenomena. Dialectic materialism was also presented as assuming historicism, anti-irrationalism, and monism (I Kongres Nauki Polskiej 1951b: 13). Stalin was presented as the most competent of Marxists, as the one who correctly determined what qualified as scientific. Thus, following Stalin, it was asserted that language was not produced by any social class and did not belong to any one class. Language, it was argued, belonged to society as a whole (i.e. to the nation). Marr, who saw language as part of the superstructure, was obviously deprecated. The historical-comparative method was seen as essential. Language was seen as evolving. Stalinist views were presented as anti-dogmatic. This was in reference to his calls for debates among linguists. The main achievements of Polish linguistics to date were seen in historical grammar, Polish dialectology, Slavic dialectology, word morphology, the study of the contemporary literary language, comparative grammar of the Slavic languages, Indo-European studies, and general linguistics. In conclusion, what was called for was a continuation of the research directions mentioned above. There was an additional call for further work on e.g. the internal laws of language development, the relationship between language and thought, and between language and social life, on idealism, dualism, Saussurism, phonology, and the possibility of applying Pavlovian concepts to linguistics, as suggested by Stalin. Further work on dialects was expected too. In the area of oriental linguistics, the earlier achievements of Kowalewski (Mongolian studies in Kazan) were recognized. In Semitic studies, more importance was expected to be given to national-liberation movements in countries of Semitic culture. In Sinology, the political context was mentioned as important. In the area of classical philology, more interest in work carried out in the Soviet Union was advised. Marxism and Stalinism were seen as being of assistance in understanding why each epoch asked different questions about classical antiquity. Ancient works were seen as demanding analysis in terms of class and social inequalities.

The report on literary studies (I Kongres Nauki Polskiej 1951a) was most probably prepared by Żółkiewski. The Polish bibliography of Karol Estreicher (1906–1984) was presented as an impressive achievement. Żółkiewski (as the presumed author) quoted other bibliographic monographs and lamented that their publication was slow as there was always a shortage of funds. In the area of dictionaries, the author saw a continuity of efforts from the nineteenth century. Żółkiewski presented a vision of a relay generations. The work of Samuel Linde was continued by Jan Karłowicz, then by Adam Antoni Kryński, Władysław Niedźwiedzki (1849–1930), and right then by Witold Doroszewski, who was working on his new dictionary. In 1927, another chain had been started by Brückner, who began work on the Etymological Dictionary of the Polish language. This project was taken over by Nitsch and his new dictionary of dialects and atlas. Work on an Old Polish dictionary, coordinated by the PAU, had taken 50 years, and as Żółkiewski announced, was only then being completed. He noted that such works had always been produced by single authors and therefore took an ordinately long time, e.g. the final volume of Estreicher's bibliography had still not been completed. It was announced that the work would be conducted in teams and that this would increase its efficiency and scale considerably (I Kongres Nauki Polskiej 1951a: 6). Thus, there was a continuity of activity in the field stretching back to the early nineteenth century. The report, and a number of other keynote speeches of the period, announced plans for a number of impressive team projects, especially historical and contemporary dictionaries. In particular, a new dictionary of the Polish language was anticipated, as were atlases, textbooks, and grammars. In addition, numerous monumental editions of collected works by historical writers, and cross-sectional editions of old Polish literature, were planned. The scale was impressive, and it was repeatedly emphasized that neither this nor the systematic nature of these works would have been possible before the war. Secondary school textbooks on the history of Polish literature, university textbooks, monographic studies, bibliographical works, lexicographical works, a dictionary of literary terms, and numerous editorial works were also planned. The systematic and planned nature of these projects was intended to lend legitimacy to the new authorities as a force that preserved and fostered Polish national culture on a scale and in a manner that was unprecedented. Urbańczyk's *Słownik staropolski* (Old Polish Dictionary) stood out among the many dictionary projects of this period. Work recommenced in 1953 as a follow-up to a series that began in 1874 pursuant to a PAU decision (Czopek-Kopciuch 2010). Also noteworthy was Mayenowa's *Słownik Polszczyzny XVI wieku* (Dictionary of Sixteenth-Century Polish), which she edited with the assistance of Franciszek Pepłowski (1921–2009). In the following decades, other impressive dictionaries were created as follow ups to these team efforts that began during the Stalinist period. These included Skorupka's two-volume phraseological dictionary (1967–1968), Doroszewski's 11-volume dictionary of Polish language (1969), and Mieczysław Szymczak's (1927–1985) three-volume *Słownik Języka Polskiego* (Dictionary of the Polish Language, 1978). The first edition of the *Słownik poprawnej polszczyny* (Dictionary of Correct Polish) by Doroszewski and Halina Kurkowska (1922–1983) was published in 1973. It was the successor to the Szober dictionary, which had been reprinted for over a quarter of a century. The *Słownik Języka Polskiego* (Polish Dictionary), edited by Mieczysła Szymczak, was first published in 1976 (Karpowicz 2006). Subsequent dictionaries were published in the following years. The remarkable continuity of this work, and the multi-generational nature of the effort that went into it (central for the field), is evident here. This was largely independent of the concurrent political and systemic changes. At its core lay the idea of nurturing or guarding the "national treasure", i.e. the Polish language and its history. Other dictionaries developed at the IBL include *Dictionary of the language of Jan Chryzostom Pasek* (Halina Koneczna, Witold Doroszewski eds.), *Adam Mickiewicz Dictionary* (Konrad Górski, Stefan Hrabuilec eds.), *Dictionary of the XVII Century and the First Half of the XVIII Century* (Kazimierz Żelazko, Krystyna Siekierska eds.), *Dictionary of the Polish of Jan Kochanowski* (Marian Kucała ed.), and *Słownik frekwencyjny współczesnej polszczyny* (Frequency Dictionary of Modern

Polish) (Zygmunt Saloni ed.). Sławiński wrote about these immense undertakings at the IBL as follows: "Yet another way of looking at the development of the IBL's output would be through the prism of - let us call them - 'mammoth' projects: great, long-term, collaborative works, whose character, scope and fundamental significance for the whole discipline meant that they were sometimes treated as the Institute's *raison d'etre*. What I have in mind is the synthesis of the history of Polish literature, as well as many other large-scale documentary, bio-bibliographical, editorial, encyclopedic and dictionary projects. The intricate fortunes of these projects, only a few of which were fortunate enough to have been completed, while the majority have remained in various stages of incompletion, like buildings beautifully begun but later neglected or abandoned by their builders, may be an extremely interesting case for the historian of science, and an instructive one for planners and managers" (Sławiński 1994: 144). Kwiryna Handke (1932–2021), on the other hand, wrote about these projects back in the modern period with a pathos typical of the intelligentsia: "After World War II, when Polish linguistics was first reactivated, the wisdom of the masters told them to continue their earlier important research ideas. And the conditions that existed in the institutes of the PAN made it possible to undertake such great undertakings" (Handke 2002: 43). It can therefore be said that the institutional and ideological efforts of this period have now been largely reinterpreted as an achievement of the intelligentsia, rather than the communist state.

But to return to 1950 and the Congress papers. An interesting observation presented in the linguistic section of the report for the First Congress of Polish Science was that folklore had been treated as a dead artistic form. However, the new generation of Polish scholars, who found inspiration in Soviet works, saw its potential as a lively and separate cultural form. Editions of sources were also considered important. Earlier literary studies were presented as incapable of explaining the dynamics of literary development and were denounced for the superficial empiricism and cognitive minimalism that typified positivist bourgeois science (I Kongres Nauki Polskiej 1951a: 14). The progressive nature of the Polish bourgeoisie was limited by their early tendencies to join forces with the aristocracy. The authors lamented the fact that no proper synthesis of romanticism based on class had ever been produced, and that Mickiewicz had never been analyzed as the author of the progressive *Trybuna Ludów*. The "incorrect aestheticization" of romantic literature was perceived to be a problem. It was argued that Brückner, as a positivist, should be commended for studying eighteenth-century literature. In general, the early critical bourgeois tendencies were seen in a rather positive light; it was the later "imperialist" bourgeois tendencies that were perceived as negative. It was suggested that Stańczyk's historians of literature be ignored as anachronistic. Conventionalism and formalism were presented negatively, as were the revival of idealism inspired by Wilhelm Dilthey (1833–1911), structuralism, and "psychology of spirit". These were all seen as stepping stones to formalism. In general, the "apology of the anti-realistic tendencies of the late bourgeois" era was condemned (I Congress of Polish Science 1951a).

The report called for the unity of linguistics and literary studies. The authors lamented that the theory of literature was practiced separately from the history of literature before the war. This supposedly reflected misguided bourgeois philosophy, with its different varieties of idealism. Nothing was supposed to have remained of these theoretical considerations. The authors distinguished three currents of the previous 30 years. The first was the conventionalist attempt to determine the content and scope of concepts. From this current, the works of Kridl and Skwarczyńska were classified as eclectic. Interestingly, they were related to the approach of sociologist Florian Znaniecki (1882–1958). The second current was the phenomenological theory of literature. This tendency to negate the servile nature of the theory of literature was viewed with disapproval. It was considered tantamount to searching for "insight into the essence of a phenomenon", and seen as extremely idealistic. According to the report, this led to "reactionary" socio-political consequences (Heidegger was mentioned as a prime example). Needless to say, this tendency was personified by Roman Ingarden in Poland. Ingarden's books were deemed inconsequential as was "repeatedly emphasized by the critics" (I Congress of Polish Science 1951a: 58). The third direction, which was seen as having derived from the Russian formal school, was the phonology, especially the Prague school. It was presented as the closest to neo-positivism. But, according to the paper, it emphasized an empirical and often vulgarly materialistic attitude. This direction, according to the report, despite its assumptions, which they considered "false", produced some positive results, as the main emphasis was on the actual linguistic material. This illustrates the more forgiving attitude to formalism in Stalinist Poland. The authors cited Jan Łoś's book as "brilliant" in view of its encyclopedic nature. Cezary Rowiński (b. 1934), Franciszek Siedlecki, Karol Wiktor Zawodziński (1890–1949) were given positive appraisals. Maria Dłuska's (1900–1992) syntheses were considered valuable. Halina Turska's (1901–1979) study of accent was regarded as sensational, as it showed that the knowledge of verse must be truly historical, and only Marxism teaches a genuine historical approach (First Congress of Polish Science 1951a: 60).

The report argued that Poland was still at the trial-and-error stage in stylistics, syntax, and the function of dialects and archaisms in literary language. Works by Klemensiewicz, Budzyk, and edited by Szober and Wędkiewicz were cited as positive examples. Kridl supposedly attracted the young by contrasting formal literary research with the "sterile meticulousness (*przyczynkarstwo*) and obscurity of the Diltheyists". But, as the report put it, "they entered the blind alley of formalism" (I Congress of Polish Science 1951a: 61). According to the report, the literary criticism of the interwar period was completely politicized, so there was no question of lasting achievements. These were supposedly the preserve of Marxist critics. Julian Brun-Bronowicz's (1886–1942) criticism of Stefan Żeromski (1964–1925) was cited as an example. The Polish Communist Party (KPP) was said to have made a significant contribution to literary criticism. Interwar aesthetics were similarly dismissed. The report proclaimed the "bankruptcy of Kantianism" and claimed that there was still no true, scientific aesthetics (I Congress of Polish Science 1951a: 48–68). In this context, reference was made to the Sixth Plenum of the PZPR at which the problem of national tradition was posed. Significantly, the problem was defined as the need to separate those traditions that were valuable and necessary in the new phase of the history of the Polish nation (i.e. necessary and valuable for creating a socialist nation), from those that were harmful. It was stated with regret that the period after 1945 should be considered transitional as idealism was still present. Works by Kleiner on Mickiewicz, Borowy on eighteenth-century poetry, and Stanisław Adamczewski's (1883–1952) reissue of Mickiewicz were proffered as examples. Wyka's pre-war book on Cyprian Kamil Norwid (1821–1883) was found to be unduly influenced by formalism. The correct distinction between critical realism and socialist realism was identified as a problem (I Congress of Polish Science 1951a: 67). The establishment of the IBL in 1948 was seen as a breakthrough in this respect, as it allowed for "a break with the anarchy and lack of professionalism (*chałupnictwo*) in Polish Studies" (First Congress of Polish Science 1951a: 68). The report noted that, despite its lack of formal powers, the institute was a center for planning and coordinating work on Polish Studies on a national scale. At the time, it had a staff of 250. Moreover, almost all the active Polish language scholars (*poloniści*) were thought to be affiliated with the IBL. As they were a mixture of young and old, Marxist methodology could be implemented in the entire field. The Mickiewicz Literary Society, formally independent, was also mentioned, but its work was already being coordinated by the IBL.

It was also argued that more work was required on the Polish language section of *Polski Słownik Biograficzny* (Polish Biographical Dictionary) (I Kongres Nauki Polskiej 1951a: 89). This project perfectly illustrates the continuity in the development of Polish culture. It is significant that its continuation was deemed important during the Stalinist period. Publication of the dictionary had begun in 1935 and 52 volumes, containing over 28,000 Polish biographies, have been published. These include biographies of figures considered historically important, starting from the legendary beginnings of Poland to the present day. Biographies are written for persons who died at least 3 years before the publication of the next volume. The dictionary can be described as a scientifically sanctioned key index of the intelligentsia elite, i.e. its state recognized pantheon, or, to put it poetically, a register of the intelligentsia's aristocracy. Although the dictionary only contains biographies of deceased persons, their genealogy is an important part of those biographies. It can therefore be used to build family trees. By way of example, the genealogist Marek Jerzy Minakowski[1] has developed a genealogical database

[1] Minakowski's data base is accessible on-line at https://wielcy.pl/. Its role in describing the contemporary Polish intelligentsia elite is discussed in a book I co-authored with Rafał Smoczyński (Smoczyński and Zarycki 2017).

of the Polish elite in which the "quality of origin" index of individuals can be calculated on the basis of ancestors who have biographical entries in the *Polski Słownik Biograficzny*. Post-war publication was managed by the PAN Institute of History. However, as the report shows, Polish philologists, and probably people from other scientific disciplines, took an interest in the dictionary, as it was important to them that there were entries for their key figures.

The literary studies report to the Congress of Polish Science ended with the following pathetic conclusion. "Literature is the song of our people, a weapon in the fight for a just society, for a free, self-determined existence, a weapon in the fight for the ultimate realization of a plan to actualize the dreams derived from the thoughts of [Andrzej Frycz] Modrzewski, [Hugo] Kołłątaj, [Adam] Mickiewicz, [Ludwik] Waryński and [Julian] Marchlewski" (I Congress of Polish Science 1951a: 91). This sentence characteristically combines different orders. The traditional classics of Polish literature and culture (the first three names), are juxtaposed with heroes of the communist movement (the last two names). At the same time, social justice slogans are combined with implicitly nationalistic pro-independence slogans. Thus, despite their aggressive tone, the papers also contain evidence of attempts to balance different orders and traditions—especially those considered traditional and those directed toward a communist future. Contemporary works reveal that realizing the postulates of these documents did not usually fulfill their most radical demands. For example, regarding the textbooks published during this period, according to Alina Witkowska, the most prominent manifestation of "vulgar Marxism" was a series of school textbooks that proposed a new interpretation of literature. Prominent among them were volumes on the "literature of the imperialist period", as it was then called, by Jan Zygmunt Jakubowski and Ewa Korzeniowska (1910–1983). Interestingly, these textbooks were severely criticized as being too simplistic and radical. The Polish Writers' Union was especially scathing (Witkowska 2002: 14).

Now to take a brief look at the report on "neophilology", which included Russian and Slavic literary studies and Western European philology (I Congress of Polish Science 1951c: 4). Criticism was leveled at the cosmopolitanism of bourgeois humanities and their subservience to everything Western. This cosmopolitanism was said to result in the literatures of the "Slavic nations" being studied far less than those of Germany, France, or Italy. Interestingly, a degree of forbearance was shown to conservative scholars and intellectuals, e.g. Wacław Lednicki, who was a professor at UC Berkeley at the time, was described as ideologically suspect, but substantively still useful. Also mentioned were Włodzimierz Spasowicz (admittedly, with references to his "justified" criticism), Alexander Brückner, and conservatives such as Marian Zdziechowski and Feliks Koneczny with his journal *Świat Słowiański* (Slavic World). It was recalled that the *Towarzystwo dla Badań Europy Wschodniej i Bliskiego Wschodu* (Society for the Study of Eastern Europe and the Near East), founded by Wacław Lednicki in 1931 to commemorate the centennial of Pushkin's death, published valuable works, including *Puszkin w polskiej krytyce i przekładach* (Pushkin in Polish Criticism and Translation) by Marian Toporowski (1901–1971). It was regretted that even less had been written about other foreign literatures than about Russian literature in the interwar period. Contempt for Ukrainian literature was discerned in interwar literary criticism, with Taras Shevchenko a special target. Arguably, Władysław Orkan (1875–1893), a progressive, was one of the few who wrote well about Shevchenko. The report evinced disgust that Stefan Kołaczkowski saw Czech literature as the literature of parvenu usurpers. There were also complaints about the lack of syntheses of Slavic literatures. Attempts were supposedly made only in "reactionary", "shallow, and obscure" chapters in works by Aleksander Brückner and Tadeusz Lehr-Spławiński. In the field of neophilology, there were complaints about the domination of extreme reactionaries, e.g. Stanisław Stroński (1882–1955), Władysław Folierski (1890–1961), and Zygmunt Łempicki (1886–1943), who were labeled "propagators of Dithleyism". It was complained that Benedetto Croce's aesthetics had also penetrated into Poland through neophilology (I Congress of Polish Science 1951c: 46). The report also expressed the opinion that neophilology was in a better position than Polish studies at the time because it could import the achievements of Soviet scholars directly. Here, too, it called for a fight against cosmopolitanism (I Congress of Polish Science 1951c: 53). Positive reference was made to Józef Chałasiński's work on the social genealogy of the Polish intelligentsia.

The model of the elite member of the intelligentsia as an ambassador of European culture in Poland was criticized, as was that of the elite member as an ambassador of Polish culture, usually in its aristocratic version, in the West. It was asserted that it was the culture of the Polish peasant or laborer that needed ambassadors abroad. It was noted that Roman Dybowski (1883–1945), otherwise seen as an excellent English scholar, glorified Cambridge and Oxford, and argued that "I would consider the aristocratization of our universities – obviously in a spiritual sense - a welcome development". This was considered unacceptable (First Congress of Polish Science 1951c: 57). The report urged neophilologists to be vigilant and not succumb to bourgeois methodologies or cosmopolitanism, and to avoid regressive cultural elements in their research. The "unmasking" of writers such as Jean-Paul Sartre (1905–1980) and Aldous Huxley (1894–1963) was also exhorted. A plan was announced to publish monographs on the classics, especially Shakespeare, Cervantes, Molière, Voltaire, Diderot, Goethe, Schiller, Heine, Byron, Dickens, Balzac, Stendhal, and Zola. Translations of classical works, dictionaries, descriptive and historical grammars, and textbooks were also called for.

3.8 The Post-1956 Era

3.8.1 The Thaw and Its Consequences

The thaw of 1955–1956 brought considerable changes to the field, and especially to the IBL. They were, however, gradual rather than abrupt. The "vulgarization of Marxism" was rejected and Żółkiewski slowly "corrected" his course. Along with many others, he presented a self-critical assessment and quickly adapted to the new period. He conceded that the fight against formalism, typological analyses, post-positivist biographism, and psychologism had gone too far. Not only Żółkiewski, but an entire generation of scholars abandoned Stalinism, including Maria Janion, Maria Żmigrodzka, Alina Witkowska, Zofia Stefanowska (1926–2007), Stefan Treugutt (1925–1991), Jerzy Ziomek (1924–1990), and Alina Brodzka (1929–2011). Their evolution was not always welcomed, e.g. Maria Rzeuska (in 1957) saw them as cynical opportunists. The thaw had several interesting moments. The conference organized in Zakopane in February 1955 by I Sector of PAN should suffice as an example. The conference was devoted to "the criteria of progressivism in science" (*Konferencja w sprawie kryterium postępowości w nauce*). According to Andrzej Czyżewski (2007), the participants fell on either side of a cleavage that was homologous to the main division within the field of power. The "dogmatic" contingent was represented by sociologist Nina Assorodobraj, who censured anyone who ascribed any progressive traits to bourgeois culture. She denied that capitalism was more progressive than feudalism. Bronisław Baczko (1924–2016), a philosopher and Stalinist at that time, rejected the ideal of national unity under the leadership of the nobility as illusory. Celina Bobińska (1913–1997) and Żanna Kormanowa (1900–1988) maintained that it was wrong to claim that there was no alternative to a noble-bourgeois alliance (Czyżewski 2007). In the "revisionist" corner were Henryk Markiewicz, Jan Kott, and Marian H. Serejski (1897–1975) (Czyżewski 2007: 30). Maria Janion berated Józef Chałasiński from a conservative Marxist perspective for being overly sympathetic toward the nineteenth-century liberal intelligentsia. As she said, "The classics of Marxism were often forced to denounce the cowardly liberalism of the European bourgeoisie, revealing its anti-national and anti-revolutionary character. Chałasiński should have highlighted the class antagonisms between liberals and revolutionary democrats. Instead, he constructed a classless model. He actually represents bourgeois views" (Czyżewski 2007: 31). Chałasiński obviously begged to differ, and quoted Stalin in his response. Wiktor Kula, a historian, appeared to a middle position. For his part Schaff was ambiguous in his warning about the dangers of "vulgarizing" Marxism. Again, the dividing line between "dogmatism" and "revisionism" was purely relational and shifted rapidly. The dogmatists were those who refused to modify their methodology in accordance with the changing political circumstances. The revisionists were those who served the ideal of science, with a special emphasis on the values of pluralism and scientific autonomy. The middle ground was to make Marxism more scientific (Czyżewski 2007). Sławiński writing about that historic moment in the early 1990s saw it as the "evolution of methodological assumptions and concepts - from the initial Marxist orthodoxy, through Marxism's opening up to previously rejected traditions of literary studies and the impulses of contemporary Western humanities, to a pluralistic acceptance of the various mutually existing ideals of literary studies, and to the recognition of the methodological multilingualism of our discipline as a proper state and one that therefore needs to be strongly maintained" (Sławiński 1994: 143). As for the development of neophilology, many of the restrictions of the Stalinist era were relaxed or lifted (Grucza 2001). After 1956, it started to grow rapidly in institutional and human resources terms. This growth was primarily in German and Romance studies, but the 1970s, particularly toward the end of the decade, also saw a significant expansion of English studies.

Curiously, Żółkiewski wrote about the Stalinist period disparagingly toward the end of his life, almost implying that he had little to do with its "perversions". Here is how he recalled the "temptations" of those years in a text dedicated to IBL director, Kazimierz Wyka: "It was all the easier to decide dogmatically what methods, what disciplines, what questions were supposedly not Marxist in relation to literature. We remember that period of dissociation from the theory of relativity, from cybernetics, and from genetics. In literary studies, there was a move away from formalism, structuralism, semiotics, and consequently, from the study of linguistic forms and the general application of linguistics in the study of literature, and from using the findings of sign theory and communications theory. The tendency toward dogmatic axiology and the aesthetics of realism to the exclusion of all else was dangerous. It was easy to impede the development of the science of literature, to cut it off from the development of world science, to overlook the difficult development of Soviet systematic, structural, semiotic research, and the Soviet use of modeling and other cybernetic inspirations in social and humanistic research" (Żółkiewski 1987: 23). Thus, Żółkiewski condemned the abandonment of structuralism and formalism, but he also identified more liberal currents in Soviet research that were worth noting. This appreciation of the importance of Soviet research at the end of the 1980s seems to distinguish him from the mainstream, which by then was almost exclusively Western-oriented.

Structuralism was relatively unhindered after the thaw of 1956. By 1958, most research restrictions had been lifted and Jerzy Kuryłowicz, Tadeusz Milewski, and Zdzisław Stieber were now free to publish and lecture (Karpowicz 2006). The turning point was the Congress of Polish Studies in 1958. Żółkiewski and Janion declared that structuralism might be acceptable as a supplementary approach to the Marxist study of literature. This signaled a shift from a Stalinist and ritual Marxism to a more open and innovative one. Translations of works from the structuralist canon began to be published in 1960. In 1961, de Saussure's *Course in Linguistics* was published, followed by a whole series of works, including Jakobson's. According to many IBL researchers, studies on structuralism became something of an oasis of neutrality. Meanwhile, Marxism was slowly losing ground, although it was still far from being marginalized. Interestingly, according to Małgorzata Gorczyńska, *Česká literatura* published an article by Maria Janion on the "contentious problems of the study of literature" in 1964. It developed the theses she presented at the 1960 Congress of Polish Studies. Janion tried to find a balanced position, one that sought common ground with structuralism without abandoning Marxism, and which emphasized the former's "Slavic origins". This apparently satisfied the needs of the Czech humanities (Gorczyńska 2009: 170–171).

3.8.2 The Golden 60s or Era of Internationalization

I would now like to take a look at the unprecedented intellectual revival of the 1960s. While this was a decade of spectacular successes for the Polish humanities, especially in the fields relevant to this work, e.g. linguistics and literary studies, it was also a decade of unfulfilled hopes. On the one hand, there was rapid and extensive development of the fields under discussion, which led to intellectual debates and international exchanges. On the other hand, Polish students of language and literature did not make a permanent entry into the global game. No specifically Polish school of language and literature studies was established, and the most active scholars did not produce any successors of a similar caliber, especially in the international arena.

Several conditions combined to pave the way for the "golden 60s". There was, as mentioned, an internal Polish political component. First, there was the political liberalization of 1956 (whose seeds were planted as early as 1953). Second, the international geopolitical context was favorable. This obviously conditioned the domestic liberalization, but it also gave a strong impetus to the development of the social sciences and humanities in Poland. This resulted in a significant interest in Poland in both the West and the East. Poland, as the most liberal communist country, gave Western scientists a point of insight into the Soviet bloc and Soviet scientists a window to the West. Much of this was due to Poland's liberal passport policy, i.e. the relative ease (by the standards of communist countries) which with Poles could travel abroad. This made Poland a meeting place for Western and Soviet scientists. The fields of interest here were particular beneficiaries of this mechanism.

The biography of Maria Renata Mayenowa is instructive in this context. Mayenowa made a key contribution to the theoretical development of structuralism and played an extremely important organizational role, especially during the 1960s. She was born into a wealthy Jewish bourgeois family in Białystok in 1908 as Rachela Gurewiczówna. Białystok was part of Russia proper (i.e. it was not in the Kingdom of Poland). It was a Russian-Jewish city in which Poles were a minority. Mayenowa listed Polish as her mother tongue on her university candidate questionnaire in 1927, but Elżbieta Janus (Janus 2021: 324) reports that she only learned it in school, that her mother tongue was Russian, and that she also knew Yiddish, although she supposedly forgot it over time. During WWI, she went to Kharkiv, where she attended secondary school in 1917. In 1922, she returned to Białystok before moving to Vilnius, where she completed her Polish language studies at the Stefan Batory University in 1927–1932. She then worked in a state Jewish high school. According to Zygmunt Saloni, Mayenowa consciously adopted Polish culture as her own (Saloni 2003). She defended her doctoral thesis in 1939 as Rachela Kapłanowa (her first husband's surname) under the supervision of Manfred Kridl. Interestingly, the defense took place in May (prior to the Soviet invasion), and the promotion in December 1939 (after the Soviet invasion and the transfer of Vilnius to Lithuania), just before the Lithuanian authorities closed the university. In the 1940–1941 academic year, Mayenowa worked at Vilnius University as a senior assistant in the Polish Studies Department, where she conducted lectures and seminars on the history of Polish literature. During the war, she went into hiding and was aided by her former teacher Tadeusz Czeżowski (1889–1981), a logical philosopher educated in Lviv who had been a professor in Vilnius since 1923. It was not uncommon for the Polonized Jewish intelligentsia to benefit from a network of community support during the German occupation. Less successful were the friends of Dawid Hopensztand, who, despite being assisted by e.g. Stefan Żółkiewski, were eventually captured and killed. During the war, Mayenowa was baptized a Catholic. Thereafter, she adopted her baptismal name as her forename and combined it with the surname of her second husband, Józef Mayen. He came from a Jewish family in Lviv that was considerably more Polonized than her own and that of her first husband, Jefim Kaplan. Immediately after the war, she worked in a library in Białystok. In 1954, she became a professor at the University of Warsaw, a position she held until the purges of March 1968. She returned to the university in 1981, thanks to the liberalization ushered in by the Solidarity movement. More importantly, she worked at the IBL, which she helped found, from day one. She established and headed the Laboratory of Theoretical Poetics and Literary Language at the IBL, and was the deputy director for scientific affairs in 1957–1969. In 1953, she was appointed editor of the series *Biblioteka Pisarzy Polskich* (The Library of Polish Writers). Apart from theoretical issues, she dealt with sixteenth-century Polish and prepared the complete edition of Jan Kochanowski's works. In 1949, she published *Poetyka opisowa* (Descriptive Poetics), which, despite having an introduction in the spirit of official Marxism, was based on the formalist school and popularized structuralist methodologies that had fallen into desuetude and were largely unknown in Poland. Within this framework, she published a selection of works by Czech literary scholars under the title *Praska szkoła strukturalna 1926–1948* (Prague Structural School 1926–1948) in 1966. In 1970, together with Zygmunt Saloni (b. 1938), she published an anthology of texts by Russian formalists under the title *Rosyjska szkoła stylistyki* (The Russian School of Stylistics). She also published a two-volume edition of Roman Jakobson's works titled

W poszukiwaniu istoty języka (In Search of the Essence of Language), and wrote the introduction under heading *Roman Jakobson—uczony i człowiek* (Roman Jakobson—Scientist and Man, 1989). In 1973, she published her ambitious book, *Poetyka teoretyczna, cz. I Zagadnienia języka* (Theoretical Poetics, Part I: Language Issues). This was a tour de force that summarized the bulk of her theoretical studies. A literary work is depicted as a semiotic construction related to the world of cultural signs and to the social context within which it functions. For this reason, Mayenowa took a special interest in text structure (a new area of research in Poland at the time), and especially in what made a text coherent. The first studies in the categories of text linguistics were published under her editorship. Especially noteworthy are *O spójności tekstu* (On the Coherence of a Text, 1971) and *Tekst i język: problemy semantyczne* (Text and Language: Semantic Issues, 1974).

In 1984, Mayenowa entered into a polemic with Tzvetan Todorov, defending structural poetics on the pages of the Polish translation of his book, to which she wrote an afterword (Mayenowa 1984). She argued that structural poetics was reborn as an independent discipline during WWII, starting with Russian formalism and later Czech structuralism. She claimed that the West saw it in the form that Roman Jakobson gave it, but insisted on recognizing another perspective that she presented as a separate school, viz. Ingarden's phenomenological poetics. This polemic, however, as mentioned, was waged in Todorov's Polish translation. Significantly, Mayenowa did not take her work to the international forum, although she did have a couple of texts published in international journals. She did, however, lecture and give talks abroad, including in Amsterdam, Los Angeles, and Stuttgart (Pszczółowska 1989). She went to the United States on a Ford Foundation fellowship and attended lectures by Roman Jakobson, Noam Chomsky, and Morris Halle. She was thus part of the crucial exchange program with the United States that was revived in 1956 and which contributed to the revival of several disciplines in the 1960s (Kilias 2021). More importantly, she organized a number of key international scientific exchanges and events in the 1960s, especially two critically important and unprecedented conferences on poetics in 1960 and 1965, and a semiotic congress in 1968. Jakobson, owing to his friendship with Mayenowa, was involved in organizing them. It was Jakobson who ensured the participation of many influential, primarily American, scholars. They were also attended by Soviet researchers, many of whom Jakobson had known from his youth, and who would not have had the chance to meet Western scientists elsewhere. Interestingly, Jakobson first met Mayenowa at a conference in Moscow in 1958. That same year, they managed to organize a large seminar in Krynica in southern Poland. The conference in 1960 attracted a large audience from both Western and communist countries. Keynote lectures were given by Roman Jakobson and Roman Ingarden, which was a symbolic configuration, a juxtaposition of two strands of structural analysis. The conferences organized by Mayenowa and Jakobson brought many other celebrities, both current and future, to Poland. Mayenowa established working relationships in the USSR. Many of the young researchers she invited to Poland went on to become important scholars. She also organized further international conferences in which eminent foreign scholars participated. One series was devoted to poetics (1960, 1961, 1964, and 1967) and the other to linguistics and semiotics (1965, 1967, 1968). The International Committee of Semantics was established through these meetings. Mayenowa joined and also became a board member of the Polish Semiotic Society (*Polskie Towarzystwo Semiotyczne*, PTS). She also organized seminars for young Polish scholars. Victor V. Gyrmunskiy, Pierre Guiraud and Roman Jakobson were among those who took part (Pszczółowska 1989). Other famous names who appeared in Warsaw through Mayenowa's efforts (usually with Żółkiewski's support) include Dmitry Likhachev (1906–1999), Roland Barthes (1915–1980), Émile Benveniste, Umberto Eco, Pierre Guiraud (1912–1983), Oswald Ducrot (b. 1930), Yuri Apresyan (b. 1930), and Boris Uspensky (b. 1937). She also collaborated with Yuri Lotman (1922–1993) in the 1960s, after meeting him during one of her many trips to the USSR. Then came March 1968. The grand Semiotic Congress, held in Warsaw in 1968, failed to meet the expectations of Mayenowa and many others. The Congress was held just after the Warsaw Pact invasion of Czechoslovakia (Poland being one of the belligerents), and was for this reason boycotted by some Westerners. This was more than symbolic, as 1968 marked the end of the "golden 60s" in the Polish social sciences and humanities. Moreover, in purely institutional terms, the failure of the Congress dashed any chance of the International Association for Semiotic Studies being founded (and headquartered) in Warsaw, thereby assigning a significant role to Poland. The Association was founded a year later—in Paris, much to Mayenowa's disappointment. The association never held a congress in Poland, although Jerzy Pelc was its second president (1984–1994) after Benveniste. Poland's contribution to the development of semiotics was modest, and Pelc was unable to achieve greater international visibility or to have his work more widely recognized. There are few traces of Poland's contribution to world literature during that period. Lucile Dumont (2018), writing on the history of the development of literary theory and semiotics internationally, only makes oblique references to Poland's contribution during the 1960s. She mentions, e.g. the international semiotic conference held in Kazimierz Dolny in 1966 at which the plan to set up the International Semiotic Association was adopted. Polish organizing committee members included Jerzy Kuryłowicz, Maria Renata Maynowa, and Stefan Żółkiewski (Dumont 2018: 61). Dumont also mentions a sub-group formed to create a semiotic journal. Its members were Algirdas Julien Greimas (1917–1992), Yuri Lotman, Thomas Sebeok (1920–2001), and—from Poland—Wojciech Skalmowski (1933–2008). Significantly, Skalmowski, who worked in the Linguistics Department at the Jagiellonian University under Kuryłowicz, left Poland and settled in Belgium in 1968.

The "golden years" of the 1960s had thus come to an end, at least in terms of the intensive international exchange of Polish linguistics and literary studies. As Dumont (2018) shows, Poland played a noticeable but now forgotten role in developing a global scientific network between 1956 and 1968. Michał Głowiński (b. 1934) recalls that well-known Western scholars suddenly stopped coming to Poland after March 1968 (Głowiński and Nasiłowska 1994: 10). Jakobson visited Poland for the last time in 1973. This symbolically drew the curtain on the unprecedented decade of the "long 1960s" in the history of Polish science. This period was not only unprecedented in the structural sense, i.e. in the sense that Poland exploited its interface location, but also in the sense of Jakobson's role. Historians of Polish linguistics recall that young Polish formalists had already established contacts with him before the war, when he was part of the Prague linguistic circle. A visible sign of these contacts was the first article by Jacobson published in Poland (Vilnius, 1937) in the volume dedicated to memory of Wójciki (mentioned above). Jacobson's intensive visits and contacts with Poland raise the question of whether he left a lasting impression. He certainly had close personal relationships, which included friendship, as in the case of Mayenowa, and love. He met his Polish third wife, Katarzyna Pomorska (1928–1986), at the 1958 conference in Krynica (mentioned above). Ulicka notes that Jakobson's two previous two wives were somewhat homologous to his life phases, as they were respectively Russian and Czech. She went so far as to suggest that his wives symbolized the three schools of structural theory in East Central Europe, viz. Russian, Czech, and Polish (Ulicka 2020a). Even in the 1960s, Jakobson provided a link between these three centers, as he was not only a frequent visitor to Poland, but also visited Moscow, Tartu, and Prague (1968 and 1969) (Gorczyńska 2009).

On the other hand, Ulicka notes that no Polish school of structuralism emerged, despite the fact that Polish scholars maintained contacts with the Russian and Czech schools throughout their existence. Maxim Waldstein (2010) offers an even pessimistic diagnosis regarding the lack of visibility of the region. He suggests that scholars like Jakobson, and others with similar career trajectories, e.g. Mikhail Bakhtin (1895–1975), Lev Vygotsky (1896–1934), Pitirim Sorokin (1889–1968), and Alexander Gerschenkron (1904–1978), are rarely associated with Eastern Europe, despite achieving global recognition. As he put it, "If we look at the ways in which they are 'remembered' (if at all) in contemporary theory textbooks and university courses and the ways that their work figures in contemporary debates (beyond second world studies), we normally see that their work on societies, cultures, literatures, and languages of the second world is least cited and least in demand. In the US, most of them have been received as generic 'European' intellectuals" (Waldstein 2010: 101). The Polish reader will additionally note that the names Waldstein mentions are mostly Russian, indicating Russia's significant advantage over Poland in the international arena.

Although contacts between Polish structuralists and the Prague School (which was almost destroyed) ended in 1968, Polish contacts with Soviet scholars continued. Bogusław Żyłko points to an interesting thread of exchanges between the Tartu School and Poland where, not surprisingly, the IBL turned out to be an important node (Żyłko 2009). For example, on June 9–10, 1973, the IBL hosted a discussion of a Tartu text entitled "Semiotics of Culture: Assumptions and Research Program". Participants included Andrzej Bogusławski, Michał Głowiński, Janusz Lalewicz, Renata Mayenowa, Janusz Sławiński, Aleksandra Okopień-Sławińska (b. 1932), Roman Zimand (1926–1992), and Stefan Żółkiewski. The text was part of a larger document entitled "Theses for Semiotic Studies of Culture (as Applied to Slavonic Texts)", prepared for the Seventh International Congress of Slavists in Warsaw in 1973. It was edited by Viacheslav Ivanov (1866–1949), Yuriy Lotman, Alexander Pyatigorskiy (1929–2009), Vladimir Toporov (1928–2005), and Boris Uspensky. According to Żyłko, the reception of the Tartu-Moscow school was particularly favorable in Poland. The geopolitical location of the country, which favored the penetration of ideas from East and West, was certainly significant. Żyłko saw a historical analogy, viz. the interwar reception of formalism in Poland. He claimed that, as a result, Poland was twenty or thirty years ahead of Western countries in this respect (Żyłko 2009: 72). Initially, the reception of the Tartu-Moscow school was led by philologists, mainly from Poland, e.g. Edward Balcerzan (b. 1937), Stanisław Barańczak (1946–2014), Włodzimierz Bolecki (b. 1952), Stanisław Dąbrowski (1930–2007), Jerzy Faryno (b. 1941), Elżbieta Janus, Henryk Markiewicz, Katarzyna Rosner, Wiktoria Śliwowska (1931–2021), and Rene Śliwowski (1930–2015). These were later joined by specialists in other fields, including historians, film scholars, ethnographers, and culturologists (Żyłko 2009: 73).

By contrast, Antoni Semczuk claims that contacts with the Soviet Union in other areas of linguistics were not so strong (Semczuk 2016). In particular, Warsaw had never had a Russian Studies tradition since the closure of the Imperial University of Warsaw in 1915. As Semczuk recalls, Wacław Lednicki, for example, was *persona non grata* in Warsaw. One exception was Samuel Fiszman (1914–1999), who worked in the East Slavonic Literatures section of the Polish-Soviet Institute and headed the Department of the History of Russian Literature at the University of Warsaw from 1957 to 1968. He then emigrated to the United States and the collaboration between literary scholars diminished considerably. According to Semczuk, Warsaw scholars began to use the collections of Russian-language literature in Prague, as they were not overly welcome in Moscow in the 1970s.

The period of relative liberalism that began in 1956 came to an end in 1968. While many of the most active participants in the March protests were students, many academics remained aloof and even gave passive support to the authorities. The repression suffered by literary scholars can hardly be considered severe; philosophers and sociologists suffered a great deal more (Tomkowski 2006: 172). Moreover, the post-March crackdown did not generally affect linguists. According to Grzegorczykowa, Doroszewski helped them avoid it (Grzegorczykowa 2016: 128). No one was laid off at the IBL, and most of the party members who worked there were not militant.

Żółkiewski and figures such as Adam Schaff were marginalized after 1968. Żółkiewski, who has openly supported the student's protests, lost his additional post at the University of Warsaw and his position as secretary of the First Department (Social Sciences) of PAN in 1968. Curiously, although he had previously only visited the Institute sporadically, he now began to work regularly in its library. Janion, Mayenowa and Żmigrodzka were also removed from their teaching positions. As the IBL had already been considered a haven for revisionists, workers employed by large enterprises, mobilized by the authorities, denounced the IBL during the events of PAN. While this was obviously stage-managed, it showed the importance of the Institute during this time. The justification for Żółkiewski's degradation was his mistakes, particularly from 1964 onward. His criticism of the merger of *Przegląd Kulturalny* and *Nowa Kultura* was mentioned in this context (Błażejowska 2018: 109). The official formulation was that "Żółkiewski blurred the ideological and worldview differences in the contemporary humanities between Marxism and bourgeois currents and succumbed to methodological concepts alien to Marxism in both literary and general cultural issues" (Błażejowska 2018: 110). Żółkiewski was also temporarily banned from being published.

3.8.3 Internal Dynamics of the Development of Polish Structuralism

There is also a domestic aspect of the development of structuralism in Poland, especially at the Institute of Literary Studies in the 1960s, to consider. According to Witkowska, structuralism became the "watchword" of the IBL in the 1960s and the first half of the 1970s (Witkowska 2002). The three major exponents of this trend at the IBL were Aleksandra Okopień-Sławińska (b. 1932), Janusz Sławiński (b. 1934), and Michał Głowiński (b. 1934). They collectively wrote their methodological manifesto in the repeatedly reissued *Zarys teorii literatury* (Outline of the Theory of Literature), first published in 1962. They had all studied under Kazimierz Budzyk. Głowiński had defended his MA thesis under the supervision of Kott in 1955. It was reviewed by Kazimierz Wyka and Czesław Zgorzelski. Budzyk and Mayenowa employed Głowiński at the IBL in 1958 and he spent his entire professional career there. He wrote a number of works with a clear theoretical orientation. He proposed, e.g. the concept of the virtual reader, which garnered a great deal of support. In the 1970s, however, like most of his generation, he abandoned structuralism and more general theoretical works. Głowiński recalled that "structuralism in the conditions of communist Poland offered a chance for independence, and this was what mattered on the ideological level. Being a Marxist in Poland was not just a scientific decision; it was a signal: 'I am loyal', or 'I am on the right side'. I wouldn't have liked to be seen that way at all. Structuralism was seemingly abstract, unrelated to the realities of the Polish People's Republic, and because of that, it offered a chance for freedom. We were treated as harmless: 'oh, they deal with form, they're crazy'" (Głowiński and Nasiłowska 1994: 9). Thus, the structuralist approach, which allowed for the study of literary work in its immanence, and which was conducive to detaching it from its philosophical or historical contexts, enabled literary research to be carried out in opposition to the then dominant, top-down Marxist paradigm. Being abstract and ostensibly unrelated to communist Poland, it was perceived as relatively harmless after 1956. It also allowed Polish scholarship to open up to the world (as discussed above). Głowiński recognized the Mayenowa's crucial role in developing structuralism in Poland. He saw his Warsaw group of structuralists as focused on the issue of literary communication (Głowiński and Wołowiec 2018). This way of treating structuralism was also, as Piotr Gierowski points out, typical of both Poland and Czechoslovakia during the 1960s (Gierowski 2018). Structuralism thus allowed for the creation of a certain neutral space, an enclave or asylum that functioned behind the frontline of the ideological struggle. Gierowski also claims that structuralism was "an antidote to the ubiquitous vulgarized Marxism" (Gierowski 2018: 54). However, he discerns a slow move away from structuralism, increasingly seen as ahistorical, static, and statistical, in the late 1960s. This promotion of structuralism as a political or tactical substitute to Marxism coincided with the disenchantment of French intellectuals with Marxism in the wake of the events of 1968. These had vitiated the status of communism as an ideology in that country (Sapiro and Dumont 2020: 50).

IBL professor Włodzimierz Bolecki (b. 1952) sheds some extreme light on the question of the political context(s) in which structuralism was used. Bolecki, a graduate of the University of Warsaw, who has been with the IBL since 1976, confirms that the language of structuralism provided literary theoreticians in Poland with a kind of political asylum, creating a hermetic (in that it was "purely" scientific) area of intellectual freedom, shielded from the attacks of communist newspeak (Bolecki 1998). According to Bolecki, structuralism was also one of the currents within which it was possible to return to the traditions of the 1930s and practice independent literary studies. Moreover, it was a source of inspiration for free scientific choices and intellectual excitement, but above all, it represented reasoning patterns that were independent of politics and which served to formulate new and strictly substantive tasks in the realm of literary theory. The views of Janusz Sławiński (1934–2014) are also topical here. Sławiński graduated in Polish philology from the University of Warsaw and joined the IBL in 1962. In 1964, he received a doctorate in humanities on the basis of his thesis *Koncepcja języka poetyckiego Awangardy Krakówskiej* (The Concept of the Poetic Language of the Kraków Avant-Garde). He received his habilitation in 1971 and was awarded the title of professor of humanities in 1983. On August 23, 1980, he was one of 64 scholars, writers, and columnists who signed a letter to the communist authorities requesting that they initiate a dialogue with the striking workers. He was married to Aleksandra Okopień-Sławińska (1932), who completed her studies and received her doctorate at the Department of Literary Theory of the University of Warsaw in 1955–1971, and worked at the IBL in 1971–2002. Thus, Sławiński also reflected on the ways in which structuralism and post-structuralism functioned at IBL. To quote his words again: "The game of quasi-independence was also played to slyly make research discourse inaccessible, to make it resistant to controlling inspections. Therefore, when considering the IBL-style literary studies of the time, its vigorous aspirations for Europeanness – its structuralist, anti-structuralist and post-structuralist achievements - one cannot regard all this simply (and exclusively) as a stage in the autonomous evolution of a normal discipline of knowledge. It was undoubtedly an important phase in the development of Polish literary studies, their modernization, but at the same time, it was a form of covert resistance to the ideological and linguistic conquest of intellectual life by the totalitarian machinery. This gave the discipline an extra significance in that it directed and integrated it, as it were, above the multitude of its orientations (from scientism to definitely anti-scientific threads). The unifying factor was a secret mission – providing an alternative language – which made it possible to recall values drowned out, forgotten or condemned in the newspeak universe. What we have here, then, is a somewhat paradoxical situation of a discipline that could thrive only on the condition of being in academic, sanctuary-like isolation" (Sławiński 1994: 151). Sławiński continued this reflection with "It was precisely from its academic character that it derived the possibility of serving non-academic purposes. Structuralism in its own modes of expression was sometimes able to substitute effectively for absent, forbidden or restricted discourses in public life. When it spoke of literature, it spoke of something more, or at least it seemed to" (Sławiński 1994: 152). Sławiński's reflection slowly turns into a criticism of the politicization of science, of being pressed into service by the anti-communist opposition.

This echoes the criticism expressed by the famous Polish historian of ideas, Andrzej Walicki (1930–2020). Walicki left Poland in 1981, i.e. during the initial triumphs of the Solidarity movement, because he found the politicization of intellectual debate unbearable (Krasowski 2002). He argued that the strategy adopted in the mid-1970s by the then opposition was a mirror image of the communist strategy in that it assumed the politicization of culture and the social sciences and appealed to the struggle for the mythical unity of the nation. From Walicki's perspective, then, only the 1960s created an academic space sufficiently autonomous from politics and the informal pressure of the intelligentsia elite. In this setting, the rise of Solidarity was a great triumph for the intelligentsia, which drowned out independent scholars reluctant to commit to unconditionally serving its political goals. Consequently, 1989 should be seen not as marking the beginning of regaining academic autonomy, as is usually assumed, but as the end of its loss in the field of the social sciences and humanities. When viewed this way, it is a moment of triumph of the politicized groupthink of the institutional intelligentsia over the free and critical spirit of independently minded intellectuals, i.e. of the identities of the intelligentsia over those of intellectuals.

Sławiński also noted that competition in producing discourses critical of the authorities began to emerge in the academic field in the 1970s. In his view, this weakened the IBL's commitment to meta-politics. As he put it, "At the end of the 1970s, with the emergence of independent centers of intellectual life and, above all, underground uncensored publishing houses, these hidden missions gradually lost their previous significance. The role of such institutionally located ideological circles at IBL began to diminish. Their forms of expression and influence, limited by the institutional requirements, subjected to various mediations enforced by censorship, ceased to be functional and gained the character of relics - in comparison with the possibilities of the unhindered word, directly naming phenomena and values, joyfully rejecting the conventions of figurative speech and the twisted Aesopian languages. The weakening of the factor that had so significantly bound the work of the IBL's community for so many years, naturally released the centrifugal force that constantly gives birth to pluralism and polycentrism of aspirations. The communicative community, which before was building over them, became unstable - and here particular schools, specialties, groups found themselves, as it were, alone with their inexorable particularisms" (Sławiński 1994: 152). He added that "In this common hustle and bustle one could feel more and more acutely the lack of a supreme order, which would fuel it with meaning, the lack of a framework for unifying experience - a project of discipline, which would attribute the multicentricity, multidirectionality and multipurposefulness of efforts to some environmentally recognized hierarchy of goals and values" (Sławiński 1994: 153). In this way, Sławiński points to major potential sources of scientific weakness at IBL, and in academia generally, at the time. These remarks might also explain why no separate theoretical school, linked by a common framework, has emerged in Poland. In its place, if we adopt Sławiński's perspective, we can see the Polish structuralism of the 1960s as a para-scientific communicative code which served not so much to generate new theoretical insights, but rather to mask meta-political positions and visions, which turned out to be strongly rooted in the context of their time. It is worth emphasizing the lack of a common theoretical position, even among the Polish structuralist of the 1960s, as this stood in stark contrast to the Prague Linguistic School. Another weakness of the Polish school of structuralism was the lack of a clear leader. Żółkiewski did not qualify, mostly for political reasons, or more broadly due to his position in the field of power. Nor did Mayenowa, although she played a major organizational role and wrote important syntheses and studies. However, she was unable to set a course for the entire group.

I now wish to return to Sławiński, who may yet play a leading role in the theory movement. As Artur Hellich has observed, "Probably no other Polish theorist of literature has had as many essays devoted to him as Sławiński" (Hellich 2020: 381). Bolecki portrays Sławiński as having anticipated a tendency which appeared much later in the West. This was fairly common for Poland and elsewhere in Central and Eastern Europe. Bolecki further argues that structuralism, as practiced in the region, especially in Poland, was far more sophisticated than standard Western structuralism. For this reason, the Western critique of structuralism, which was the basis of post-structuralism, could not take hold in Poland. As Bolecki argues, "The image of literary structuralism that emerges from the attacks of Western post-structuralists, mainly deconstructionists, is an unknown animal to literary scholars in Russia, the former Czechoslovakia, Hungary, the former Yugoslavia, and Poland, who refer to the tradition of the formalist school and the Prague circle. I have no hesitation in saying that these two research schools, and here we can also add the inspiration of Mikhail Bakhtin's writings, which had a decisive influence on the development of literary studies in Central Europe, turned out to be precursors to many considerations of Western European or American post-structuralists which emerged decades later" (Bolecki 1998: 17). Bolecki produces a long list of issues that demonstrate the pioneering nature of Polish linguistics, including the role of the reader and the viewer, the limits of the literary text, and the historical limitations of literatures. Among other things, Bolecki argues that "Eastern European post-structuralism developed out of literary constructivism, linguism (distrust of language!), surrealism, symbolism, and a rejection of realism and the tradition of transparent language" (Bolecki 1998: 42). Bolecki seems to be turning Polish "backwardness" on its head. He writes that "In Poland, this post-structuralist methodological perspective - although it is called'structuralist ' by the researchers themselves - has for many years been the basic determinant of the work of Janusz Sławiński, Michał Głowiński, Edward Balcerzan, Zdzisław Łapiński and several generations of other researchers associated with them" (Bolecki 1998: 18). Interestingly, Bolecki even writes about the "… confrontation between two schools (Eastern and Western) of the same current, viz. post-structuralism. The difference between them is that Western post-structuralists reject structuralism en bloc" (Bolecki 1998: 19). According to Bolecki, Eastern and Western European (and American) post-structuralism also differ in their attitude toward Marxism. For Eastern European post-structuralists, Marxism denoted a totalitarian utopia and a totalitarian state, while for Westerners, it was a philosophy of freedom and denitrification of state and social structures. He therefore describes the paradoxical rejection of structuralism in West "where it was seen as hermetic, orthodox, dogmatic, and totalistic, while in Central-Eastern Europe it was a symbol of freedom that was considered to be anti-ideological and anti-totalitarian" (Bolecki 1998: 56). Gierowski in turn quotes Milan Kundera (b. 1929), for whom the similar Prague structuralism was comprehensible and deepened his understanding of literature. By contrast, he considered the French structuralists to be illegible, esoteric, and far removed from the reality of contemporary literature (Gierowski 2018: 57). Żyłko also points to the different strands of the Tartu-Moscow school of semiotics, which was close to structuralism. It tended to distance itself from the dominant Marxism, and was sympathetic to scientism, which was seen as opposed to Marxism. It was thus a current related to the Warsaw-Lviv school and which found favor with conservatives. Żyłko further notes that structuralism and semiotics were strongly associated with the political left in the West during this time (Żyłko 2009: 78).

Sławiński can now be appraised as an emblematic figure for the intelligentsia elite. The way Bolecki writes about him is telling. Bolecki notes that Sławiński "published little and reluctantly, even keeping works in a drawer for ten or more years. But whatever he presented publicly was always highly regarded" (Bolecki 1998: 6). Bolecki goes on to say that "Sławiński, this insatiable and never satisfied seeker of new and original ideas, is at the same time a curator, not to say a gourmet, of continuity in science and a master of the art of adapting new theoretical ideas to earlier legacies" (Bolecki 1998: 21). He later opines that Sławiński was not attached to structuralism per se, but was fascinated by the "structuring of the objects" he described. Bolecki claims that, for Sławiński, structure was always the result of the researcher's creation (Bolecki 1998: 22). It should be noted that this describes a classic member of the Polish intelligentsia rather than a professional scholar focused on acquiring academic recognition. The intelligentsia does not have to prove anything in a purely professional sense; it is mostly organized around the informal hierarchies of its social circles (*środowiska*). By his lifestyle, his independence, his originality, and his courage to depart from canons, political expectations, and intellectual fashions, Sławiński established excellent intelligentsia credentials—at least in Bolecki's estimation. The point of reference for his symbolic games, rather than purely scientific criteria, are wider social concerns and political trends. An elite member of the intelligentsia can throw out theoretical (and other ideas), but it is up to others to evaluate and apply them. A lack of theoretical ideas can be attributed to his perception of theory as an insufficiently challenging subject matter. This was, e.g. Ulicka's strategy (Ulicka and Wierzejska 2021: 8). The greatness of a member of the intelligentsia thus rests on his or her personality, which can be revealed through his or her interpretations of national culture. It falls on the commentariat, especially obituary writers, to associate the great Polish intelligentsia members with various theoretical frameworks, usually of Western origin, and to demonstrate their pioneering status. Thus, Bolecki praises e.g. Ryszard Nycz (b. 1951) for being the only one to notice Sławiński and recognize him as a post-structuralist, i.e. Nycz is extolled as a worthy successor to Sławiński. An indication that Sławiński, as a classic member of the intelligentsia, did not feel the pressure of purely academic competition, especially with the West, was his rather optimistic opinion that the quality of research conducted at the IBL was world-class. He wrote that "The literary studies issues that the IBL dealt with in the 1970s were not all that different from those that were popular at Western universities" (Sławiński 1994: 150). Any such assessment is obviously relative, but Sławiński does not seem overly concerned that the IBL's foreign impact was never particularly impressive and that it never developed an internationally recognized intellectual school beyond its own language. Nor does Bolecki judge Sławiński in terms of his international recognition, which was rather limited. An intelligentsia member is primarily great in his or her own country. He or she can become a global figure by serving that country, but this is normally the fruit of a position in the field of power, not just an academic career.

3.9 Post-1968 Developments in the Field: Toward Anti-communism

3.9.1 The Growing Significance of the IBL in the Field of Power

This part of the book uses the files that the Security Services (*Służba Bezpieczeństwa*, SB) and the PZPR kept on academics in the 1970s and 1980s, and especially the two-volume collection edited by Patryk Pleskot and Tadeusz Rutkowski (Pleskot and Rutkowski 2009, 2012). They give a unique insight into the relations between the communist state and selected fields of the social sciences and humanities. Not surprisingly, the IBL was constantly in the cross-hairs of both the PZPR and the SB. Of interest here are the files on Żółkiewski. A 1984 report describes him as follows: "At the beginning of the 1960s, Prof. Stefan Żółkiewski began to clearly demonstrate views that were unambiguously revisionist and Zionist. At this time, he began to gather IBL employees with similar views. Taking advantage of his position, he facilitated the employment of such people as Roman Zimand, Samuel Sandler, Halina Adalińska, Aleksander Zatorski, and David Shugar, all of whom are of Jewish origin. As IBL director, Prof. Żółkiewski worked closely with the leading Zionist figures of the political opposition at home and abroad (Leszek Kołakowski, Krzysztof Pomian, Michał Głowiński, and Roman Zimand). A good example of mutual assistance in this milieu is the support that Jan Tyranowski gave to Barbara Engelking, the daughter of the famous mathematician Prof. Ryszard Engelking, in her research on anti-Semitism in Poland. In the 1970s, members of Zionist circles in the PAN took advantage of the liberal policy of the state to establish contacts with Zionist and Jewish organizations and in the West. During numerous trips and scholarships, mostly on personal invitations, they were trained and instructed to fight against the system of the Polish People's Republic" (Pleskot and Rutkowski 2009: 478). It should be noted that the research of Barbara Engelking, who works at the Institute of Philosophy and Sociology of the PAN, continues to be controversial and draws a good deal of criticism from the current authorities for her work on Polish antisemitism. But to return to Żółkiewski. According to SB analyses from 1980s, he was already becoming politically alienated from the PZPR in the early 1960s. His programmatic and political ideas betray his cooperation with leading opposition figures including Jan Józef Lipski, Michał Głowiński, Roman Zimand, Leszek Kołakowski, and Krzysztof Pomian (Błażejowska 2018: 111).

A critical attitude toward the dominant forces in the PZPR prevailed in the IBL after 1968. On March 13, 1968, the IBL passed a resolution condemning antisemitism (Błażejowska 2018). Moreover, there were no antisemitic sentiments within the IBL itself, as Michał Głowiński asserted (Błażejowska 2018: 117). However, several IBL scholars left Poland after 1968. One of the most prominent was Jan Kott, who had worked at IBL until 1965. In 1968, he was a visiting professor at Yale University. He was one of several American intellectuals who signed a letter to the Polish Embassy that protested the 1968 anti-Semitic hate campaign in Poland. A New York Times obituary penned by Eric Pace described Kott as a "Polish-American theater critic and expert on Shakespeare whose theories influenced some of the most innovative of modern theater directors" (Pace 2002). Pace went on to say that "Mr. Kott was one of a handful of theater critics who have changed the perception of masterpieces. His main strong point as a critic lay in his skill at showing 'the way in which the history is part of the drama and the drama is part of the history,' as he put it in a 1985 interview" (Pace 2002). One of the Kott's most influential books was "Shakespeare, Our Contemporary" (Doubleday, 1964). Andrzej Lam (b. 1929) reported that Kott also enjoyed wide recognition during his post-war years in Poland. His seminars attracted most of the ambitious students at the University of Warsaw. Interestingly, Lam mentions that Kott wrote his essays on the works of French writers, which were published in 1946, in the "Kuźnica library" (*Biblioteka Kuźnicy*) in his spare time, (Lam 2016: 205). Samuel Sandler (1925–2020), a researcher of Aleksander Świętochowski's work, emigrated in 1969. On the other hand, Janusz Maciejewski (1930–2011) was dismissed from the University of Warsaw in 1968, but was then employed at the IBL (Tomkowski 2006).

The Institute was headed by Kazimierz Wyka from 1953 to 1970. Obviously, Żółkiewski was looming in the background until at least 1968. He wielded enormous influence on the Institute and rescued many "enemies of the people" there (Błażejowska 2018: 596). Żółkiewski invited Konrad Górski (and others), who had been banned from lecturing in 1968. This tradition was continued also by Kazimierz Wyka, who employed several people who were considered "unreliable" by the PZPR and the SB. A characteristic assessment of Wyka's personnel policy at the IBL was provided by Jan Józef Lipski (1926–1991)—a well-known member of the anti-communist opposition. Lipski in a letter to Wyka dated May 8, 1965 wrote that the IBL "is a rare reserve for liberals, revisionists and Catholics in an era of partisan rule. No one looks into our souls here" (Błażejowska 2018: 616). Among these liberals were many former communists who had fallen out of favor, as well as conservatives. Many of these people were outstanding intellectuals, and above all, important actors in the field of the intelligentsia elite. One example was Zdzisław Najder (1930–2021), who was employed at the IBL as an assistant. In 1958, Romand Zimand (1926–1992), who had been expelled from the PZPR and publicly branded a "renegade" by Gomułka, joined the staff in 1958 (Błażejowska 2018: 601). Zimand's case is an interesting one. During the initial period of the thaw (1956–1957), he was a columnist for *Po prostu*. He was a well-known academic and active public intellectual, but evidently not a very systematic scholar, as he had not published any major works. In this, he was similar to many intelligentsia elite members who straddled the border between the fields of academia and politics, and who therefore could not commit themselves systematically and wholeheartedly to the former. Zimand was a militant communist during the Stalinist period. He later leaned toward liberalism, and became an internal, highly reflexive critic of the anti-communist opposition. Some of his former colleagues, who had adopted more conservative positions, believed that Zimand was the only former Stalinist in the IBL who expressed regret and analyzed his involvement critically (Błażejowska 2018: 225). It is also interesting that Zimand was a link between the Polish and Russian intellectual fields. Thus, for example, on April 25, 1981, he delivered a lecture titled "Notes to Zinoviev" (Błażejowska 2018: 226). He also published in the Paris-based Russian émigré magazine *Kontinent* at roughly the same time (Błażejowska 2018: 227).

The IBL's liberal employment policy made it a safe haven for several nationally recognized members of the intelligentsia elite. This contributed to the visibility and political importance of the institute, particularly in the 1960s. It can be argued that some of these individuals were de facto in a field of power that had been expanded—primarily in 1956. As discussed, social gatherings were a crucial form of activity for the intelligentsia elite. The post-1956 intelligentsia elites outside the strict political field were extremely active in this regard. As mentioned in the previous chapter, one of the important figures around whom the liberal part of the field of power met in the 1960s was Jan Józef Lipski. His name days were known as "the President's Ball". They constituted platforms where the elite of the IBL could interact with other factions of the intelligentsia elite. Justyna Błażejowska lists the following circles whose members were regular guests: comrades-in-arms from his "*Baszta*" Home Army regiment; friends from university; former colleagues from the *Polski Instytut Wydawniczy* Publishers (PIW); and IBL colleagues.

Another important group were friends from the Crooked Circle Club, from 1977 from KSS KOR, and from 1980, from "Solidarity". Using a slightly different classification, we can say that Lipski's guests included: (i) well-known writers and poets, e.g. Antoni Słonimski (1895–1976), Aniela Steinbergowa (1896–1988), Zbigniew Herbert (1924–1998), Janusz Szpotański (1929–2001) and Jerzy Ficowski (1924–2006); and (ii) well-known opposition activists, e.g. Adam Michnik, Jan Olszewski (1930–2019), Jacek Kuroń, Zbigniew Bujak (b. 1954), and Maciej Jankowski (b. 1945). "Who wasn't there?", as Janusz Maciejewski put it (Maciejewski 1996: 85).

Another prominent IBL member active in the opposition movement was the poet Jarosław Marek Rymkiewicz (1935–2022). He initiated several protest letters, including one in 1971 that condemned the verdict against the "Ruch" underground anti-communist organization. Signing protest letters to the authorities was a commonplace among opposition intellectuals, although they were often persecuted in retaliation. The party archives contain several summary lists of names, e.g. a document prepared for the Sejm of the People's Republic of Poland by the Department of Science and Education of the Central Committee of PZPR lists the signatories to such letters in 1976–1977 (Błażejowska 2018: 152–153). These include the Letter of 59, the Student Petition, the Letter of 101, and the Letter of 25 (which protested proposals to limit the sovereignty and rights of the citizenry. These were all signed by a number of IBL employees, including Tomasz Burek, Maria Dernałowicz (1928–2009), Marek Gumkowski (b. 1948), Maryla Hopfinger (b. 1942), Jan Józef Lipski, Zdzisław Łapiński (b. 1930), Janusz Pawłowski, Jarosław Marek Rymkiewicz, Zofia Stefanowska, Jacek Trznadel (1930–2022), Helena Zaworska (b. 1930), Krzysztof Zalewski (1948–2008), Antoni Libera (b. 1949), and Janusz Maciejewski (Błażejowska 2018: 153). When Lipski was imprisoned in May 1977, there were letters demanding his release. It is worth mentioning in this context that Lipski was not dismissed from his job while in prison. This must have been due to a party-level decision. Jan Kielanowski (1910–1989), who was also involved in opposition activities, similarly retained his PAN membership (Pleskot and Rutkowski 2009: 22). Błażejowska reports that there was a rupture among IBL staff around 1975 (Błażejowska 2018: 618).

This probably had to do with the disintegration of the alliance between a major part of the IBL around March 1968. Two camps of intelligentsia emerged at the IBL in the second half of the 1970s. One was increasingly involved in opposition activities, while the other resented them for harming the interests of the institute.

3.9.2 Toward Political Activism: Committee for the Defense of Workers, Society for Scientific Courses, and "Solidarity" Trade Union

Politically engaged IBL employees were also among the founders of the Workers' Defence Committee (KOR). Antoni Libera (b. 1949) of the IBL was among those present at the founding meeting of the KOR in Edward Lipiński's apartment on September 12, 1976 (Pleskot and Rutkowski 2009: 184–185). Lipski, although initially opposed to establishing the KOR, joined the initiative to establish it. Lipski compared his behavior to his earlier opposition to the Warsaw Uprising in 1944, which he also joined once it had broken out. Hanna Maciarewicz (b. 1948), the wife of Antoni Maciarewicz (b. 1948), a KOR co-founder and later a high-profile right-wing politician, also worked at the IBL. She was employed in the nineteenth- and twentieth-century History of Polish Literature Laboratory. Another KOR collaborator was soon-to-be politician, Jarosław Kaczyński (b. 1949), then a senior assistant at the Institute for Scientific Policy and Higher Education.

As mentioned, an important event for the development of the opposition in the late 1970s was the establishment of the Flying University and the Society for Scientific Courses (*Towarzystwo Kursów Naukowych*, TKN). The signatories of the founding of the Society for Scientific Courses (in 1977) included 6 IBL employees: Alina Brodzka, Tomasz Burek, Michał Głowiński, Maria Janion, Jan Józef Lipski, and Andrzej Werner (b. 1940). Zdzisław Łapiński joined in November 1980. Interestingly, Maria Janion was expelled from the PZPR for joining the TKN. Żółkiewski was also considered as a potential lecturer. In 1978, the inaugural TKN lecture, entitled "*Język jako rozkład komunikacji—o problemach propagandy*" (Language as a decomposition of communication—on the problems of propaganda), was delivered by Michał Głowiński. He then conducted a course titled "*Nowomowa—język propagandy*" (Newspeak—the language of propaganda). Other TKN courses presented by IBL employees included "*Jakiej historii literatury potrzebujemy*" (What kind of literary history do we need) by Tomasz Burek, "*Bohater narodowy—mit i rzeczywistość*" (The national hero—myth and reality) by Andrzej Kijowski (1928–1985), and "*Dylemat zasady narodowej—zasada narodowa a zasada historyczna*" (The dilemma of the national principle—the national principle versus the historical principle) by Roman Zimand (Błażejowska 2018: 248). On December 10, 1978, Głowiński gave a lecture titled "*Życie społeczne a scentralizowana struktura władzy*" (Social Life and the Centralized Structure of Power) (Błażejowska 2018: 251). Maria Janion, Tomasz Burek, Wiktor Woroszylski, Alina Brodzka, and Michał Głowiński lectured in 1979, and Zdzisław Łapinski volunteered to teach seminars on Czesław Miłosz. Zdzisław Łapiński also offered a course titled "*Literatura polska na emigracji od roku 1939*" (Polish Literature in Exile since 1939). This was also when the booklet *Zeszyt TKN* devoted to the language of propaganda, and edited by Stefan Amsterdamski, Aldona Jawłowska, and Tadeusz Kowalik, was published by the underground publisher NOWA. Interestingly, Błażejowska notes that the TKN had 3 PZPR members employed at the IBL (Maria Janion, Jan Strzelecki, and Krystyna Kersten [1931–2008]) in 1978. All were expelled from the party. Altogether there were 6 IBL scholars active in the TKN. Four people from the Institute of Philosophy and Sociology (IFIS) were in the TKN, as were 3 people from KUL. By comparison, there were only 6 from the University of Warsaw (Błażejowska 2018: 265). This shows an above-average level of involvement of IBL scholars in the TKN. Mieczysław Klimowicz (1919–2008) (IBL Director, 1975–1981) warned that this could have led to the liquidation of the institution (Błażejowska 2018: 263). Rumors that the IBL was to be moved from the prestigious and centrally located Staszic Palace (*Pałac Staszica*) to barracks on the outskirts of the city were also circulating (Błażejowska 2018: 264). Although it was only a rumor, it illustrates the importance of the homology between the rank of an institution and its location in the city space. Krakowskie Przedmieście St., where the magnificent Staszic Palace (which was built for the Warsaw Scientific Society and which still houses the IBL and several other PAN institutes) is located, is undoubtedly a prestigious location from both a historical and a contemporary perspective. Several government institutions and key scientific and cultural institutions, including the University of Warsaw, are on this street. The role of "accidental meetings" on Krakowskie Przedmieście St. is a recurring theme in the biographies of intelligentsia members. For example, Antoni Kukliński (1927–2015), a well-known geographer about whom I have written extensively elsewhere (Zarycki 2020), claimed to have owed his professorship at the University of Warsaw to an accidental meeting with the rector of the university after a mass he attended at the Church of the Nuns of the Visitation (*Kościół Wizytek*) in Krakowskie Przedmieście (Kukliński 2007). The New York Times obituary of renowned Polish sculptor Magdalena Abakanowicz (1930–2017) is another example. William Girmes suggested that Abakanowicz owed her international career to the fact that another famous artist happened to be walking along Krakowskie Przedmieście St. and saw an exhibition of Abakanowicz's work in the small but prestigious Kordegarda Gallery. Even though the exhibition was never officially opened, this was a sufficient impulse for Abakonowicz's career. As Grimes wrote "In 1960, a show of her work at the Kordegarda Gallery in Warsaw was not allowed to open after a cultural official deemed it formalist. As luck would have it, the eminent tapestry artist Maria Laszkiewicz peeked inside and, seeing the fiber-art works that Ms. Abakanowicz had included in the show, added her name to a list of artists to be included in the first *Biennale de la Tapisserie* in Lausanne in 1962" (Grimes 2017). But to return to the IBL, the threat of closing the institute was occasionally raised by PZPR loyalists.

Researchers joined the TKN for a variety of reasons. According to SB documents, Teresa Kostkiewicz (b. 1936) of the IBL joined the TKN because government censors had prevented three books by former students from being printed (Błażejowska 2018: 266). Those who participated in illegal courses suffered various forms of repression, e.g. Maria Janion, Michał Głowiński, and Andrzej Werner had their passports revoked and their promotions blocked. Moreover, as an SB document explains: "An evaluation of the academic work performance of Janion and Werner was demanded, resulting in a greater time commitment at the workplace" (Błażejowska 2018: 276). In 1984, there was an attempt to politically purge the institute under the guise of a "personnel review". Again, academic performance was the criterion used as a pretext to dismiss or demote staff.

As mentioned above, Maria Janion was expelled from the PZPR as punishment for her TKN activities. Stefan Żółkiewski, Hanna Kirchner (b. 1930), Maria Żmigrodzka, and Janina Żurawicka (1914–2006) spoke in her defense at the party meeting. On June 15, 1978, Janion appealed the decision, claiming that she had assiduously employed Marxist methodology in the study of literature since her student days (Błażejowska 2018: 282). She was obviously fence sitting; she did not want to abandon Marxism or leave the PZPR. However, the decision was made for her. She was then pushed to adopt increasingly radical positions. The 1979 SB list of IBL employees involved in opposition activities included Alina Brodzka, Tomasz Burek, Maria Dernałowicz, Michał Głowiński, Marek Gumkowski, Maryla Hopfinger-Amsterdamska (wife of Stefan Amsterdamski), Maria Janion, Jan Józef Lipski, Zdzisław Łapiński, Janusz Maciejewski, Agnieszka Pliszkiewicz, Janusz Pawłowski, Jarosław Marek Rymkiewicz, Jacek Trznadel, Andrzej Werner, Bożena Wojnowska, Helena Zaworska, Roman Zimand, and Roman Loth (1931–2019) (Błażejowska 2018: 289–299).

On November 24, 1980, more than 150 IBL employees joined the Solidarity trade union. The scientific council of the IBL elected on July 15, 1981 had 21 Solidarity members out of a total of 27 (Błażejowska 2018: 309). They were Stefan Żółkiewski (president of the council), Michał Głowiński, Maria Janion, Edmund Jankowski, Renata Maria Mayenowa, Zofia Stefańska, Alina Witkowska, Maria Żmigrodzka, Alina Brodzka, Zbigniew Gliński, Ryszard Górski, Teresa Kostkiewiczowa (b. 1936), Roman Loth, Jerzy Łojek (1932–1986), Teresa Michałowska (b. 1932), Ryszard Przybylski (1928–2016), Lucylla Pszczółkowska, Janusz Sławiński, Jadwiga Czachowska (1922–2013), Krzysztof Dmitruk (1939–2020), Aleksandra Okopień-Sławińska, and Ryszard Nycz. The six who declined to join Solidarity (none of whom had a comparable profile to those who did) were Roman Kaleta (1924–1989),

Elżbieta Aleksandrowska (1928–2018), Maria Karpluk (1925–2016), Franciszek Pepłowski (1921–2009), Jerzy Woronczak (1923–2003), and Krystyna Tokarz. Lipski was elected by the IBL Solidarity members to represent the PAN at the inaugural general assembly of Solidarity of Mazovia. Jerzy Łojek of the IBL became a member of the Program Council of the Solidarity Social Research Centre for the Mazovia Region (*Ośrodek Badań Społecznych przy Regionie Mazowsze NSZZ „Solidarność", OBS*). The council also included such prominent intelligentsia members as Zbigniew Bogusławski (1929–1983), Wiesław Chrzanowski (1923–2012), Ludwik Dorn (1954–2022), Jan Falewicz (1927–1995), Andrzej Grabiński (1922–2006), Jerzy Jasiński (1939–1998), Jarosław Kaczyński, Jacek Kurczewski (b. 1943), Stefan Kurowski (1923–2011), Antoni Maciarewicz (scientific secretary), Piotr Naimski (b. 1951), and Jan Olszewski.

On October 9, 1980, Czesław Miłosz was awarded the Nobel Prize in Literature. The very next day, Roman Zimand proposed to organize a session dedicated to Miłosz. It was held at the PAN on December 15–16, and discussed the significance of the poet's work for the literary tradition of Poland and the world. Participants included Jerzy Kwiatkowski, Andrzej Werner, Tomasz Burek, Stefan Treugutt, Roman Zimand, Jan Błonski, Krzysztof Dybciak (b. 1948), Aleksander Fiut (b. 1945), and Zdzisław Łapiński (Błażejowska 2018: 356). During the Solidarity period, a number of conferences with more or less oppositional overtones were organized at the IBL. These included a session titled "*Literatura źle obecna*" (Literature Badly Present) on October 27–30, 1981, which was devoted to Polish literature in exile (Błażejowska 2018: 358), and the "citizen's" "*Ku niepodległości*" (Toward Independence) conference held on November 14, 1981. On December 11, 1982, Maria Janion delivered a lecture titled "*Słowo i symbol w miesiącach przełomu*" (Word and Symbol in the Months of Breakthrough) at the Congress of Polish Culture. The congress was interrupted by the declaration of martial law. Significantly, she spoke about the role of the Polish Romantic tradition in the "post-August" (1980) culture, i.e. culture in opposition to communism. This example illustrates the importance of inscribing political breakthroughs in distinct historical narratives and cultural traditions to the Polish intelligentsia. The role of intellectuals from the IBL in this respect cannot be overstated—neither during the Stalinist period nor during the slow collapse of communism.

The following IBL employees were interned when martial law was declared (on December 13, 1981): Maryla Hopfinger, Marek Karpiński (b. 1948), Eugeniusz Kloc (1947–1994), Jerzy Łojek, Hanna Maciarewicz, Jan Walc (1948–1993), Andrzej Werner, Roman Zimand, and Tadeusz Witkowski (b. 1946). By this time, being imprisoned was a badge of honor for the intelligentsia elite. Although it could be traumatic, due to the almost complete loss of symbolic capital by the communists, it bestowed high moral ground status. Most oppositionists were aware of this, in particular Adam Michnik, who ostentatiously refused the interior minister's offer to go into exile in the 1980s (Zarycki 2009). Meanwhile, the politicization of academia continued unabated.

On October 23, 1987, a memorandum entitled "The Situation of the Social Sciences and the Humanities at the PAN" was released. It was written by Michał Głowiński and Paweł Czartoryski (1924–1999), together with Juliusz Domański (b. 1927), Aleksander Gieysztor, Stefan Marczewski, and Emmanuel Rostworowski (1923–1989) (Błażejowska 2018: 523). The authors argued that "No ideology or political orientation can be imposed on science" and criticized "extra-ideological criteria", implicitly political, in the management of academia. According to the authors, all power at the PAN was in the hands of officials. On November 23, 1987, PAN president Zdzisław Kaczmarek (1928–2008) decried the memorandum as slanderous (Błażejowska 2018: 524). Kaczmarek asked "Is it questionable that the Scientific Secretary of PAN has the right to privilege, in accordance with the resolutions of the Ninth and Tenth Congresses of the PZPR, a Marxist methodological orientation in the scientific institutions of the Academy?" (Błażejowska 2018: 525). Thus, as can be seen, by the second half of the 1980s Marxism had become a mere slogan; one that was treated in a dismissive manner, and defended with difficulty by the authorities.

3.9.3 Social Science as Seen Through the Prism of Communist Secret Service Documents

Internal PZPR documents obtained by Patryk Pleskot and Tadeusz Rutkowski (2009) offer some insight into the way the institute and its circles functioned. In particular, the deep roots of the intelligentsia elite's multi-positionality are evident. This has always been one of its important aspects. For example, according to SB analyses conducted in the early 1980s, around 10–15% of PAN employees held two jobs, including PAN president Aleksander Gieysztor, who also held the position of director of the Royal Castle in Warsaw (Pleskot and Rutkowski 2009: 337). Another characteristic trait of Polish academia, both in the modern and communist periods, is the opacity and often arbitrariness of the salary system in public institutions. Thus, the SB noted that the remuneration system at the IBL had been completely non-transparent for years (Błażejowska 2018: 322). In 1971, the SB suggested that the Polish science system was uncoordinated, and that there were no centers linking it to the needs of the economy. It was argued that: "At the PAN, scientific policy is pursued by individual researchers following their own goals and views. They are under anyone's supervision and are not responsible to anyone for anything" (Pleskot and Rutkowski 2009: 71). This observation evidences the significant degree of autonomy of science in the PRL. On the other hand, the intelligentsia members themselves usually either underestimated their autonomy, failed to notice it, or took it for granted. In any case, they always thought it insufficient. Intelligentsia members employed at public institutions invariably complained of excessive supervision and bureaucracy. Sławiński, for example, recalled his years working at the IBL as agonizing due to the ever-present party oversight and the multiple inspection of texts. He additionally complained about "the immensely elaborate, meticulous, and burdensome system of planning and reporting, imposed on the realm of research (which is a universe of creative work) – this system adapted from factories, a factory-like system, one might say, which undoubtedly constituted an element of the sociotechnical training to which people engaged in intellectual work were subjected" (Sławiński 1994: 141). This picture is diametrically opposed to the one drawn by the SB, who further complained about researchers' "unwillingness to integrate the research potential and research plans of the Comecon countries" (Pleskot and Rutkowski 2009: 72). That complaint was well-founded; most Polish scientists, contrary to party slogans, were more interested in working with Western countries than communist ones after 1956. It is therefore appropriate to examine this cooperation and its role in the evolution of the field.

The point of interest is foreign travel. Only 39 trips from the PAN to the West were recorded in 1954. This increased to 349 in 1956 and 549 in 1959 (Pleskot and Rutkowski 2009: 17). In a 1960 report, the Science Department of the Central Committee noted a sharp increase in the number of trips abroad in 1956–1957, i.e. during a period of political difficulties and a weakening of the party leadership over science. The report argued that the lack of a unified organizational and political concept, and almost complete decentralization, made scientific exchanges random and spontaneous (Pleskot and Rutkowski 2009: 17–18). Tighter control of foreign trips was therefore ordered. One example of punitive measures was the removal of Józef Chałasiński from his posts as deputy scientific secretary of the Polish Academy of Sciences (PAN), as a delegate of the People's Republic of Poland at UNESCO, and as the chair at the University of Łódź, in 1960 for his critical evaluation of the situation of Polish sociology at the Sociological Congress in Stresa, Italy, in September 1956 (Pleskot and Rutkowski 2009). SB documents also provide detailed information on the foreign cooperation of PAN scientists. One memo, dated May 16, 1972, concerned cooperation with Section VI of the *École pratique des hautes études* (headed at that time by Fernand Braudel (1902–1985), leader of the *Annales* School, which in 1975 gained independence to became the *École des hautes études en sciences sociales,* EHESS) (Pleskot and Rutkowski 2009: 104). The SB clearly believed that contacts with Section VI "had a certain positive impact on the development of social sciences in Poland. The fellowships at EHESS created a considerable chance to improve their scientific skills, especially for younger scholars" (Pleskot and Rutkowski 2009: 106). However, they enumerated several negative aspects of this cooperation. For one thing, Fernand Braudel, Raymond Aron (1905–1983), Lucien Goldman (1913–1970), Georges Friedman, Claude Lévi-Strauss (1908–2009), and André Marchal were able to politically influence Polish scholars. The SB also pointed out interesting asymmetries in the public recognition of some scholars. Some of the French intellectuals engaged in exchanges, SB officers believed, enjoyed significantly more recognition in Poland than in France, an example being Jacques Le Goff (1924–2014). For many Poles, it was argued, their stay in France was a chance to make a name for themselves in their own country. The report also argued that several other French institutions were critical of Section VI for political reasons, and that it monopolized contacts with Poland. It was therefore recommended to increase the range of contacts with other French research centers.

A separate thread in the SB analyses published by Pleskot and Rutkowski is the analysis of the social networks created between Polish and Western researchers. SB officers frequently found "Zionist elements" in them, i.e. they saw those networks as mostly consisting of people of Jewish origin. This could have been a legacy of the post-1968 period in which the authorities, and particularly the SB, saw the liberal opposition factions through the prism of overt anti-Semitic schemes. This also applied to circles critical of the communist authorities at Western universities. According to SB analysts, the "managerial core" of Section VI was "strongly linked personally and ideologically with Zionist organizations". On the Polish side, in the Institute of History of the PAN, the IBL, and Polish Studies departments, "there was a significant percentage of people with that background and record of extreme political activity (Zionist revisionism), e.g. Aleksander Gieysztor, Henryk Samsonowicz, Włodzimierz Wesołowski, Andrzej Werblan, Jadwiga Puzynina, Edward Lipiński, Leszek Nowak, Janusz Maciejewski, Bronisław Geremek, Aldona Jawłowska, and Jan Józef Lipski" (Pleskot and Rutkowski 2009: 350). In a report dated June 22, 1984, it was stated that "the most numerous group of Jewish scholars is present in social and political sciences, i.e. philosophy, sociology, history, economics, and political science" (Pleskot and Rutkowski 2009: 476). Polish circles perceived as "Zionist" or "Jewish" were considered to be exclusively Western-oriented, primarily because of their social contacts. It was further argued that "scholars of Jewish origin are primarily oriented towards Western countries, in which they have family, social and other contacts. As a result, these people have few contacts with their peers in socialist countries.

They mostly maintain contacts with others who are of Jewish origin and who hold Zionist views. Zionist circles operating in the area of Polish science exert a destructive influence on the development of science in the name of their particular group interests" (Pleskot and Rutkowski 2009: 481–480).

Putting the anti-Semitism to one side, what really stands out is the informal nature of the networks that the SB officers observed. This typified elite intelligentsia circles, but was also common in the scientific world generally. To quote once more from the report: "The above-mentioned persons, by capitalizing on their positions in science, their social connections, and above all, their opportunities to obtain scholarships and visit capitalist countries, have subjugated part of the scientific community in order to propagate anti-socialist views and attitudes. It is noteworthy that the Zionist circle inspires various activities and participates in the initial stage of their development and then withdraws in the next stage in order to observe the course of events and the behavior of the authorities. However that may be, the Zionist circle exerts a wide influence on the scientific community, and maintains a vast network of contacts with scholars, social and political activists, journalists, and often the special services of Western European countries" (Pleskot and Rutkowski 2009: 377). The reports also list the key nodes of the network. It is worth noting that these include some top Western universities, as well as Polish émigré institutions, starting with the Paris-based *Kultura* monthly. I now wish to examine an SB report dated October 20, 1983 entitled "Information about the current socio-political situation in the scientific circles of PAN". It discusses "political players" from the West, and names such institutions as *Kultura*, the Council for International Research and Exchange (headed by Samuel Kassow from Trinity College, Hartford, CT), Berkeley, Harvard, Yale, Oxford, the European Foundation for Mutual Intellectual Assistance in Paris, Section VI of the EPHE, the Humboldt Foundation in Germany, and the Jurzykowski Foundation in New York. Their "main direction of diversionary activities", according to the SB, "was directed at the socio-political sciences, i.e. sociology, history, philosophy, economics, psychology, Polish studies (literary research), and was concentrated in the

Institute of Philosophy and Sociology, the Institute of History, the Institute of Literary Research (IBL), and the Institute of Economic Sciences. These institutes employ a significant percentage of scholars with bourgeois, revisionist, Zionist and Masonic backgrounds, often engaging in open struggle against the socialist relations of the Polish People's Republic" (Pleskot and Rutkowski 2009: 415). In an SB on the social science institutes at the PAN, dated December 28, 1983, the authors demand that these institutes "ruthlessly program and implement the assumptions and ideals of the socialist system. Those institutes which do not fulfil these tasks should be subjected to a detailed inspection of their activities to date, and in the event of that they are found to have propagated foreign ideological models, severe staff cuts should be applied, including the temporary liquidation of certain institutes. Scientific councils in institutes should fulfil their functions for the benefit of the management and not the other way round, as is currently the case" (Pleskot and Rutkowski 2009: 446). Apart from the asymmetry between the de facto self-governing scientists and the party-appointed directors who had little influence on them, there was also mention of the asymmetry between Poland and the West, in favor of the latter, when scholars left on scholarships. For example, a report dated September 15, 1985 argued: "Those leaving on scholarships funded by Western countries implement various components of complex research plans, the goals of which they are seldom apprised, at local scientific institutions" (Pleskot and Rutkowski 2009: 497). The SB also conducted class analyses of the scholars they were interested in, e.g. a report dated September 11, 1982 states that "The social and political sciences have traditionally been studied by the bourgeois class (landowners), the interwar clerical strata, which had close links with the authorities, freemasons, etc. After liberation, they found refuge in the humanities at the University of Warsaw, the Jagiellonian University, and other tertiary institutions. After the establishment of the PAN institutes in the social sciences and humanities, and the removal of the abovementioned from teaching positions at universities, they found refuge in the institutes of the PAN. Over the years, they have gained considerable influence on the development of younger scholars and their political attitudes" (Pleskot and Rutkowski 2009: 349).

3.9.4 The Twilight of Marxism Through the Prism of Party Documents

The Central Committee Science Department materials published by Pleskot and Rutkowski reveal the details of functioning of Marxism in the last decade of the communist Poland (Pleskot and Rutkowski 2012). In a report on a meeting of the Party Task Force of the General Assembly of the Polish Academy of Sciences dated June 22, 1981, i.e. during the "Solidarity carnival", or short period of liberalization, Żółkiewski is quoted as saying that this was necessary to defend the "openness of Marxism" and condemning the old dogmatic Marxism (Pleskot and Rutkowski 2012: 700). After the imposition of martial law, the "conservative" factors dominating the party made a last attempt to defend Marxism. In June 1982, a PZPR document titled "Information on the Situation in the Social Sciences" (Pleskot and Rutkowski 2012: 758) stated that "The primary problem of literary studies is the analysis of the ideological content of literature, the development of a Marxist science of literature. This is all the more important as various non-Marxist theoretical positions influence the consciousness and research processes of numerous Polish literary scholars". Interestingly enough, there also appears an open attack on structuralism, probably one of the last in the history of communist Poland. It was argued that "structuralist tendencies have the widest currency, while they are irreconcilable with Marxism because of their overly narrow definition of the essence of literature and their misidentification of its developmental regularities" (Pleskot and Rutkowski 2012: 763). The authors of the document, however, conceded that this was a minority position, despite formally representing the ruling party. They pointed out that "The circle of Marxist scholars of literature today is extremely thin. Studies of contemporary literature, especially over the last quarter century, and of literature and young adult literature, are in an unsatisfactory state. The position of the party is particularly weak among literary scholars. And the political difficulties are exacerbated by the considerable influence of the anti-socialist opposition. The situation in this field requires the introduction of classes on the culture of the Polish language at all university faculties and the resumption of lectures on pronunciation. This will help better prepare university graduates linguistically. The party's initiative in this field will help rebuild and strengthen the party's position in this milieu, and weaken the influence of the political opposition, which has been using the slogan 'newspeak' in the past period" (Pleskot and Rutkowski 2012: 764). Apart from criticism of disciplines such as literature studies, which were seen as disloyal to the party, the document also criticizes political science, which was definitely loyal to the communist authorities but which, in the opinion of the SB analysts, failed to meet basic professional and scientific standards. Political science was described as too descriptive and normative, and detached from reality. Some political scientists were accused of violating the professional ethics of a scientist and "… abandoning the principles of scientific integrity in favor of apologetics for the current decisions of the authorities and the practice of journalism (*publicystyka*) in a normative convention" (Pleskot and Rutkowski 2012: 766). The PZPR executive closely monitored the percentage of party members in each PAN institute. According to a table on party membership statistics (*upartyjnienie*), for 1979, there were only 8.3% at the IBL (down from 12.6% in 1972), 22.8% at the Institute of History, and 46.4% at the Institute of Socialist Countries—the highest figure among PAN institutes (Pleskot and Rutkowski 2012: 657).

3.9.5 Two Visions of the IBL

The legacy of the IBL that emerged in the post-communist period was seen both critically (and from a decidedly anti-communist standpoint) and as unequivocally positive. These two perspectives homologically correspond to the two main poles of the field of power in their 1990s configuration. Staunch anti-communists were opposed to post-communists, which included both former PZPR members and part of liberals who balked at a reckoning with the party leadership. Sławiński gives an extensive and comprehensively outlined account of the IBL's history from an anti-communist perspective (Sławiński 1994: 146). His litany of communist period sins is unsparing. Sławiński claims the IBL was "established as a strictly ideological institution, although the justifications that accompanied it were masked in scientistic rhetoric (of pre-war origin). It was created in order to balance the ideologically unstable university Polish Studies Departments. Its program was simply and unambiguously to gather and train a cadre of Marxist scholars capable of completely reinterpreting the history of Polish literature, using the ideological language that would be produced at the IBL. This language should explain itself and define its identity as a specialized variety of Communist Party discourse. In the field of theoretical-creative work, the primary task consisted in a 'creative' exegesis - and adaptation - of the concepts and formulas that Soviet scholars had previously derived from the writings of the classics of Marxism-Leninism, i.e. the theory that reality is reflected in art, the theory that there are two currents in culture, the theory of base and superstructure, the theory of socialist realism, etc. Its constitutive component remained the uninterrupted struggle against delusions, falsehoods, and in general, the anti-scientific nature of bourgeois knowledge of literature - domestic and foreign. This was the main task in the field of interpretative (historical-literary) work" (Sławiński 1994: 146). Sławiński goes on to say that this task also involved selecting "progressive traditions within the storehouse of the literary past, by which is meant traditions which could - after appropriate adjustments - be treated as the genealogy of the communist present: plebeian, anti-religious, anti-noble, materialistic (from "*Biernat z Lublina*" to nineteenth-century novel realism). The output of theoretical work and the models of interpretative action were intended to be - and in fact were - forcibly disseminated in university literary teaching, in school Polish studies, and in normative literary criticism. The total combination was called the Marxist breakthrough in literary studies" (Sławiński 1994: 147). Sławiński noted that "The IBL's gradual disengagement from the rigid confines of an ideological institution, which began with the 1956 thaw (because this had to occur, as it did everywhere, at about that time), emphatically does not mean that it unequivocally ceased to be one. Official justifications and activity programs continued to emphasize fidelity to the principles of the doctrine that had underpinned the Institute from the very outset. However, such declarations were increasingly rhetorical – as evidenced by the transformations that actually took place.

Thus, they fulfilled tasks that were analogous (although contradictory) to the earlier scientistic rhetoric in that they masked the domination of ideological content" (Sławiński 1994: 147). As regards the role of Marxism, Sławiński writes, "The will to refine Marxism through successive 'openings' sometimes led to errors and deviations that were not tolerated by the Party overseers. The final years of the 1960s finally made this trend irrelevant" (Sławiński 1994: 148). He considered the IBL of the 1960s and 1970s to be a polycentric, multi-directional and multi-specialist institution. Interestingly, he draws attention to the autonomy that the humanities received during this period, arguing that "During the Gomułka and Gierek eras, academic quasi-independence was tolerated in the humanities. This was the niche in which the liberal science of literature developed. How should this academic quasi-independence be understood? It meant, first and foremost, the recognition of the right of researchers to take independent problem-solving initiatives without having to relativize them to the expectations of their political supervisors (although such expectations were obviously expressed and a positive response could result in certain gratifications for the researcher or the institution). Someone professionally involved in the history or theory of literature was not forced to orient his or her actions towards the thematic and ideological preferences of the party authorities. The field of his investigations could be delineated arbitrarily, finding inspiration for concepts or methods of interpretation wherever his or her own tastes and inclinations directed him" (Sławiński 1994: 149).

The opposing view of the IBL, best represented by Michał Głowiński and Alina Witkowska, is much more neutral. Głowiński sees the IBL as a relatively normal scientific institute. The term "normal" is obviously judged by the standards of the "golden" 1960s. In this context, Głowiński speaks of a "miracle" that made the IBL not only a "normal" institute which maintained high academic standards, but a haven for many of the intelligentsia who were opposed to the regime and could not countenance its ideology, the best example being Lipski. Głowiński marvels at the freedom of, and timeframes for, research at the IBL. He sees the Stalinist legacy as a rather superficial and unthreatening ideological layer, which was almost entirely discarded after 1956.

Witkowska also protests the "martyrdom" paradigm of the IBL's history, especially in so far as it depicts scientists as being unrelentingly persecuted by the SB throughout the communist period, and in this regard, names a film on the history of the Institute. She claims that the film showed "… that we were mainly engaged in anti-state and conspiratorial activities, and were persecuted and even imprisoned. There is no place for scientific research in an image of the past that attaches superfluous flowers to the anti-regime militant's coat" (Witkowska 2002: 22). Thus, the image of a "normal" institute, focused on research rather than politics, also emerges from Witkowska's words. The pressures of communism, ideology, and the SB were not overbearing because, as Witkowska observes, the communists "… were primitive and clumsy, and few took up their invitation to follow a common path" (Witkowska 2002: 14). She writes about the "IBL language" with some recognition as "the fashion of the times" (Witkowska 2002: 15). In this context, "IBLspeak" was probably also a product of a particular intellectual milieu and an important tool of distinction (in Bourdieu's terms). If we follow Sławiński and examine this language critically, it becomes apparent that this distinctive IBL language or discourse reflected aspects of the intelligentsia's domination over purely academic circles, i.e. circles in which the realization of scientific ambitions, as measured by internationally recognized progress, was more important than snobbish ostentation and covert power plays.

These two images of the IBL involve a historical dispute and a contemporary conflict over how that dispute should be interpreted. The protagonists and their positions only partially overlap, as the axes of conflict have been redefined and the positions have evolved. The former conflict was between orthodox Marxists and those critical of communism. The contemporary conflict concerns the interpretation of the communist period. Some of those who were heavily involved in constructing the communist system took more critical positions.

3.10 The Late 1970s and the 1980s: Toward the End of Theorization and Communism

3.10.1 The Institutional Infrastructure of the Field

The 1970s and 1980s saw a continuation of most of the major long-term projects in the field, especially dictionaries, encyclopedias, and textbooks. These included the *Encyklopedia wiedzy o języku polskim* (Encyclopedia of Knowledge of the Polish Language) and the *Encyklopedia języka polskiego* (Encyclopedia of the Polish Language), both edited by Stanisław Urbańczyk and published respectively in 1978 and 1994, as well as the *Encyklopedia językoznawstwa ogólnego* (Encyclopedia of General Linguistics, 1992), edited by Kazimierz Polański. Implementing these programs was facilitated by the growth of the institutional infrastructure of language and literature research, especially the establishment of the Polish Language Institute (*Instytut Języka Polskiego*, IJP) at the PAN in 1973 (Czopek-Kopciuch 2010; Polański 2002). Its directors were (in succession) Stanisław Urbańczyk, Władysław Lubaś (1932–2014), Kazimierz Rymut (1935–2006), and Ireneusz Bobrowski (b. 1954). The institute originally consisted of four departments: the Department of Linguistics (located in Warsaw and Kraków, and based on a unit that had been part of the IBL in 1969–1973); the Department of Modern Polish Language and Language Theory; the Department of History of Polish Language; and the Department of Polish Onomastics. Employment at the IJP peaked in 1976 (as it did at the IBL). This was partly due to its having to coordinate the interdepartmental "Polish language - its structure and development tendencies" program (1976–1980). This basic research program consisted of 7 groups of problems: contemporary Polish grammar and the theory of language description; the development of contemporary Polish vocabulary and language culture; past developmental directions of the Polish language; the mutual integration of literary language and folk dialects; the language of urban environments; folk dialects; and Polish toponyms and anthroponyms (Czopek-Kopciuch 2010: 63). Thus, it was clearly a traditional program, although it included some research on contemporary language. IJP director Władysław Lubaś introduced sociolinguistic research, which included expanding and enriching the *Słownik polskich leksemów potocznych* (Dictionary of Polish Colloquial Lexemes). The same Lubaś, however, declared that IJP PAN "remained faithful to age-old traditions" (Czopek-Kopciuch 2010: 66). This made it plain that the key stake for him was for the IJP to become part of the intelligentsia tradition, an important aspect of which was to build a scientific infrastructure for the Polish language.

The evolution of the main IBL research directions in the late 1970s is also worth noting. As Jan Tomkowski (2006) points out, the 1970s saw the full rehabilitation of the Baroque, while the Enlightenment came to be seen as a diverse mosaic of currents and styles that were not necessarily harmonious. At the beginning of the 1970s, the poet Cyprian Kamil Norwid was the most widely read and quoted classicist, while interest in structuralism steadily waned during the decade. In 1972, for example, Renata Mayenowa's work "Theoretical Poetics" was published, and one of Poland's most prominent Polish linguists, Anna Wierzbicka (b. 1938), best known for her work in semantics, pragmatics, and cross-cultural linguistics (especially for the natural semantic metalanguage and the concept of semantic primitives), emigrated to Australia. Wierzbicka published her seminal *Semantic Primitives* in Germany the same year. As mentioned earlier, she can be considered an indirect pupil of Kuryłowicz. One of the more prominent linguists-theoreticians who remained in Poland, Andrzej Bogusławski (b. 1931), published his *Problems of the Thematic-Rhematic Structure of Sentences* in 1977. According to Karpowicz (2006), Renata Grzegorczykowa's *Zarys słowotwórstwa polskiego* (An Outline of Polish Wordsmithing, 1972) and Grzegorczykowa and Puzynina's monograph *Słowotwórstwo współczesnej języka polskiego* (Contemporary Polish Word Formation, 1979) sum up this trend in the study of vocabulary, which was influenced by the Prague School. Several works that drew directly on Prague structuralism were also published in the first half of the 1970s, Jan Tokarski's *Fleksja polska* (Polish inflection, 1973) and Krystyna Pisarkowa's (1932–2010) *Historia składni polskiej* (History of Polish Syntax, 1974) being prime examples.

In the second half of the 1970s, however, historicism triumphed, as did the dream of "action", i.e. political or even armed action. There was something of a sentimental longing to return to tradition, which included militarily resisting occupiers. According to Tomkowski, it is hardly surprising that Romanticism became the most fashionable epoch at that time, and poets such as Juliusz Słowacki, Zygmunt Krasiński (1812–1859), and even the mystical visionary Andrzej Towiański (1799–1878) were fully rehabilitated. This change seems to be closely associated with the re-traditionalization trend, described in the previous section, that began in the 1970s. After the election of Karol Wojtyła as Pope John Paul II in 1978, the role of the Church and religion in cultural life increased even more. Hermeneutics, which Tomkowski claims can be seen as supplanting structuralism, was increasingly popular (Tomkowski 2006). In the final years of communism, when censorship was relaxed, previously taboo topics appeared, such as those related to the culture of regions, and historical regions that were annexed by the USSR after World War II. These former eastern regions of Poland are often called "*Kresy*" (borderlands), especially in more conservative discourses, and their main cultural centers were Lviv and Vilnius. Any mention of the Polishness of these cities had been prohibited by the communist authorities, so when such discussion was finally permitted in the 1980s, a number of publications, including linguistic ones, appeared. This research can be described as a current of the "nostalgic" works that appeared at this time (Karpowicz 2006). In 1997, the IJP set up the "*Pracownia Polszczyzny Kresowej*" (Polishness of the Kresy Region Workshop), and as early as 1989, field recordings of native Polish speakers were made in Ukraine. Several institutions received grants to host researchers from Belarus, Lithuania, and Ukraine (Czopek-Kopciuch 2010: 64).

3.10.2 Configuration of the Field in the Late Communist Period

The geometry of the field in the personal dimension was also interesting. The main cleavage during the 1970s and 1980s indicates an increasing homology with political cleavages. Some scholars were loyal to the authorities, while others were openly involved in opposition activities. Michał Głowiński, obviously in the latter camp, called his opponents "hard-headed". He noticed that there were far more of them at the Faculty of Polish Studies at the University of Warsaw than at the IBL (Głowiński and Nasiłowska 1994). The main figure in the "hardliners" camp is usually considered to be Jan Zygmunt Jakubowski (1909–1975), Dean of the Polish Language Studies Department at the University of Warsaw. Jakubowski graduated in Polish Language Studies at the University of Warsaw in 1934 and later worked as a teacher. During the war, he joined the Polish Workers' Party (PPR). In 1946, he received his doctorate for a thesis written under Juliusz Kleiner. Lam mentions that he considered Józef Ujejski (1883–1937) to be his patron (Lam 2016: 207). Jakubowski was one of the founders of the IBL in 1948, and headed the Laboratory of the School Textbook of the History of Polish Language until 1956. From 1951, he taught at the University of Warsaw. He was a party member and also a member of the communist veterans' organization, ZBOWiD. After 1968, he was one of those who allegedly demanded that the IBL be closed on account of its having succumbed to "Western influences", as evidenced by its highly internationalized library. Głowiński called this circle "Marxist obscurants" (Głowiński and Nasiłowska 1994), and "the worst of the Moczar party pig ignorance" (Głowiński, Wołowiec 2018). Głowiński also dismissed Jakubowski's work as "the humanities of least effort", and reproached him for being able to edit five journals simultaneously. According to Lam, Jakubowski had little interest in methodological disputes and complained of a terminological and conceptual hermeticism. Lam construed this prejudice against the "IBL school" as overly scientistic (Lam 2016: 208). However, Lam believes that it is difficult to speak of a single IBL school, or even several schools, as most scholars had highly individual research styles. Jakubowski, however, was not completely incognizant of theoretical or even structuralist issues. He translated Wolfgang Kayser's (1906–1960) treatise on the modern novel, which became the basis for a structurally oriented narratology (Lam 2016: 210). The confrontation between the two camps additionally involved something of a dispute as to whether literary studies should target dedicated professionals or typical members of the intelligentsia. Lam states that the IBL had little "essayists", i.e. scholars writing accessible texts for a wider public, in the 1960s. Their numbers increased considerably in the 1970s. He further states that writers began to be invited to give lectures at the IBL, which suggests a weakening of the strict boundary between literature and literary studies. Among those invited were Jarosław Iwaszkiewicz (1894–1980), Mieczysław Jastrun, Teodor Parnicki (1908–1988) and Melchior Wańkowicz. In contrast to Głowiński, Lam does not demonize the University of Warsaw Polish Language Department, pointing to its internal diversity. He also notes that even during the Stalinist period, there was no administrative interference in the content of classes at the department and that pre-war professors were not removed from teaching.

The PZPR documents revealed by Pleskot and Rutkowski show that the authorities consciously promoted Jakubowski's career, although not always effectively (Pleskot and Rutkowski 2012: 348). For example, in March 1969, an attempt was made to have Jakubowski elected a member of the PAN, but the election was won by Konrad Górski, who had been nominated by Kazimierz Wyka, Stefan Żółkiewski, and Henryk Markiewicz. This demonstrates that, even when the authorities were most aggressively attacking the autonomy of science, their options were limited. Nor did this situation change significantly, despite statutory changes. In the period of liberalization following the strikes of August 1980, the independently minded and highly respected Aleksander Gieysztor was elected President of PAN, in December 1980. Gieysztor refused to join the "Patriotic Movement for National Rebirth" (*Patriotyczny Front Odrodzenia Narodowego*, PRON), an organization set up to publicly support the authorities. He also confronted the authorities over the many oppositionists that had been arrested after December 1981. In February 1982, SB reports mentioned the general "capitulation and acquiescence [towards the opposition] of some of the communist party circles in PAN". The Institute of History, Institute of Philosophy and Sociology (IFiS), the Institute of Fundamental Technological Research or Institute of Chemistry, and of course, the IBL were mentioned in this context as the most problematic (Pleskot and Rutkowski 2009: 337). The communist government therefore had little control over much of academia at this time. Karol Janicki contrasts the situation in Poland, where linguists loyal to the authorities were in a significant minority in the 1980s, with other communist countries as follows: "In Poland, as opposed to other former East European countries (e.g. East Germany), it was not necessary to write pseudo-philosophical introductions to papers and books in which the 'only true' philosophy and its political exponents would be glorified and extolled. Neither was it necessary or appropriate (again, as in the former East Germany) to make politically tainted conference opening speeches. The few contributions which included salient adherence to the communist ideology were largely ignored by the academic community. The authors were perceived as opportunists and they were severely stigmatized" (Janicki 1995: 169).

As for other members of the loyalist camp, Bazyli Białokozowicz (1932–2010), a scholar specializing in Russian literature, was considered an ally of Jakubowski. Białokozowicz, who worked at the Institute of Slavic Studies of PAN, was vehemently opposed to the IBL. His anti-communist opponents castigated him not only for his party affiliation, but for his contacts with the Soviet embassy and his ties to the USSR (Błażejowska 2018: 135). Apart from specific figures, some works also became points of reference in the configuration of the field. For instance, "The Dictionary of the Polish Language", edited by Mieczysław Szymczak, which appeared in 1978, reveals an interesting tension. In particular, Karpowicz argues that "Polish society perceived this dictionary as a manifestation of the propaganda language and newspeak known from the media of the Gierek era. The value-laden formulations in the individual definitions were glaring" (Karpowicz 2006: 203).

At least until the mid-1980s, the IBL was subjected to political pressure in an attempt to restrain opposition scientists, and possibly encourage them to champion the "Marxist" themes that were still expected by the most conservative members of the communist authorities. For example, Michał Głowiński notes that, in May 1984, "It was alleged that the institute was not conducting Marxist research, that no one was dealing with the problems of Marxist aesthetics, and that the theory of literature practiced at IBL was alien to Marxism" (Głowiński 1996: 207). Głowiński considered this criticism to be unfounded, because according to him, Marxism was still very much alive at the Institute, although usually in more sophisticated rather than ostentatious forms. Another accusation against the IBL was that "Slavonic comparativism is not practiced at the Institute". As Głowiński stated, this meant that Polish historians of literature were insufficiently interested in Russian literature, which he saw as an attack from Białokozowicz (Głowiński 1996: 208).

As mentioned earlier, Żółkiewski was somewhere between the clashing camps in 1980–1981. Although Żółkiewski declared his readiness to defend Marxism, the PZPR found many "revisionist" features in his position (e.g. he denied the need to introduce martial law in 1981) (Pleskot and Rutkowski 2009: 338). Kazimierz Wyka, can also be considered a "man of the middle". He tried to remain loyal to the authorities, but gave the institute a lot of autonomy and hired numerous people who were viewed with suspicion, and even aversion, by the authorities. In the final years of communism, the directors were state officials rather than scientists. Witold Nawrocki was appointed IBL director after the imposition of martial law. He was a professor at the University of Silesia and the Academy of Social Sciences (ANS), a PZPR school where he headed the Department of Fundamental Cultural Problems (Błażejowska 2018: 475). Previously, he had been PZPR secretary of the Katowice region and head of the Culture Department in the PZPR Central Committee (1983–1986). In other words, he was a quintessential member of the nomenklatura. Not surprisingly, Bolecki called him a "political commissar" (Błażejowska 2018: 476). In his defense, his actions were informally presented to IBL staff as caused by "the necessity to defend the IBL as an indispensable substance creating national culture" (Błażejowska 2018: 620). In any case, the possible liquidation of the IBL was an ever-present and constantly discussed threat. However, there were no personnel purges at the IBL. Tomasz Burek was one of the few to lose his position. He was dismissed in 1985, the same year that Bronisław Geremek was dismissed from the Institute of History of the PAN (Pleskot and Rutkowski 2009). The PZPR's attempts to discipline the PZPR were supposedly thwarted by the historian Jarema Maciszewski (1930–2006). Maciszewski, a senior state and party official, was the rector of the Academy of Social Sciences (ANS). In his official capacity as head of the Science Department of PZPR, he presented views and positions in line with the party stance, but personally never interfered in the internal affairs of the institute, even when the management disregarded party directives (Błażejowska 2018: 157). The authorities ceased imposing IBL directors in 1989. Since then, they have been elected by the institute's community.

3.10.3 Re-traditionalization in the Field of Linguistics and Literary Studies

I now wish to discuss the key intellectual breakthrough of the 1970s and the backdrop against which it occurred, viz. a gradual abandonment of the earlier interest in structuralism in favor of humanistic approaches with political undertones. Karpowicz describes this slow but deep intellectual turn, for which the IBL was a particularly clear lens, as a "turn toward the human". Within linguistics, this turn would lead to a paradigm in which "the description of language took into account the broader social context" (Karpowicz 2006: 204). Two important currents can be distinguished in that wide new wave of "human-oriented" studies. The more conservative of them was oriented toward the world of values, i.e. the traditional intellectual narrative. This was strongly bound by the canons of the Warsaw-Lviv school. The other was the newspeak paradigm. Jadwiga Puzynina (b. 1928) was a prominent representative of the former current. Karpowicz argues that Puzytnina best expressed the essence of the wider "turn to man" when she wrote that "man was now at the center - a social being who uses language, not language itself" (Karpowicz 2006: 205). Puzynina graduated from the University of Warsaw in 1951. Norwid was one of her main interests. Among her other functions, she headed the "Dictionary of Norwid's Language Lab" at the University of Warsaw. She became politically active in the 1970s. In 1976, she signed the Memorial of 101 against changes to the Polish constitution and was consequently persecuted. The university senate had voted to offer her a professorship in 1975, but this was blocked for signing the letter. This ban was not rescinded until 1987. The 1970s, she declared, saw a shift in her interests from intra-linguistics, mainly vocabulary issues, to interdisciplinary relations between linguistics, axiology, ethics, and pedagogy (Puzynina 2016). She studied lying and manipulation, with the implicit assumption that they typified the language of the Party. After August 1980, she also wrote about the concept of solidarity

(Puzynina 2016: 337). Puzynina also fought for the autonomy of science, and especially the autonomy of university departments. This commitment ensured her election as Dean of the Faculty of Polish Studies at the University of Warsaw during the "Solidarity revolution" of 1980–1981. However, she soon resigned after suffering a nervous breakdown brought on by political pressure and conflicting expectations. After the fall of communism, she worked on subjects that were considered the domain of anti-communists or even conservatives, publishing works on the Katyń massacre, Jerzy Giedroyc (1906–2000), and Pope John Paul II. She also published *Język wartości* (The Language of Values, 1992) and a number of textbooks. Puzynina is a longstanding member of the Council of the Polish Language (discussed in detail below), where she heads the "Ethics of the Word Unit" (*Zespół Etyki Słowa*). The unit issued a number of statements and publications, one example being a comprehensive volume titled *Etyka słowa: wybór opracowań* (Ethics of Words: A Selection of Studies, Lublin 2017) and edited by Jerzy Bartmiński (1939–2022), Stanisława Niebrzegowska-Bartmińska (b. 1962), Marta Nowosad-Bakalarczyk and Jadwiga Puzynina. This volume of almost 600 pages contains a selection of texts by many Polish linguists. Most are highly normative in nature, and contain numerous postulates toward Polish language users, and even specific "decalogues" of language use, usually with reference to morality and culture. One of the texts, written by Jadwiga Puzynina and Anna Pajdzińska (b. 1953), is particularly significant in this context (Puzynina and Pajdzińska 1995). It refers to the classics theorists of linguistic pragmatics, in particular Herbert Paul Grice (1913–1988), John Langshaw Austin (1911–1960), John R. Searle (b. 1932), and linguistic politeness theorists, e.g. Penelope Brown (b. 1944) and Stephen C. Levinson (b. 1947). Although the introduction briefly mentions the principle of implicature, i.e. the mechanism of generating meanings by breaking the rules of "ideal communication", the remainder of the text considers these rules primarily as moral norms. Moreover, they link them to certain concepts of man, including the religious vision of man as a child of God and language as a gift from God (although they emphasize that non-believers may not recognize this). Central to the authors' considerations is the principle of telling the truth, and they consider its violation to be a moral evil. They link the issue of linguistic politeness to Maria Ossowska's considerations on the norms of good manners, which they regard as ethical postulates. This way of interpreting the classics of pragmatics is continued by e.g. Anna Cegieła (2012), and can be seen as turning the entire field of pragmatics on its head. After all, it is based on breaking the rules of ideal communication to create new meanings. In the interpretation of the Polish authors enumerated above, it is transformed into an ethical treatise on the moral culpability of breaking those rules. Such a normative use of theories taken from the global circulation was quite typical during this period. This is noted in the second part of this book in reference to an article I wrote with Tomasz Warczok on normative interpretations of Bourdieu's critical approaches (Warczok and Zarycki 2014). This interpretation of the classics of pragmatics is continued by e.g. Anna Cegiela (2012).

3.10.4 From Structuralism to Theory of Newspeak

As mentioned, the second current was research on "newspeak", or more broadly, the language of propaganda. The development of these trends can be connected homologically with the changes taking place in the field of power at the same time, i.e. with the consolidation of the anti-communist opposition in the second half of the 1970s. As the IBL gathered many politically active intelligentsia members, it was no coincidence that newspeak was studied there. The most famous and active researcher in this area has always been Michał Głowiński, and it is his paradigm that is presented here. It should be borne in mind, however, other linguists conducted similar research on the discourse of the communist leaders and media, especially Walery Pisarek (1931–2017), the first chairman of the Polish Language Council, and Jerzy Bralczyk (b. 1947), who wrote about the "language of propaganda". Janicki reported that Bralczyk's book *O języku polskiej propagandy politycznej lat siedemdziesiątych* (On the language of 1970s Polish political propaganda) was initially rejected by Ossolineum. Only in 1986 was the Ośrodek Badań Prasoznawczych in Kraków allowed to print 100 copies (Bralczyk 1987). Moreover, neither Bralczyk's 1976 article for *Polityka* weekly nor a 1987 conference paper ever saw the light of day.

Both were on the "language of political propaganda" (Janicki 1995: 170). Głowiński did not even try to publish his studies on newspeak officially. As mentioned, his first essay on the topic was published by the underground publishing press "NOWA" in 1979. Now to summarize the basic tenets of Głowiński's approach on the basis of that and his later publications. One of the key features of newspeak in Głowiński's model is the clear valuation it imposes, i.e. a dichotomous view of the world with a clearly identified enemy. This dichotomous division into "we" and "they" is meant to make dialogue impossible, as "we cannot talk to enemies" (Głowiński 2009: 225). Głowiński further argues that all newspeak words carry an expressive component of evaluation, and that the attitude toward an opponent is of crucial importance. He also contends that newspeak is characterized by unanimity, magic, and a ritualistic and conspiratorial worldview. As for its wider structural features, Głowiński claims that newspeak is monopolistic by definition (Głowiński 2009: 180). Its use "disintegrates communication", "devastates language", and "destroys the world". It "devastates language, it also devastates tradition, for example the tradition of revolutionary language, the tradition of patriotic language" (Głowiński 2009: 30). Głowiński's *Marcowe gadanie* (March Talking) (Głowiński 1991), on the language of the media during the political crisis of March 1968, is one of his major works. He claims that it was written "for the drawer" during the events themselves, as publication was out of the question. However, Głowiński considered that writing it was a "moral duty" and a way of "freeing" himself from the traumatic feelings connected with the political situation. This declaration can be interpreted as a manifestation of the intelligentsia's sense of being responsible for society. In the preface, written after the fall of communism (because it was only then that the work was published), Głowiński states that he was using the language of "revisionism" in 1968, i.e. he wrote the book from political standpoints that were soon to become outdated. Thus, the book presents the language of the 1968 press as a reactivation of Stalinism. An important thread of the book is Głowiński's criticism of the intelligentsia status (or lack thereof) of his opponents. In particular, he refers to the notorious party-affiliated columnist Władysław Machejek (1920–1991) as a "semi-intelligentsia member" (*pół-inteligent*). Głowiński's emotive critique demonstrates the stakes of belonging to the intelligentsia and the contested limits of the intelligentsia as a field. As a member of the liberal, and radically anti-communist intelligentsia faction, his linguistic analyses pointedly exclude representatives of the authorities from the intelligentsia field. This follows from his implicit assumption that the language of newspeak is incompatible with intelligentsia status. Głowiński's notes on the language of the state media in the 1980s discuss the intelligentsia extensively. Głowiński was outraged that party journalists portrayed the opposition intelligentsia as self-interested sellouts. Any criticism of the opposition intelligentsia in the state media is presented by Głowiński as an attack on the intelligentsia per se. He even goes so far as to state that the "anti-intelligentsia campaign is now one of the main components of the Sovietization process" (Głowiński 1996: 232). He was also outraged that intelligentsia status was attributed to the functionaries of the Ministry of Internal Affairs. When talking about the agents who murdered the famous priest Jerzy Popiełuszko (1947–1984), Głowiński writes: "Attributing intelligentsia background to gangsters whose parents were also gangsters strikes me as ghastly" (Głowiński 1996: 278).

Głowiński claims that one of the principal adversaries of newspeak was Pope John Paul II. Głowiński analyses the pontiff's speech during his first visit to Poland in 1979 (Głowiński 2009: 124). According to Głowiński, the Pope "was fighting over language". Although John Paul II uses formulations used by the state media, he "often restores their conventional meanings". Głowiński praises the Pope for "daring to speak about patriotism, national history, moral issues, political and social problems with words that have been corrupted" (Głowiński 2009: 124). Głowiński sees John Paul II's homilies as "one of the great acts of rebuilding language" (Głowiński 2009: 126). Głowiński is similarly complimentary about the language of John Paul II during his subsequent pilgrimages during the communist period, especially in 1987. He is scathing about General Zbigniew Pudysz (1931–2010), the Deputy Minister of Internal Affairs, for creating "a false 'we'" (Głowiński 2009: 158). This use of "we/us", he claims, was mendaciously inclusive, as it was predicated on the police and security services being in communion with society at large, whereas the main axis of social conflict demarcated the two, or more broadly "society" (implicitly represented by the anti-communist intelligentsia) and "the police and the authorities" (which only maintained power through force.)

Głowiński's texts were accompanied by a transcript of the discussion held after his TKN lecture in the underground edition of his first booklet on newspeak (published by *Niezależna Oficyna Wydawnicza* NOWA in 1979). All the comments are anonymized and some are critical of Głowiński. For example, he is accused of being ahistorical, and the term "newspeak" is challenged on the grounds that the same features were present in fascist and Nazi propaganda, monarchical language, and contemporary ecclesiastical rhetoric. It was noted that Leszek Kołakowski had previously labeled Catholic theology and then Party "theology" obscurantism. One of those present dated the origins of newspeak as far back as the early 1840s. Interestingly, the transcript of that discussion was not included in any of the volumes of Głowiński's work published after the fall of communism. Here, I would like to add another critical remark concerning his model. It is unduly normative and the way it ascribes monological nature and anti-intelligentsia attitudes to specific texts seems arbitrary at best, and conditioned by the political position of the author at worst. Meanwhile, a look at just one speech given by Gen. Jaruzelski from the 1980s can complicate this picture. Consider Gen. Jaruzelski's speech at the Second Congress of the Culture of the Polish Language in December 1988. Jaruzelski criticized "bureaucratic and colloquial newspeak", which "grows on a background of real social problems" (Jaruzelski 1989: 341). Jaruzelski therefore used the term "newspeak" with reference to "bureaucrats"—not the opposition. This indicates that he was aware of the term and sought to challenge its use by anti-communist intellectuals. The fact that Jaruzelski used the same concept as Głowiński can be seen as an attempt to enter into a dialogue with his opponents. This possibility, however, has never been considered by any of the theoreticians of newspeak, who invariably see it as monological.

The persistence of newspeak during the post-communist period and its revival in subsequent political contexts is a recurring theme in Głowiński's more recent works. He wrote e.g. that "Amazingly similar, similar semantic mechanisms and closely related rhetorical processes can also be identified today" (Głowiński 2009: 7). The newspeak paradigm can therefore be used as a tool for critiquing political discourses even after the fall of communism. Conservative, Euroskeptic forces, with Jarosław Kaczyński's Law and Justice party (which first came to power in 2005) in the vanguard, were especially harshly criticized by Głowiński. The discourse of this political milieu strikes Głowiński as largely a reincarnation of communist discourse at its most nationalistic, e.g. in March 1968. A number of other linguists, e.g. the eminent discourse analyst Anna Duszak (1950–2015), made similar claims (Duszak 2006). It can be argued that Duszak's integration of the "newspeak paradigm" into the modern critical discourse analysis framework confirms the frequency with which the latter is strongly embedded in more commonsensical (and dominant) modes of political thinking. This is due to the normative nature of part of its theoretical framework. I have argued this point at length in a previous paper (Zarycki 2017). In any case, these linguists have been joined in their anti-newspeak campaign by researchers from other disciplines, especially sociologists. Sociologists have been studying "newspeak" almost as long as linguists. The works of Jakub Karpiński (1940–2003) (Karpiński 1984) and Mirosława Marody (b. 1947) (Marody 1984) are prime examples. Sociologist Ireneusz Krzemiński (b. 1949) argues that the language of the communists, especially ca. March 1968, is very similar to that of the Law and Justice (PiS) party and its supporters (Krzemiński 2006). These authors often additionally discuss a variety of deeper, non-linguistic similarities between these two political groups. Newspeak theorists partly attribute this to a similar "authoritarian mindset" and an innate dislike of the intelligentsia.

To conclude this discussion of Głowiński's "newspeak paradigm", it is worth emphasizing how deeply it is embedded in the identity of the intelligentsia. It serves to illustrate the trend under discussion in the Polish social sciences. This began in the second half of the 1970s and has continued to grow during the fall of communism. It involves a gradual shift from theoretical work, especially in international cooperation, and a growing involvement in social and political life, which is strongly connected with the intelligentsia identity of the researcher. The assignment of intelligentsia status has also become a key stake in social games. Although Głowiński wrote exclusively in Polish, several of his works on newspeak were collected and translated into English and published under the title "Totalitarian Speech" (Głowiński 2014). The volume was published by the German publishing house, Peter Lang. Bibliographic databases show that it is rarely cited. This is the case with most Polish authors published in large numbers by Peter Lang. Most are financed by the Polish government under the National Programme for the Development of the Humanities (*Narodowy Program Rozwoju Humanistyki*, NPRH). The NPRH program was an attempt to alleviate the frustration of the humanities. It was oriented toward Polish, and some less internationalized, parts of the social sciences, and followed the establishment of the National Science Center (*Narodowe Centrum Nauki*, NCN) and the introduction of a system of assessment of scientific units based primarily on publications in peer-reviewed international journals listed in databases such as Web of Science. Most of these circles protested, believing that their main mission is to "serve Polish culture" and, more broadly, the Polish nation, so it was difficult to expect them to publish internationally at the same time. The government was also called upon to help them disseminate their research results internationally. This would supposedly have served to promote Polish culture as well. The NPRH was set up to meet these expectations by making it possible to receive research grants without having any international publications (which is almost impossible with the NCN) and to receive grants for translating and publishing books in English. Peter Lang is most amenable to such translation and publishing projects. The publisher has opened an office in Warsaw and offers publishing services to applicants for NPRH grants. As a result, significant public funds are allocated to publishing translations of works by Polish humanists and social scientists that seldom have much, which usually have little international resonance. However, the system makes it possible to fulfill the formal expectation of "internationalization" of Polish science. There is nothing new in the expectation that translations arranged by Polish institutions will bring Polish scholars into international circulation. As early as 1975, Michał Głowiński declared in an article translated by Ewa Thompson (b. 1937) for "Books Abroad" that "Due to the language barrier, the works of Polish structuralists are little known abroad. Recently a number of German translations have appeared, both in book-length collections of essays and in periodicals such as *Sprache im technischen Zeitalter* and *Poetica*. Several German anthologies of Polish literary scholarship are in preparation. In addition, a large anthology of Polish structuralist studies is being readied for publication by Mouton. This makes one hopeful that the works under discussion will soon become available to those who do not read Polish" (Głowiński 1975). Only some of these projects were realized, and none of them brought Polish structuralism to wider prominence.

This form of institutionally enforced internationalization has largely replaced the much more profound internationalization that many Polish social science disciplines boasted, particularly in the 1960s. As noted in a work I wrote with Tomasz Warczok (Warczok and Zarycki 2018), Once the share of papers by Polish researchers listed in international journals in the Web of Science database is taken into account, it dropped noticeably in many social science disciplines after the fall of communism (the physical sciences mostly retained their much higher positions). This can be interpreted as a result of the weakening of the position of science (described in a previous chapter), and the shift from an ethos of professional excellence (which was especially strong in the 1950s and 1960s) to one of social and political engagement. Moreover, the publishing policy of Polish institutions in the 1960s was more ambitious in certain respects. For example, starting in 1959, *Polskie Wydawnictwo Naukowe* (Polish Scientific Publishers, PWN) attempted to enter the global market in collaboration with e.g. Pergamon Press. The apogee of this partnership was the publication (in English) of the great *Atlas of the World* elaborated by Polish cartographers. This undertaking was not entirely successful, but it triggered a global momentum unknown to Polish scientific publishers today (Rutkowski 2010). Nowadays, Western publishers are entering the Polish academic book market. This is especially visible in the legal sciences, where the Polish branches of Wolters Kluwer and C. H. Beck hold a large market share. In the 1960s, Poland was able to introduce academic journals to the international market. A good example was the *International Review of Sport Sociology*, founded as a 1966 yearbook under the auspices of the International Sociological Association (ISA) and published in Poland until the early 1970s (Kilias 2017:130).

3.11 The Post-communist Era

3.11.1 The Key Cleavage in the Field

After the fall of communism, the "turn to the human", which is primarily defined by a move away from structuralism, and arguably theory altogether, gradually intensified. Paweł Bohuszewicz (2012) argues that an important milestone of the changes that were taking place in Polish literary studies was a survey published in the journal *Teksty Drugie* in 1990 (issue 2). It clearly defined the direction of literary studies, while demonstrating its polyphony. Bohuszewicz concluded that Henryk Markiewicz and Michał Głowiński appeared to be the most conservative in their understanding of literature, although Głowiński revealed a sympathy for gender, feminist and post-colonial approaches. Bohuszewicz considered Danuta Ulicka, who proposed an "archival turn," and Ewa Domańska (b. 1963), who sought alternatives to post-structuralism, to be particularly important. He also singles out Grażyna Borkowska (b. 1956), Dorota Głowacka (b. 1960), Inga Iwasiów (b. 1963), Krzysztof Kłosiński (b. 1954), Michał Paweł Markowski (b. 1962), Zofia Mitosek (b. 1943), and Andrzej Szahaj (b. 1958), who all deemed theory to be of secondary importance in literary studies. Theory thus becomes mostly a diverse repertoire of references rather a goal in itself. Still, the use of theory was not entirely incidental.

More generally, a number of other themes have gained traction since 1989. These are arranged homologously on either side of the field of literary studies, along lines that are visible to the fields of power and politics. On one side of the field, there are religious and "patriotic" themes. The latter include the *Kresy*, and anti-communist writers and intellectuals whose work was banned during the communist period. The other side of the field features topics such as regional and national minorities, and their languages and culture, with Jewish issues and the Holocaust at the forefront, followed by feminist and gender studies, and finally (especially recently), studies re-evaluating the legacy of communism and other leftist traditions. Also common on this side of the field are critical analyses of the legacy of Polish nationalism and colonial ambitions. Post-colonial theory is interpreted in parallel at two opposite poles of not only the literary studies field, but also the Polish social sciences and humanities fields. This duality in the application of post-colonial theory is discussed in an earlier book (Zarycki 2014). The field of literary studies is characterized by a prevalence of left-liberal variants of the use of post-colonial theory, and these are critical of conservative narratives of Polish culture (e.g. Kołodziejczyk 2011). However, conservative applications of post-colonial theory are also present in Polish literary studies. This is best exemplified by the work of Dariusz Skórczewski (b. 1967) of the KUL (Skórczewski 2020), and earlier, Ewa Thompson (b. 1937), who has long worked at Rice University in Houston, and is also present in the Polish literary studies field, and even in the Polish conservative mass media. Thompson is a graduate of the University of Warsaw, and after emigrating to the United States in the mid-1960s, she received her Ph.D. from Vanderbilt University in 1967 for a dissertation on "Russian Formalism and Anglo-American New Criticism". Her intellectual trajectory seems thus to mirror that of Michał Głowiński. Following a period of fascination with structuralism in the 1970s, they moved toward "humanist" topics with a clear anti-communist slant. For example, Thompson published a book on Witold Gombrowicz (1904–1969), an emigree writer banned in communist Poland, in 1979. However, their paths diverged in the 1990s, and they found themselves at opposite poles of the main Polish political cleavage. As for the left pole of the field, Maria Janion's observations on post-colonial theory, especially those in her *Niesamowita Słowiańszczyzna* (The Amazing Slavonic World, 2006), are pertinent here. Although Janion adopted a left-liberal position, she did so in a manner typical of the mainstream Polish intelligentsia, i.e. she took upon herself the role of gatekeeper for Polish cultural traditions and spirituality. For example, in the book under discussion, she defends the pre-Christian Slavic traditions in the territory of present-day Poland. This position is as far removed from national-Catholic orthodoxy as you can get. However, the villain in Janion's account is the Teutonic Order, i.e. the colonizer is German. This is consonant with the common stereotype of Poland as the eternal victim of Germany and adds weight to the imperative for the intelligentsia to search for the nation's primordial and indigenous cultural roots. In any case, scholars in these areas spend much of their time fulfilling an intelligentsia mission, which they see as a duty to society. These duties often have a clear political aspect, which is distributed in those fields of linguistics and literary studies homologous to the main political divisions. Ulicka uses the notion of homology similarly, but her understanding of the concept is more metaphorical than that proposed by Bourdieu (Ulicka 2020c). She divides Polish literary scholars into "homologues" (the majority) and "analogues". The latter ascribe metaphysical and existential tasks to literature, while the former see its primary purpose as reflecting reality. For "analogues", literature expands the sphere of individual freedom and offers a revelation, while for "homologues", it attempts to change the world, even by revolution. Ulicka views "homologues", i.e. those who assume a similarity between an artistic work and socio-political reality, as the dominant camp. Moreover, she distinguishes between "deformed" homology, which was characteristic of Stalinism, and "straight" homology, which was characteristic of the radical anti-communist period from the second half of the 1970s through the 1980s. At that time, literature was considered to be a direct reflection of "objective" social and political issues, and literary studies scholars assumed a direct responsibility for the ethics of their analyses. This was especially the case with e.g. Stanisław Barańczak and Czesław Miłosz. Finally, "inverted" homology is intended to treat literature as a tool for changing, or at least influencing, the world. This is how Ulicka sees literary interpretations that problematize the relations between culture, society, and the individual and power, especially the issues of the body, gender, and sexuality studied by feminist literary scholars such as Grażyna Borkowska, Anna Nasiłowska (b. 1958), Ewa Kraskowska (b. 1954), and Inga Iwasiów (Ulicka 2020c: 563), studies in the paradigm of direct representation of reality, studies stemming from post-colonial programs, including the "post-dependence" paradigm proposed by Hanna Gosk (b. 1952), and studies on protecting minorities (Gosk 2010). According to Ulicka, the cornerstone of these studies is the cultural theory of literature propounded by Michał Paweł Markowski and Ryszard Nycz (Markowski and Nycz 2006). This mainly refers to the principles of reverse homologous referentiality (Ulicka 2020c: 564). Markowski and Nycz's work is also considered crucial for the contemporary Polish field by those working on synthesizing the history of literary studies in Central Europe at the University of Tübingen (Schahadat et al. 2015).

Now to examine linguistic pragmatics. This is an area where Poland has been relatively successful in internationalization. Not surprisingly, most works in that sub-field have been written by researchers at English language institutes or who are English majors. Examples include Anna Duszak (1950–2015) and Piotr Cap (b. 1970), who have both published in such journals as *Poetics* and the *Journal of Pragmatics*. However, weak cooperation between linguists and sociologists can be disappointing. Although there have been attempts to develop a sociology of language (e.g. Bokaszański et al. 1977), they can hardly be said to have resulted in original studies which would gain international recognition. As mentioned above, a number of sociologists interested in language have also been involved in developing the "newspeak" paradigm and a critique of "language manipulation". A well-known example of this is *Technologie Intelektu* (Technologies of the Intellect, 1987) by Mirosława Marody. This book examines the effectiveness of "language manipulation". Typically for that period, it contains more references to psychology (e.g. Eliot Aronson b. 1932) and Basil Bernstein (1924–2000) than to sociology. Sociologists and linguists have attempted to work together on critical discourse analysis over the past decade. Much of this has been sponsored by Anna Duszak and Marek Czyżewski (in particular the Consortium of Discourse Analysis Workshops WAD). For her part, Duszak has been working on text linguistics and has had articles on the subject published in international journals. She has also written monographs in Polish e.g. *Tekst, dyskurs, komunikacja międzykulturowa* (Text, Discourse, Intercultural Communication, 1998). It is quite telling, however, that these works contain few references to the achievements of Polish structuralists, such as Siedlecki, Hopensztadt, or even Mayenowa. Her main points of reference are contemporary Western scholars of discourse analysis, such as Teun van Dijk (b. 1943) and Norman Fairclough (b. 1941). This seems to evidence a typical peripheral discontinuity in the development of theoretical schools and the virtual disappearance of the Warsaw structuralist tradition. This state of affairs also corresponds with Danuta Ulicka's observation that the legacy of Polish structuralism and formalism is not only poorly known around the world, but also in Poland

(Ulicka 2020b). Significantly, an impressive volume edited by Koerner and Szwedek (2001) is also silent about the Warsaw-Vilnius formalist school of the 1930s.

3.11.2 Language Standardization and Polish Language Council

I will now briefly discuss some other tendencies evident in recent decades, in particular in the study of dialects and policies of language standardization. Interestingly, previous debates, some of them going back to the nineteenth century, have been revived since 1989. The key debate concerns the status of the languages spoken in Upper Silesia and Kashubia. Some see them as independent languages, while others consider them to be dialects of Polish. The strong post-war political pressure to maintain cultural and linguistic unity ended with communism. Measures aimed at cultural homogenization and language standardization likewise abated. As a result, there was a sudden outburst of associative activity aimed at renewing regional and linguistic identities in the early 1990s. As mentioned above, this sort of activity was particularly intense in Upper Silesia and Kashubia (*Kaszuby*). These regions were historically situated on the Polish-German cultural border and have retained a considerable degree of cultural autonomy due to their interface periphery status. During the communist period, the distinctiveness of Upper Silesian culture was politically controversial, and its ambiguous relations to Polish culture could not be discussed in the public sphere. After the fall of communism, the Silesian Autonomy Movement (*Ruch Autonomii Śląska*, RAŚ) was set up to restore the region's pre-war autonomy, and another group (whose aims partly overlapped) lobbied to have Silesians recognized as a national minority and Silesian as a (regional) language. Obviously, linguists became involved and adopted the entire spectrum of possible positions, from seeing Silesian as a dialect of Polish to seeing it as a full-fledged language in its own right. The first position is taken by e.g. the renowned Polish philologist Jan Miodek (b. 1946), who argued that the Silesian dialect has too few words and that there is no tradition of using it outside private life.

Those who support this view, including Bogusław Wyderka (b. 1951), also point to the historical closeness of the Silesian dialect to other variants of Polish. As already mentioned, this was the argument that Baudouin de Courtenay raised to support having Kashubian classified as a dialect. Wyderka noted, however, that there is a process of "linguistic liberalization". This especially concerns the increasing media tolerance toward regional languages. However, for him, this did not suffice to justify having Silesian recognized as a separate language (Wyderka 2014). Jolanta Tambor (b. 1958) adopted the opposite viewpoint and argued in favor of assigning Silesian regional language status for political and social reasons (Tambor 2006). She was also involved in standardizing Silesian. Tomasz Kamusella (b. 1967), currently a professor at the University of St Andrews, strongly supports the thesis that Silesians are a distinct ethnic and linguistic group. He contends that, as political division in Poland has intensified over recent years, the Silesian and Kashubian languages have become increasingly politicized issues. Kamusella makes his own political position unambiguous when he states that the ruling conservatives in Poland rely on ideals of ethno-linguistic that are undemocratic and ipso facto unacceptable (Kamusella 2016). The situation in Kashubia is not as tense, as the language issue is not accompanied by a debate on whether Kashubians are a distinct ethnic group. The Act on National and Ethnic Minorities recognized Kashubian, but not Silesian, as a regional language in 2005. However, Kwiryna Handke points out that, despite the huge amount of material compiled on the Kashubian language for over 100 years, "the conclusions drawn from it have not yet convinced everyone, so the discussion on the status of Kashubian continues" (Handke 2002: 52).

Although the dispute over regional languages is becoming increasingly polarized, and positions more diverse, there is another aspect of language homogenization at work in Poland, viz. the normative definition of "correct Polish" laid down by the intelligentsia elite. In this respect, Polish linguistics generally follows a historically shaped paradigm. It implicitly assumes that Polish linguists, as members of the intelligentsia, are guardians of the Polish language and have been charged with safeguarding its norms. This is usually legitimized, not by direct nationalistic arguments, but by classical intelligentsia arguments on the necessity to maintain the norms of high culture, good manners, politeness, and knowledge of language. Thus, there is still a large group of public linguists serving as advisors on and guardians of the correct Polish language. Some of them have their own radio or TV programs in which they discuss examples of "good" and "bad" Polish. By definition "good Polish" language is the language of the intelligentsia elite.

An important tool of the public role of linguists, especially in formulating and monitoring language policy, is the Polish Language Council, which was set up in 1996. Its first chairman was Walery Pisarek (1931–2017) and the second was Andrzej Markowski (b. 1948). The council has been headed by Katarzyna Kłosińska (b. 1965) since 2019. Several council members are "evangelists of correct Polish". The scholarly activity of many is directed almost exclusively at the domestic market. They hardly publish abroad, and their works are usually prescriptive in nature. This accords with the spirit of the council, whose official website defines its aims as follows: "The Council's objectives include fostering and protecting the Polish language as a national cultural treasure and part of our national heritage, disseminating knowledge about the Polish language in the media, at scientific and popular conferences, and through its own publications. The Council expresses its opinion on all important issues concerning contemporary Polish language, contributes to the government's language policy, and analyzes and critically evaluates the Polish language used in public. In particular, the Council issues opinions and formulates expert opinions concerning the use of the Polish language in public and in legal circulation, and especially in advertising, press, radio, television and public institutions".[2] Council members frequently issue appeals to observe the norms and culture of the language, and to teach and observe its patterns. For example, former chairman Andrzej Markowski argued that "Certain communicative situations and certain genres of speech require that only linguistic means approved by the model language norm be used. The model norm of the Polish language should be taught in at school […] and promoted by the media

[2] https://rjp.pan.pl/index.php?option=com_content&view=article&id=214&catid=36, accessed November 21, 2021.

[…] Creating a zone of good Polish, i.e. Polish language based on the model norm, will enable the identity of our language to be maintained in this day and age of globalization, and will ensure its continuity of development, linking it to past epochs" (Markowski 2012: 61). Tadeusz Polański in turn made a characteristic statement in response to comments critical of the council. This position seems to be typical for the intelligentsia's elite stance: "From time to time, criticism is voiced against the Council, accusing it of excessively purist or conservative tendencies towards the Polish language. Such accusations are unfounded and result from a misunderstanding of the Council's assumptions and goals. The Council sees its goal, and indeed its raison d'être, in promoting the maintenance of the language and setting the directions for its development" (Polański 2002: 39). The Council has held a number of conferences and popularization events, and sponsors many publications. The Council's regulation of the Polish language is a continuation of an activity that has not changed for many decades. The Committee for Linguistics (est. 1952) of the PAN previously performed a similar role. The Committee was chaired by Kazimierz Nitsch (1952–1958), Jerzy Kuryłowicz (1958–1971), Jan Safarewicz (1972–1974), Mieczysław Karaś (1975–1977), Mieczysław Szymczak (1978–1983), Stanisław Urbańczyk (1984–1986), Witold Śmiech (1987–1990), Kazimierz Polański (1990–1998), and Stanisław Gajda (1999–). Since its inception, the Committee has codified linguistic norms, in particular those of pronunciation, spelling, grammatical categories, and vocabulary. It also contributed to the Act on the Polish Language. This active involvement of Polish linguists in Polish language regulation policies contrasts with the trends described in France by Vincent Dubois (2014). Dubois argues that, until the 1980s, few French linguists were involved in active language regulation policies and most were opposed to linguistic purism. This was probably due to their dominated position in the social sciences field and homologically in the field of power. Dubois believes that the increased involvement of linguists in formulating state language policies until the 1980s was due to the increasing importance of linguistics and to changes in the French political field. One aspect of changes after 1980s is the departure of the French authorities from their ambitions of strict language regulation and the withdrawal of most defenders of linguistic purism. In

Poland, the uninterrupted engagement of linguists in regulating language and preserving the hierarchy of "language culture" is due to the strong position of the intelligentsia in the field of power.

References

Adamska-Sałaciak, Arleta. 2001. Jan Baudouin de Courtenay's Contribution to General Linguistics. In *Towards a History of Linguistics in Poland: From the Early Beginnings to the End of the Twentieth Century*, ed. E.F.K. Koerner and A. Szwedek, 175–208. Amsterdam, Philadelphia: John Benjamins Publishing Company.

Axer, Jerzy. 2013. Kazimierz Kumaniecki and the Evolution of Classical Studies in the People's Republic of Poland. In *Classics and Communism: Greek and Latin Behind the Iron Curtain*, ed. G. Karsai, G. Klaniczay, D. Movrin, and E. Olechowska, 187–221. Budapest: CEU Press.

Bajerowa, Irena. 1999. Językoznawstwo polonistyczne. In *Historia nauki polskiej wiek XX: Nauki filologiczne*, ed. A. Śródka. 201–228. Warszawa: Wydawnictwo Instytutu Historii Nauki PAN.

Bajerowa, Irena. 2002. Dwie szkoły językoznawstwa polonistycznego XX w. - konflikty& współpraca. *Analecta: Studia i materiały z dziejów nauki* XI: 9–20.

Baluch, Jacek. 1998. Czeskie inspiracje polskiego językoznawstwa. In *Czechy i Polska na szlakach ich kulturalnego rozwoju*, ed. J. Wyrozumski, 207–213. Kraków: Międzynarodowe Centrum Kultury w Krakowie.

Bańczerowski, Jerzy. 2001a. Aspects of Ludwik Zabrocki's Linguistic World. In *Towards a History of Linguistics in Poland: From the Early Beginnings to the End of the Twentieth Century*, ed. E.F.K. Koerner and A. Szwedek, 273–312. Amsterdam, Philadelphia: John Benjamins Publishing Company.

Bańczerowski, Jerzy. 2001b. Mikołaj Rudnicki's General Linguistic Conceptions. In *Towards a History of Linguistics in Poland: From the Early Beginnings to the End of the Twentieth Century*, ed. E.F.K. Koerner and A. Szwedek, 232–254. Amsterdam, Philadelphia: John Benjamins Publishing Company.

Behr, Valentin. 2017. *Science du passé et politique du présent en Pologne: L'histoire du temps présent (1939–1989), de la genèse à l'Institut de la Mémoire Nationale*. Strasbourg: Thèse de doctorat pour obtenir le grade de Docteur de l'Université de Strasbourg, soutenue le 18 octobre 2017.

Behr, Valentin. 2021. *Powojenna historiografia polska jako pole walki: studium z socjologii wiedzy i polityki*. Warszawa: Wydawnictwa Uniwersytetu Warszawskiego.

Berezin, Fedor M. 2001. Mikolaj Kruszewski and 20th-Century Linguistics. In *Towards a History of Linguistics in Poland: From the Early Beginnings to the End of the Twentieth Century*, ed. E.F.K. Koerner and A. Szwedek, 209–231. Amsterdam, Philadelphia: John Benjamins Publishing Company.

Błażejowska, Justyna. 2018. *Opozycja antyreżimowa w Instytucie Badań Literackich Polskiej Akademii Nauk w latach 1956–1989*. Warszawa: Oficyna Wydawnicza Volumen.

Bogusławski, Andrzej. 2014. Leon Zawadowski: Szkic z perspektywy osobistej. *Linguistica Copernicana* 11: 15–28.

Bralczyk, Jerzy. 1987. *O jezyku polskiej propagandy politycznej lat siedemdziesiatych*. Kraków: Ośrodek Badań Prasoznawczych.

Budzyk, Kazimierz. 1936. Gwara a utwór literacki. *Język Polski* XXI: 97—105.

Budzyk, Kazimierz. 1946. O Franciszku Siedleckim wspomnienie. *Pamiętnik Literacki* 36 (3–4): 375–386.

Bohuszewicz, Paweł. 2012. Teoria literatury doby „zwrotów": o jubileuszowej ankiecie „Tekstów Drugich. *Przestrzenie Teorii* 17: 1–20.

Bokszański, Zbigniew, Andrzej Piotrowski, and Marek Ziółkowski. 1977. *Socjologia języka*. Warszawa: Wiedza Powszechna.

Bogusławski, Andrzej. 1977. *Problems of the Thematic-Rhematic Structure of Sentences*. Warszawa: Państwowe Wydawnictwo Naukowe.

Bolecki, Włodzimierz. 1998. Janusz Sławiński: U źródeł polskiego poststrukturalizmu. In *Koncepcja języka poetyckiego awangardy krakowskiej*, ed. J. Sławiński, 5–43. Kraków: Universitas.

Borzym, Stanisław. 2006. Filozofia w Polsce w latach 1945–1990. In *Humanistyka polska w latach 1945–1990*, ed. U. Jakubowska and J. Myśliński, 67–95. Warszawa: Fundacja Akademia Humanistyczna; Instytut Badań Literackich PAN.

Budzyńska, Katarzyna, Michał Araszkiewicz, Barbara Bogołębska, Piotr Cap, Tadeusz Ciecierski, Kamila Dębowska-Kozlowska, Barbara Dunin-Kęplicz, Marcin Dziubiński, Michał Federowicz, Anna Gomolińska, Andrzej Grabowski, Teresa Hołówka, Łukasz Jochemczyk, Magdalena Kacprzak, Paweł Kawalec, Maciej Kielar, Andrzej Kisielewicz, Marcin Koszowy, Robert Kublikowski, Piotr Kulicki, Anna Kuzio, Piotr Lewiński, Jakub Z. Lichański, Jacek Malinowski, Witold Marciszewski, Edward Nieznański, Janina Pietrzak, Jerzy Pogonowski, Tomasz A. Puczyłowski, Jolanta Rytel, Anna Sawicka, Marcin Selinger, Andrzej Skowron, Joanna Skulska, Marek

Smolak, Ma.łgorzata Sokół, Agnieszka Sowińska, Piotr Stalmaszczyk, Tomasz Stawecki, Jarosław Stepaniuk, Alina Strachocka, Wojciech Suchoń, Krzysztof Szymanek, Justyna Tomczyk, Robert Trypuz, Kazimierz Trzęsicki, Mariusz Urbański, Ewa Wasilewska-Kamińska, Krzysztof A. Wieczorek, Maciej Witek, Urszula Wybraniec-Skardowska, Olena Yaskorska, Maria Załęska, Konrad Zdanowski, and Tomasz Żurek. 2014. The Polish School of Argumentation: A Manifesto. *Argumentation* 28: 267–282.

Cegieła, Anna. 2012. O retoryce pogardy i wykluczenia w polskim dyskursie publicznym. *Poradnik Językowy* 9: 14–25.

Chałasiński, Józef. 1946. *Społeczna genealogia inteligencji polskiej*. Warszawa: Czytelnik.

Connelly, John. 2000. *Captive University: The Sovietization of East German, Czech, and Polish Higher Education, 1945–1956*. Chapel Hill and London: The University of North Carolina Press.

Czelakowska, Anna and Mirosław Skarżyński. 2011. *Materiały do dziejów polskiego językoznawstwa: Listy Jana Niecisława Baudouina de Courtenay, Jana Łosia, Kazimierza Nitscha, Jana Michała Rozwadowskiego, Henryka Ułaszyna*. Kraków: Uniwersytet Jagielloński, Wydział Polonistyki, Księgarnia Akademicka.

Czopek-Kopciuch, Barbara. 2010. Instytut Języka Polskiego PAN i jego dyrektorzy. In *Język polski - wczoraj, dziś, jutro…*, ed. B. Czepek-Kopciuch and P. Żmigrodzki, 61–73. Kraków: Wydawnictwo Lexis, Instytut Języka Polskiego PAN.

Czyżewski, Andrzej. 2007. *Proces destalinizacji polskiej nauki historycznej w drugiej połowie lat 50. XX wieku*. Warszawa: Wydawnictwo Naukowe PWN.

Dębska, Hanna. 2019. Idea i struktura – społeczne warunki kształtowania się filozofii i teorii prawa w Polsce. *Państwo i Prawo* 6: 73–90.

Doroszewski, Witold. 2016 (1974). Jan Niecisław Baudouin de Courtenay. In *Portrerty uczonych: Profesorowie Uniwersytetu Warszawskiego 1915–1945 (A-Ł)*, ed. P. Salwa and A.K. Wróblewski, 82–93. Warszawa: Wydawnictwa Uniwersytetu Warszawskiego.

Dubois, Vincent. 2014. Le rôle des linguistes dans les politiques de la langue française (1960–1990): éléments pour une analyse socio-politique. *Dossiers d'HEL, SHESL, 2014, Linguistiques d'intervention. Des usages socio-politiques des savoirs sur le langage et les langues*. http://dossierhel.hypotheses.org/.

Dumont, Lucile. 2018. Théories sans frontières? Les théories littéraires en France et la construction d'un espace transnational, années 1960-années 1970. *Actes de la recherche en sciences sociales* 4: 46–63.

Duszak A. 2006. Why 'New' Newspeak?: Axiological Insights into Language Ideologies and Practices in Poland. In *Language Ideologies, Policies and Practices. Language and Globalization*, ed. C. Mar-Molinero and P. Stevenson, 91–103. London: Palgrave Macmillan.

Duszak, Anna. 1998. *Tekst, dyskurs, komunikacja międzykulturowa*. Warszawa: Wydawnictwo Naukowe PWN.

Gawrak, Wit. 1968. Metamorfozy Adama Schaffa. *Prawo i Życie, 22*.

Gierowski, Piotr. 2013. *Struktury historii: O czeskim projekcie dziejów literatury na tle recepcji praskiego strukturalizmu w Polsce*. Kraków: Wydawnictwo Uniwersytetu Jagiellońskiego.

Gierowski, Piotr. 2018. *Praski strukturalizm: przekroje, miejsca, tożsamości*. Kraków: Wydawnictwo "scriptum".

Głowiński, Michał. 1975. Polish Structuralism. *Books Abroad* 49: 239–243.

Głowiński, Michał. 1991. *Marcowe gadanie: komentarze do słów 1966–1971*. Warszawa: Wydawnictwo PoMOST.

Głowiński, Michał. 1996. *Mowa w stanie oblężenia 1982–1985*. Warszawa: Open.

Głowiński, Michał. 2009. *Nowomowa i ciągi dalsze: szkice dawne i nowe*. Kraków: Universitas.

Głowiński, Michał. 2014. *Totalitarian Speech*. Frankfurt am Main: Peter Lang.

Głowiński, Michał and Anna Nasiłowska. 1994. Strukturalizm, modernizm i cień marksizmu (Z Michałem Głowińskim rozmawia Anna Nasiłowska). *Teksty Drugie* 29 (30): 5–16.

Głowiński, Michał and Grzegorz Wołowiec. 2018. *Czas nieprzewidziany: Długa rozprawa bez pana, wójta i plebana*. Warszawa: Wielka Litera.

Gogolewski, Stanisław. 2001. Dialectology in Poland, 1873–1997. In *Towards a History of Linguistics in Poland: From the Early Beginnings to the End of the Twentieth Century*, ed. E.F.K. Koerner and A. Szwedek, 123–146. Amsterdam, Philadelphia: John Benjamins Publishing Company.

Gorczyńska, Małgorzata. 2009. "Dobra tradycja". Praskiego strukturalizmu ciągi dalsze. *Postscriptum Polonistyczne* 2: 167–184.

Górny, Maciej. 2017. *Kreślarze ojczyzn. Geografowie i granice międzywojennej Europy*. Warszawa: Instytut Historii PAN.

Gorzelak, Grzegorz. 1996. *The Regional Dimension of Transformation in Central Europe*. London and Bristol: J. Kingsley, Regional Studies Association.

Gosk, Hanna. 1985. *W kręgu „Kuźnicy": Dyskusje krytycznoliterackie lat 1945–1948*. Warszawa: Państwowe Wydawnictwo Naukowe.

Gosk, Hanna. 2010. *Opowieści "skolonizowanego/kolonizatora": w kręgu studiów postzależnościowych nad literaturą polską XX i XXI wieku*. Kraków: Universitas.

Grimes, William. 2017. Magdalena Abakanowicz, Sculptor of Brooding Forms, Dies at 86. *New York Times*, April 21.

Grucza, Franciszek. 2001. Origins and Development of Applied Linguistics in Poland. In *Towards a History of Linguistics in Poland: From the Early Beginnings to the End of the Twentieth Century*, ed. E.F.K. Koerner and A. Szwedek, 53–100. Amsterdam, Philadelphia: John Benjamins Publishing Company.

Grzegorczykowa, Renata. 1972. *Zarys słowotwórstwa polskiego*. Warszawa: Państwowe Wydawnictwo Naukowe.

Grzegorczykowa, Renata. 2016. Z dziejów warszawskiej polonistyki w latach 1949–2000. In *Nasz Uniwersytet: Wspomnienia pracowników Uniwersytetu Warszawskiego*, ed. Z. Boglewska-Hulanicka, A. Kędziorek, and B. Jankowiak-Konik, 123–131. Warszawa: Wydawnictwa Uniwersytetu Warszawskiego.

Grzegorczykowa, Renata and Jadwiga Puzynina. 1979. *Słowotwórstwo współczesnej języka polskiego*. Warszawa: Wydawnictwo Naukowe PWN.

Handke, Kwiryna. 2002. Mistrzowie i uczniowie - ważniejsze dokonania polskich językoznawców. In *Uczeni i ich badania*, ed. J. Tazbir, 41–53. Warszawa: Kancelaria Polskiej Akademii Nauk.

Hellich, Artur. 2020. Heterologie. Między literaturoznawstwem a literaturą. In *Wiek teorii: Sto lat nowoczesnego literaturoznawstwa polskiego*, ed. D. Ulicka, 369–424. Warszawa: Instytut Badań Literackich PAN.

Hutnikiewicz, Artur. 1999. Rozwój badań literackich. In *Historia nauki polskiej wiek XX: Nauki filologiczne*, ed. A. Śródka. 7–36. Warszawa: Wydawnictwo Instytutu Historii Nauki PAN.

I Kongres Nauki Polskiej. 1951a. *Referat podsekcji badań literackich*. Warszawa: Sekcja nauk społecznych & humanistycznych.

I Kongres Nauki Polskiej. 1951b. *Referat podsekcji językoznawstwa*. Warszawa: Sekcja nauk społecznych & humanistycznych.

I Kongres Nauki Polskiej. 1951c. *Referat podsekcji neofilologii*. Warszawa: Sekcja nauk społecznych & humanistycznych.

Ingarden, Roman. 1937 *O poznawaniu dzieła literackiego*. Lwów: Zakład Narodowy im. Ossolińskich.

Ingarden, Roman. 1931. *Das Literarische Kunstwerk*. Halle: Max Niemeyer Verlag.

Jakobson, Roman. 1971. *Selected Writings. Vol.11: Word and Language*. The Hague: Mouton.
Jakobson, Roman. 1989. *W poszukiwaniu istoty języka: wybór pism", t. 1–2, wybór, redakcja naukowa i wstęp Maria Renata Mayenowa*. Warszawa: Państwowy Instytut Wydawniczy.
Janicki, Karol. 1995. Sociolinguistics in Poland: History and Prospects (a Socio-Political Analysis). In *When East Met West: Sociolinguistics in the Former Socialist Bloc*, ed. J. Harlig and C. Plé, 165–184. Berlin: Mouton de Gruyter.
Janion, Maria. 2006. *Niesamowita słowiańszczyzna. Fantazmaty literatury*. Kraków: Wydawnictwo Literackie.
Janus, Elżbieta. 2021. A Page from the History of the University of Vilnius: Selected Biographical Threads. *Vilnius University* Open Series, 314–327.
Jaruzelski, Wojciech. 1989. Przemówienie na II Kongresie Kultury Języka Polskiego. Wygłoszone 7 grudnia 1988 r. In: *Przemówienia 1988*, 340–342. Warszawa: Wydawnictwo Książka i Wiedza.
Kamusella, Tomasz. 2016. Silesian: From Gwara to Language After 1989. *Rocznik Polsko-Niemiecki/Deutsch-Polnisches Jahrbuch* 24: 78–119.
Karcz, Andrzej. 2002. *The Polish Formalist School and Russian Formalism*. Rocherster and Kraków: University of Rocherster Press and Jagiellonian University Press.
Karpiński, Jakub. 1984: *Mowa do ludu: szkice o języku polityki*. London: Puls Publication.
Karpiński, Jakub. 2001. *Wykres gorączki. Polska pod rządami komunistycznymi*. Lublin: Wydawnictwo UMCS.
Karpowicz, Tomasz. 2006. Polskie językoznawstwo w latach 1945–1990. In *Humanistyka polska w latach 1945–1990*, ed. U. Jakubowska and J. Myśliński, 183–205. Warszawa: Fundacja Akademia Humanistyczna; Instytut Badań Literackich PAN.
Kilias, Jarosław. 2017. *Goście ze wschodu: socjologia polska lat sześdziesiątych XX wieku a nauka światowa*. Kraków: Nomos.
Kilias, Jarosław. 2021. Not Only Scholarships: The Ford Foundation, Its Material Support, and the Rise of Social Research in Poland. *Serendipities. Journal for the Sociology and History of the Social Sciences* 5: 33-46.
Koerner, Ernst Frideryk Konrad and Aleksander Szwedek (eds.). 2001. *Towards a History of Linguistics in Poland: From the Early Beginnings to the End of the Twentieth Century*. Amsterdam, Philadelphia: John Benjamins Publishing Company.
Kołaczkowski, Stefan. 1937. Bilans 'estetyzmu.' *Marchołt* 3: 149–185.

Kołodziejczyk, Dorota. 2011. The uncanny space of "lesser" Europe: transborder corpses and transnational ghosts in post-1989 Eastern European fiction. *Postcolonial Text* 6 (2).
Kornat, Marek. 2000. Instytut Naukowo-Badawczy Europy Wschodniej w Wilnie (1930–1939) i jego wkład w rozwój polskiej sowietologii. *Kwartalnik Historyczny* 107: 43–89.
Krasowski, Robert. 2002. Czekając na odwilż, czyli umysł nadal zniewolony. *Życie* Junem 29, 2002.
Krąpiec, Mieczysław A. 1987. Czy językoznawstwo warunkuje filozofowanie? *Roczniki Filozoficzne* XXXV: 125–136.
Kridl, Manfred. 1936. *Wstęp do badań nad dziełem literackim*. Wilno: Z zasiłku Funduszu Kultury Narodowej.
Kridl, Manfred. 1945. *Literatura polska (na tle rozwoju kultury)*. New York: Roy Publishers.
Krzemiński, Ireneusz. 2006. Republika zawistników. *Przegląd Polityczny* 78: 21–25.
Kukliński, Antoni. 2007. Curriculum Vitae (unpublished manuscript). Warszawa.
Kupiszewski, Władysław. 2016. Moja praca na Uniwersytecie: Katedra Języka Polskiego. In *Nasz Uniwersytet: Wspomnienia pracowników Uniwersytetu Warszawskiego*, ed. Z. Boglewska-Hulanicka, A. Kędziorek, and B. Jankowiak-Konik, 187–193. Warszawa: Wydawnictwa Uniwersytetu Warszawskiego.
Kuryłowicz, Jerzy. 1977. *Problèmes de linguistique indo-européenne*. Wrocław: Ossolineum.
Labbé, Morgane. 2019. *La nationalité, une histoire de chiffres: Politique et statistiques en Europe Centrale (1848–1919)*. Paris: Les Presses de Sciences Po.
Lam, Andrzej. 2016. Polonistyka warszawska po wojnie. In *Nasz Uniwersytet: Wspomnienia pracowników Uniwersytetu Warszawskiego*, ed. Z. Boglewska-Hulanicka, Anna Kędziorek and Beta Jankowiak-Konik, 199–216. Warszawa: Wydawnictwa Uniwersytetu Warszawskiego.
Leszczyński, Adam. 2020. *Ludowa historia Polski*. Warszawa: Wydawnictwo WAB.
Maciejewski, Janusz. 1996. Heroizm uzwyczajniony. In *Jan Józef – spotkania i spojrzenia. Książka o Janie Józefie Lipskim*, ed. A. Brodzka, 75–77. Warszawa: Państwowy Instytut Wydawniczy.
Markiewicz, Henryk. 1979. Teoria literatury i badań literackich w latach 1918–1939. *Pamiętnik Literacki* LXX: 49–60.

Markowski, Andrzej. 2012. Norma wzorcowa. In *Oblicza polszczyzny*, ed. A. Markowski and R. Pawelec, 57–63. Warszawa: Narodowe Centrum Kultury.

Markowski, Michał Paweł and Ryszard Nycz. 2006. *Kulturowa teoria literatury. Główne pojęcia i problemy*. Kraków: Universitas.

Marody, Mirosława. 1984. *Język propagandy i typy jego odbioru*. Warszawa: Uniwersytet Warszawski.

Marody, Mirosława. 1987. *Technologie intelektu. Językowe determinanty wiedzy potocznej i działania*. Warszawa: Państwowe Wydawnictwo Naukowe.

Maslov, Vitalyi N. 2016. Sovetskaya akademicheskaya nauka i formirovanie isstorichieskoi pamiati pri pereimienovanii naselionnykh punktov Kaliningradskoi oblasti. *Istoricheskaya Expertiza/The Historical Expertis*e 3: 100–107.

Mayenowa, Maria Renata. 1949. *Poetyka opisowa. Opis utworu literackiego: książka pomocnicza dla nauczyciela*. Warszawa: Państwowe Zakłady Wydawnictw Szkolnych.

Mayenowa, Maria Renata. 1955. *Walka o język w życiu i literaturze staropolskiej*. Warszawa: Państwowy Instytut Wydawniczy.

Mayenowa, Maria Renata (ed). 1966. *Praska Szkoła Strukturalna w latach 1926–1948*. Warszawa: Państwowe Wydawnictwo Naukowe.

Mayenowa, Maria Renata and Zygmunt Saloni (eds). 1970. *Rosyjska szkoła stylistyki*. Warszawa: Państwowy Instytut Wydawniczy.

Mayenowa, Maria Renata. 1971. *O spójności tekstu*. Wrocław: Zakład Narodowy im. Ossolińskich.

Mayenowa, Maria Renata. 1973. *Poetyka teoretyczna: zagadnienia języka*. Wrocław; Warszawa, Kraków, Gdańsk: Zakład Narodowy im. Ossolińskich.

Mayenowa, Maria Renata. 1974. *Tekst i język: problemy semantyczne* Wrocław: Zakład Narodowy im. Ossolińskich.

Mayenowa, Maria Renata. 1984. O perspektywie poetyki inaczej. In *Poetyka*, ed. T. Todorov, 107–152. Warszawa: Wiedza Powszechna.

Milewski, Tadeusz. 1972. *Językoznawstwo*. Warszawa: Państwowe Wydawnictwo Naukowe.

Nitsch, Kazimierz. 1948. *Studia z historii polskiego słownictwa*. Kraków: Polska Akademia Umiejętności.

Okoniowa, Joanna. 2011. Szkoła dialektologiczna Kazimierza Nitscha. In *Kazimierz Nitsch 1874–1958: Materiały z posiedzenia naukowego w dniu 20 czerwca 2008 r*, ed. R. Makowska and T. Filip, 19–29. Kraków: Polska Akademia Umiejętności.

Otrębski, Jan. 1951. Na drodze ku realizacji językoznawstwa materialistycznego. *Lingua Posnaniensis* III: 1–12.

Pace, Eric. 2002. Jan Kott, 87, Critic and Shakespeare Scholar. *New York Times* January 4, 2002: 10.

Pelc, Jerzy. 1961. *O pojęciu tematu*. Wrocław: Zakład Narodowy im. Ossolińskich.

Pelc, Jerzy. 1979. *Semiotics in Poland 1894–1969*. Warsaw, Dordrecht: Polish Scientific Publishers; D.Reidel Publishing Company.

Pleskot, Patryk and Tadeusz Paweł Rutkowski. 2009. *Spętana Akademia: Polska Akademia Nauk w dokumentach władz PRL. Materiały Służby Bezpieczeństwa (1967–1987)*. Warszawa: Instytut Pamięci Narodowej.

Pleskot, Patryk and Tadeusz Paweł Rutkowski. 2012. *Spętana Akademia: Polska Akademia Nauk w dokumentach władz PRL. Materiały partyjne (1950–1986)*. Warszawa: Instytut Pamięci Narodowej.

Pobłocki, Kacper. 2021. *Chamstwo*. Wołowiec: Wydawnictwo Czarne.

Polański, Tadeusz. 2002. Badania nad językiem polskim. In *Uczeni i ich badania*, ed. J. Tazbir, 27–40. Warszawa: Kancelaria Polskiej Akademii Nauk.

Pisarkowa, Krystyna. 1974. *Historia składni polskiej*. Wrocław: Zakład Narodowy im. Ossolińskich.

Pszczółowska, Lucylla. 1989. Maria Renata Mayenowa. *Biuletyn Polonistyczny* 32 (113): 168–174.

Puzynina, Jadwiga. 1992. *Język wartości*. Warszawa: Państwowe Wydawnictwo Naukowe PWN.

Puzynina, Jadwiga and Anna Pajdzińska. 1995. Etyka słowa. In *O zagrożeniach i bogactwie polszczyzny: Formy kultury słowa*, ed. J. Miodek, 26–35. Wrocław: Towarzystwo Przyjaciół Polonistyki Wrocławskiej.

Puzynina, Jadwiga. 2016. Wspomnienia. In *Nasz Uniwersytet: Wspomnienia pracowników Uniwersytetu Warszawskiego*, ed. Z. Boglewska-Hulanicka, A. Kędziorek, and B. Jankowiak-Konik, 333–346. Warszawa: Wydawnictwa Uniwersytetu Warszawskiego.

Rauszer, Michał. 2020. *Bękarty pańszczyzny: Historia buntów chłopskich*. Warszawa: Wydawnictwo RM.

Robins, Robert Henry. 1997. *A Short History of Linguistics*. London: Longman.

Romer, Eugeniusz. 2010. *Pamiętnik Paryski 1918–1919: Tom I*. Wrocław: Ossolineum.

Róziewicz, Jerzy. 1984. *Polsko-rosyjskie powiązania naukowe (1725–1918)*. Wrocław: Zakład Narodowy im. Ossolińskich.

Rutkowski, Tadeusz Paweł. 2010. *Adam Bromberg i "encyklopedyści". Kartka z dziejów inteligencji w PRL*. Warszawa: Wydawnictwa Uniwersytetu Warszawskiego.

Saloni, Zygmunt. 2003. Wspomnienie o Marii Renacie Mayenowej. In *Okna pamięci: Maria Renata Mayenowa (1910–1988)*, ed. B. Chodźko, E. Feliksiak, and E. Konończuk, 57–71. Białystok: Towarzystwo Literackie im. Adama Mickiewicza, Oddział Białostocki; Uniwersytet w Białymstoku, Instytut Filologii Polskiej.

Sapiro, Gisèle and Lucile Dumont. 2020. The International Circulation of Structuralism: Between Appropriations and Rejections. In *Ideas on the Move in the Social Sciences and Humanities*, ed. G. Sapiro, M. Santoro, and P. Baert, 31–54. Cham: Palgrave Macmilian.

Saussure, Ferdinand de. 1983. *Cours de Linguistique Générale*. London: Duckworth.

Schahadat, Schamma, Michał Mrugalski, and Irina Wutsdorff. 2015. Modern Literary Theory in the Cultures of Central and Eastern Europe as an Entangled Intellectual History Beginning in the 20th Century to the Present: A Handbook Project at the University of Tübingen. *Slovo a Smysl—Word & Sense* 12: 231–238.

Seegel, Steven. 2018. *Map Men: Transnational Lives and Deaths of Geographers in the Making of East Central Europe*. Chicago: The University of Chicago Press.

Semczuk, Antoni. 2016. Kartki o powstaniu & rozwoju rusycystyki na Uniwersytecie Warszawskim. In *Nasz Uniwersytet: Wspomnienia pracowników Uniwersytetu Warszawskiego*, ed. Z. Boglewska-Hulanicka, A. Kędziorek, and B. Jankowiak-Konik, 377–387. Warszawa: Wydawnictwa Uniwersytetu Warszawskiego.

Siedlecki, Franciszek. 1937. *Studia z metryki polskiej*. Wilno: Fundusz Kultury Narodowej.

Skórczewski, Dariusz. 2020. *Polish Literature and National Identity: A Postcolonial Perspective*. Rochester: University of Rochester Press.

Sławiński, Janusz. 1994. Jakiej historii IBL-u nam potrzeba? *Teksty Drugie* 28: 137–154.

Smoczyński, Wojciech. 2001. Jerzy Kuryłowicz as Indo-Europeanist and Theorist of Language. In *Towards a History of Linguistics in Poland: From the Early Beginnings to the End of the Twentieth Century*, ed. E.F.K. Koerner and Meneses A. Szwedek, 255–272, Amsterdam, Philadelphia: John Benjamins Publishing Company.

Smoczyński, Rafał and Tomasz Zarycki. 2017. *Totem inteligencki: Arystokracja, szlachta i ziemiaństwo w polskiej przestrzeni społecznej*. Warszawa: Wydawnictwo Naukowe "Scholar".

Stieber, Zdzisław. 1952. *Rozwój fonologiczny języka polskiego*. Warszawa: Państwowe Wydawnictwo Naukowe PWN.

Stieber, Zdzisław. 1982. Pisownia i poprawna wymowa w Polsce niepodległej. In *Język i językoznawstwo polskie w sześćdziesięcioleciu niepodległości (1918–1978)*, ed. J. Rieger and M. Szymczak, 245–248. Wrocław: Zakład Narodowy im. Ossolińskich.

Szober, Stanisław. 1924. *Zarys językoznawstwa ogólnego*, vol. 1. Warszawa: Kasa im. J.Mianowskego.

Szymański, Wiesław. 2000. Stefan Kołaczkowski (1887–1940). In *Złota księga Wydziału Filologicznego*, ed. J. Michalik and W. Walecki. 362–376. Kraków: Księgarnia Akademicka.

Śródka, Andrzej (ed.) 1999. *Historia nauki polskiej wiek XX: Nauki filologiczne*. Warszawa: Wydawnictwo Instytutu Historii Nauki PAN.

Tambor, Jolanta. 2006. *Mowa Górnoślązaków oraz ich świadomość językowa i etniczna*. Katowice: Wydawnictwo Uniwersytetu Śląskiego.

Taszycki, Witold. 1946. *Słowiańskie nazwy miejscowe*. Kraków: Polska Akademii Umiejętności.

Tihanov, Galin. 2004. Why Did Modern Literary Theory Originate in Central and Eastern Europe? (And Why Is It Now Dead?). *Common Knowledge* 10: 61–81.

Tihanov, Galin. 2019. *The Birth and Death of Literary Theory: Regimes of Relevance in Russia and Beyond*. Stanford, CA: Stanford University Press.

Tomkowski, Jan. 2006. Nauka o literaturze w Polsce (1945–1990). In *Humanistyka polska w latach 1945–1990*, ed. U. Jakubowska and J. Myśliński, 156–182. Warszawa: Fundacja Akademia Humanistyczna; Instytut Badań Literackich PAN.

Tokarski, Jan. 1973. *Fleksja polska*. Warszawa: Państwowe Wydawnictwi Naukowe.

tp. 2018. Prof. Leon Zawadowski – wybitny filolog & teoretyk języka. https://bejsment.com/index.php/2018/08/09/zmarl-prof-leon-zawadowski/. Accessed January 21, 2021.

Turkowski, Andrzej. 2018. Polish Intelligentsia Totems in Elites' Struggles for Legitimization: The Case of Jerzy Giedroyc and Poland's Eastern Policy. *East European Politics and Societies: and Cultures* 33: 66-88.

Ulicka, Danuta. 2007. *Literaturoznawcze dyskursy możliwe: Studia z dziejów nowoczesnej teorii literatury w Europie Środkowo-Wschodniej*. Kraków: Universitas.

Ulicka, Danuta. 2018. Przegapiona okazja. *Teksty Drugie* 6: 217–230.

Ulicka, Danuta (ed.). 2020a. *Wiek teorii: Sto lat nowoczesnego literaturoznawstwa polskiego.* Warszawa: Instytut Badań Literackich PAN.

Ulicka, Danuta. 2020b. Rzut oka na nowoczesne polskie literaturoznawstwo teoretyczne. In *Wiek teorii: Sto lat nowoczesnego literaturoznawstwa polskiego*, ed. D. Ulicka, 9–160. Warszawa: Instytut Badań Literackich PAN.

Ulicka, Danuta. 2020c. Między światami. Rzeczywistość w literaturze – literatura w rzeczywistości – rzeczywistość literatury. In *Wiek teorii: Sto lat nowoczesnego literaturoznawstwa polskiego*, ed. D. Ulicka, 541–602. Warszawa: Instytut Badań Literackich PAN.

Ulicka, Danuta, and Jagoda Wierzejska. 2021. Jest po to, by uczyć myślenia: Z profesor Danutą Ulicką rozmawia Jagoda Wierzejska. *Nowe Książki* 3: 4–9.

Urbańczyk, Stanisław. 1993. *Dwieście lat polskiego językoznawstwa: 1751–1950*, vol. 69: Secesja.

Waldstein, Maxim. 2010. Theorizing the Second World. *Ab Imperio* 1: 98–117.

Warczok, Tomasz, and Tomasz Zarycki. 2014. Bourdieu Recontextualized: Redefinitions of Western Critical Thought in the Periphery. *Current Sociology* 62 (3): 334–351.

Warczok, Tomasz and Tomasz Zarycki. 2018. Polish Political Sciences in a Global Context. In *Oxford Research Encyclopedia of Politics*, ed. W.R. Thompson. Oxford University Press. https://doi.org/10.1093/acrefore/9780190228637.001.0001/acrefore-9780190228637-e-872.

Wąsik, Zdzisław. 1999. Językoznawstwo ogólne. In *Historia nauki polskiej wiek XX: Nauki filologiczne*, ed. A. Śródka, 37–63. Warszawa: Wydawnictwo Instytutu Historii Nauki PAN.

Wąsik, Zdzisław. 2001. General Linguistics in the History of the Language Sciences in Poland: Late 1860s—Late 1960s. In *Towards a History of Linguistics in Poland: From the Early Beginnings to the End of the Twentieth Century*, ed. E.F.K. Koerner and A. Szwedek, 3–52. Amsterdam, Philadelphia: John Benjamins Publishing Company.

Wellek, René and Austin Warren. 1949. *Literary history*. New York: Harcourt, Brace.

Wieczorkiewicz, Bronisław. 1959. Życie i działalność naukowa Stanisława Szobera. In *Stanisław Szober: Wybór pism*, ed. B. Wieczorkiewicz, 7–22 Warszawa: Państwowe Wydawnictwo Naukowe.

Wierzbicka, Anna. 1972. *Semantic Primitives*. Frankfurt/M: Athenäum-Verlag.

Wierzbicka, Anna. 1992. Semantic Primitives and Semantic Fields. In *Frames, Fields, and Contrasts: New Essays in Semantic and Lexical Organization*, ed. A. Lehrer and E. Feder Kittay, 209–227. London: Routledge.

Witkowska, Alina. 2002. Czas przypomnieć ojców dzieje. In: J. Tazbir (ed.) *Uczeni i ich badania.* 11–26. Warszawa: Kancelaria Polskiej Akademii Nauk.
Woleński, Jan. 2014. Szkoła Lwowsko-Warszawska z perspektywy historycznej. *Przegląd Pedagogiczny* 1: 9–18.
Wójcik, Bogusław. 1993. Spór o filozofię lingwistyczną w Polsce. *Zagadnienia Filozoficzne w Nauce* XV:59–68.
Wyderka, Bogusław. 2014. O rozwoju polskich dialektów. *Poznańskie Studia Polonistyczne: Seria Językoznawcza* 21 (41): 103–113.
Yurchak, A. 2003. Soviet Hegemony of Form: Everything Was Forever, Until It Was No More. *Comparative Study of Society and History* 45: 480–510.
Zarycki, Tomasz. 1998. Towards the Theory of Structures of Maps in Their Deep Concept Dimension. *Cartographic Journal* 35: 71–78.
Zarycki, Tomasz. 2002. Four Dimensions of the Centre-Periphery Conflict in the Polish Electoral Geography. In *Social Change: Adaptation and Resistance*, ed. T. Klonowicz and G. Wieczorkowska, 19–38. Warszawa: Wydawnictwa Instytutu Studiów Społecznch UW.
Zarycki, Tomasz. 2004. Uses of Russia: The Role of Russia in the Modern Polish National Identity. *East European Politics and Societies* 18: 595–627.
Zarycki, Tomasz. 2007a. History and Regional Development. A Controversy over the 'Right' Interpretation of the Role of History in the Development of the Polish Regions. *Geoforum* 38: 485–493.
Zarycki, Tomasz. 2007b. An Interdisciplinary Model of Centre-Periphery Relations: A Theoretical Proposition. *Regional and Local Studies* Special Issue 2007b: 110–130.
Zarycki, Tomasz. 2009. The Power of the Intelligentsia. The Rywin Affair and the Challenge of Applying the Concept of Cultural Capital to Analyze Poland's Elites. *Theory and Society* 38 (6): 613–648.
Zarycki, Tomasz. 2011. From Soviet to a Western-Dominated Political Scene: The Geopolitical Context of Politics in Poland. In *Cultural Transformations After Communism: Central and Eastern Europe in Focus*, ed. B. Törnquist-Plewa and K. Stala, 82–108. Lund: Nordic Academic Press.
Zarycki, Tomasz. 2012. The Embarrassing Russian Connection: Selective Memory of the Russian Heritage in Contemporary Poland. In *Russia's Identity in International Relations. Images, Perceptions, Misperceptions*, ed. R. Taras , 133–148 Routledge, London.
Zarycki, Tomasz. 2013. Aleksander Lednicki i los jego środowiska Polonii w Rosji na przełomie XIX i XX wieku jako zwierciadło przemian polskiego pola władzy *Sprawy Narodowościowe* 42: 67–83.

Zarycki, Tomasz. 2014. *Ideologies of Eastness in Central and Eastern Europe*. London: Routledge.
Zarycki, Tomasz. 2017. For a Relational Critical Discourse Analysis. *Stan Rzeczy/State of Affairs* 1 (12): 303–328.
Zarycki, Tomasz. 2020 Why Geography in Poland Has Never Radicalized: Political and International Entanglements of Polish Geography Seen Through the Prism of Antoni Kukliński's Professional Trajectory, *Serendipities. Journal for the Sociology and History of the Social Sciences*.
Zawadowski, Leon. 1949. Rzeczywisty i pozorny wpływ kontekstu na znaczenie: przedstawione na posiedzeniu Wydziału Nauk Filologicznych dnia 1 lipca 1949 roku. *Sprawozdania Wrocławskiego Towarzystwa Naukowego* 4 (2).
Zawadowski, Leon. 1950. On the Elements of Semantic Systems. *Biuletyn Polskiego Towarzystwa Językoznawczego* X: 77–97.
Zawadowski, Leon. 1952. *Zagadnienia teorii zdań względnych*. Wrocław: Wrocławskie Towarzystwo Naukowe.
Zawadowski, Leon. 1966. *Lingwistyczna teoria języka*. Warszawa: Państwowe Wydawnictwo Naukowe.
Zysiak, Agata. 2016. *Punkty za pochodzenie: Powojenna modernizacja& uniwersytet w robotniczym mieście*. Kraków: NOMOS.
Żabicki, Zbigniew. 1966. *„Kuźnica" i jej program literacki*. Kraków: Wydawnictwo Literackie.
Żółkiewski, Stefan. 1937. O podstawach metodologii badan literackich. *Życie Literackie* 1: 6–9.
Żółkiewski, Stefan. 1950. *Stare i nowe literaturoznawstwo: szkice krytyczno-naukowe*. Wrocław: Wydawnictwo Zakładu Narodowego im. Ossolińskich.
Żółkiewski, Stefan. 1951. *Badania nad literaturą polską: dorobek, stan i potrzeby*. Warszawa: Państwowy Instytut Wydawniczy.
Żółkiewski, Stefan. 1952. *Spór o Mickiewicza*. Wrocław: Zakład Narodowy im. Ossolińskich.
Żółkiewski, Stefan. 1987. Kazimierz Wyka - dyrektor Instytutu Badań Literackich. *Pamiętnik Literacki* 78: 21–24.
Żółkiewski, Stefan and Janusz Stradecki. 1955. *Rozwój badań literatury polskiej w latach 1944–1954*. Warszawa: Państwowy Instytut Wydawniczy.
Żyłko, Bogusław. 2009. *Semiotyka kultury: Szkoła tartusko-moskiewska*. Gdańsk: słowo/obraz terytoria.

4

Conclusion

4.1 Inevitable Entanglements in Polish Contexts

As mentioned above, this book was written as an attempt to go beyond the hitherto dominant interpretative schemes, particularly those concerning Polish history and the history of the social sciences and humanities in Poland. However, it inevitably has a purely Polish, or intra-Polish, dimension as well. In that sense, it is an attempt to contribute to intra-Polish debates. This strong connection with a purely Polish perspective can be seen in terms of the homology mentioned above. That is, the idea that departing from familiar patterns, especially those that create the basic senses of opposition, is only possible to a limited extent. If this were not the case, our texts, and social behavior in general, would not be intelligible. Thus, although this book attempts to rise above local Polish disputes, including political ones, it inevitably has to refer to them, and to some extent, remains a part of them. It will probably be read from this angle by some of my Polish readers.

I will mention here some of the more important of these contexts. First, there is the dispute about the role of communism and the Soviet

T. Zarycki, *The Polish Elite and Language Sciences*,
https://doi.org/10.1007/978-3-031-07345-8_4

Union in Polish history. It was essentially settled politically and ended with a condemnation of the role of both. This book has no intention of debating the pros and cons of communism. However, by offering a closer look at its role in the construction of the post-war social sciences and humanities in Poland, it can be read as a voice in this dispute. Related to this Polish dispute is a broader and deeper dispute over the role of Russia in Polish history. It, too, seems to be fundamentally resolved in favor of a general acceptance of a critical interpretation of Russia as an unambiguously negative factor throughout most of Polish history. This work, meanwhile, offers a much more ambivalent view of Russia's role in the development of Polish social sciences and humanities. In my view, the Polish–Russian relationship is crucial to understanding the historical development of Polish culture and science. Studying it is not only necessary for drawing a complete picture of the social sciences and humanities fields in Poland. Being a key axis of creative tension homologously linked to political and related disputes, the axis of Polish–Russian relations can be presented as one of the most interesting dimensions of intellectual dynamics in Central and Eastern Europe. As Maxim Waldstein suggested in the work discussed in the first part of this book (Waldstein 2010), it can even be universalized as a paradigmatic axis of creative tensions in the broader area of European civilization. It is hard not to notice that such a reading of the history of Polish–Russian relations after Russia's invasion of Ukraine may prove even more difficult than it has been so far. The third Polish dispute to which this book refers is the dispute over the country's relations with the West. In the Polish political field, relations between the parties to this dispute currently appear evenly balanced. However, the unusually high emotional tension that accompanies this dispute does nothing to facilitate a calm analysis of the intellectual dilemmas surrounding Poland's dependence on the broader Western core. Nonetheless, I hope that this book will also find its audience in Poland; one that will find it useful for analyzing the dilemmas of the development of the field of social sciences in a way which goes beyond the abovementioned contexts of current disputes with a strong political aspect. For this reason, some of the more general implications of the issues considered in this book are mentioned below.

4.2 Hegemony of the Intelligentsia as a Key Factor Shaping the Polish Social Sciences Field

One of the key arguments of this book is that the relative weakness of the Polish social sciences and humanities, especially in terms of their international status and influence, is a corollary of the special role of the intelligentsia in Poland. The intelligentsia is a cultural elite by definition, and in Poland, it is a dominant elite. This strengthens the position of culture in the Polish field of power, and consequently, in Polish society. However, I contend that after the 1980s the intelligentsia's privileged position has weakened the field of science on a purely academic level and proved detrimental to its position in the hierarchies of the global system of social sciences. This is because the privileged position of the intelligentsia elite in the field of power necessarily corrupts, restricts autonomy of the cultural field, and suppresses its creative forces. The power of the intelligentsia translates into numerous pressures, but above all, the pressure to assume the role of creator, promoter, and guardian of national culture, usually in its dominant variants. This culture, however, is largely a constitutive dimension of the political community. As mentioned above, this can be linked to the weakness of the political and (especially) economic elites, whose capacity to generate the framework of the national community is limited. The privileged position of the intelligentsia is a constant source of temptation to occupy privileged roles in the field of power and other fields close to it, like bureaucratic field for example. This applies to both purely political and less formalized roles, including social ones, which nevertheless entail certain privileges and sometimes have leadership aspects. However, they are a distraction from academic pursuits, and they demand a huge commitment, not least in terms of time. These pressures translate into the multi-positioning that is so typical of the intelligentsia, especially its elite. This feature of intelligentsia status manifests itself clearly in the biographies of many of the researchers mentioned in this book. They often feel a structural calling to assume a typically intelligentsia attitude of moral responsibility for the nation or society (sometimes for all of humanity, as in the case of communists or radical Catholics). For humanists, this pressure imposes a universal sense of obligation to set moral guideposts for social development and to assume responsibility for national culture. For conservatives, it is a matter of "protecting" or "saving" the national heritage. For progressives, it is about development and modernization, but is almost always built on historical foundations. An intelligentsia member is therefore something of a "secular priest", instilled with a strong moral drive to fulfill not only the duties of a scientist (for those in the academy), but also those of a guide, and a protector of the nation. These impulses are evident both in the consciousness of individual scholars, in their life paths, and in the structural features of the state created under their influence, and especially in individual institutions.

It bears repeating that the structurally dominant position of the intelligentsia in the field of power (at least since 1918) has been the key force conditioning a given tendency. The year 1918 witnessed the full institutionalization of a given domination in the sovereign Polish state. However, it should also be recalled that the Polish proto-field of power had existed much earlier, and the powerful position of the intelligentsia in it became apparent in the second half of the nineteenth century. The issue is therefore the position of the cultural elite in the dominant part of the field of power. Conditions favorable to the development of the social sciences emerge when they are generously funded by the state, while the fields of science and culture simultaneously enjoy considerable autonomy. Science in such a system tries to defend its independence from the dominant factions of the field of power, but is also concerned for the legitimacy of its position and anxious to secure adequate funding. This is largely based on a professional ethos and an ability to ensure international recognition defined by the standard of the global academic fields. Given the peripherality of Poland and the hegemony of its intelligentsia, academic autonomy of science is not all that highly valued. However, more important than international recognition for the legitimacy of the intelligentsia elite, and by extension, the scientific elite, are its moral, and often political, leadership. Thus, in Poland, national culture not only legitimizes the domination of the upper class, but to a large extent its very essence, since the dominant faction of the upper class is the intelligentsia. In most other countries (including Russia), the intelligentsia, or simply the cultural capital elite, are factions of the dominated factions of the elite. This makes them more likely to adopt strongly critical attitudes, including their attitudes toward the canons of national culture, and to create autonomous cultural, intellectual, and academic niches. This creates "creative margins"—at least when viewed from the fields of power of countries such as Russia or France, but also to some extent the Czech Republic (where the intelligentsia is also not as powerful as in Poland). Such creative margins were especially the intellectual circles envied by the Poles to the Russians or the Czechs, not to mention the French. It can therefore ironically be said that the Polish mainstream eats away at these margins the most, and weakens them to such an extent that they are unable to develop on a larger scale, least of all one that would bring international recognition. This help explain the imposing careers of Jan Baudouin de Courtenay and Leon Petrażycki, who were active on the creative margins of the Russian Empire, and were also present on the margins of the fields of power of other countries. Since becoming independent, Poland has not managed to create such permanent margins, and not only because of its average size and limited resources. In the case of linguistics, the Warsaw-Vilnius formalist circle was the beginning of such a margin, but it failed to develop into a permanent structure. The history of the IBL shows that not even the Stalinist system imposed by the Soviet Union could overthrow the hegemony of the intelligentsia. If anything, the margins in question widened during the 1960s. This was even more the case with history, where there was a clear division between the communist-subordinated recent history faction and the faction that studied earlier periods. In literary and linguistic studies, not even Stalin's sloganeering, as the above discussion of the papers of the First Congress of Polish Science amply demonstrates, could overturn the crucial role of the traditional canons of national culture or prevent most of the intelligentsia from working on them. This is obviously not to say that some were not temporarily marginalized. The IBL, especially after 1956, proved to be relatively inclusive and attracted a larger portion of the field's elite to work on national culture. Its relatively rebellious factions never contested the dominant role of the intelligentsia; they merely called for changes in the factional hierarchy. Hence, most major debates in the field were centered around current politics. As such, they were not conducive to the deeper intellectual schisms that would have established inspiring new intellectual paradigms. Rather they can be seen as a homologous transfer of cleavages from the field of power, which are essentially internal-intelligentsia divisions. Thus, despite their often emotional nature, they rarely went beyond rather detailed disputes about the reinterpretations of the canons of the national culture in terms of political divisions at any given moment and their reference to dynamically changing global political trends.

Thus, the intelligentsia elite is under a lot of pressure to work hard on topics that are not likely to lead to internationally recognized academic discoveries, but which will be of service in domestic intellectual debates and even political disputes. This is eloquently illustrated by Maria Janion, one of the most important literary scholars associated with the IBL. In an interview with Kazimiera Szczuka (Janion and Szczuka 2012), she said that she had always felt a broader, public responsibility as a literary historian, adding that, "In the Polish tradition, the study of literature had the status of an anointed domain, a national good. We were to be the servants of the sacred canon, the very core of which was Romanticism" (Janion and Szczuka 2012: 45).

4.3 Polish Structuralism and Interpretations of Its Failure: From Trajectories of Selected Scholars to the Global Scale

On an individual level, a similar hypothesis with specific reference to the history of Polish structuralism can be put forward with regard to the biography of Maria Renata Mayenowa, who is usually presented as someone who worked tirelessly on behalf of Polish culture and science (Janus 2021). Michał Głowiński wrote that she appeared far more concerned about the fate of the *Dictionary of Sixteenth Century Polish* and an edition of Jan Kochanowski's works that she was coordinating than her own fame and career (Janus 2021: 325). In her dedication to these titanic projects aimed at strengthening the edifice of Polish culture, Mayenowa's biography reveals her intelligentsia nature. In a similar vein,

Artur Hellich writes of "a reflex of intelligentsia duty [in which] literary scholars also published strictly autobiographical works, usually motivated by the need to bear witness. Their authors presented themselves as witnesses of the turbulent history of the twentieth century" (Hellich 2020: 377–378).

Michał Łuczewski, in turn, presents a similar thesis, although set in a slightly different theoretical framework in his biographical study of the renowned Polish sociologist Maria Ossowska (Łuczewski 2020). Łuczewski argues that, despite her ambitions to take to the world stage, Ossowska failed to promote her "sociology of morality" project internationally. While it was ostensibly an empirical research project, i.e. a purely academic exercise, Ossowska was perceived in the West as the "priestess of an unknown religion". The "unknown religion" was in fact the "religion of the intelligentsia". Her academic works, possibly because of their strong grounding in intelligentsia culture, were read primarily as commentaries on contemporary Poland. Thus, Łuczewski suggests that her status as an intelligentsia celebrity, and in time even an intelligentsia totem, ensured her greatness in Poland but hindered her international career. However, two aspects of intelligentsia membership need to be distinguished in this context. What most hampers the international careers of the elite of the Polish intelligentsia is their dominant position in the Polish field of power. It is this that presses them into service as the guardians of national culture. They might occasionally lean toward irony or criticism, but they nevertheless take this role very seriously. Intelligentsia opinions that are decidedly critical of the intelligentsia are normally only directed at competing factions. This dominant position, based as it is on structural mechanisms, makes them *nolens volens* priests of national culture, and prevents them from adopting fully revolutionary or even autonomous positions, as they feel a structural responsibility toward the nation. In contrast, the intelligentsia in other countries do not have, and seldom feel, such a strong responsibility, because they occupy dominated positions in the field of power. As a result, the intelligentsia can develop into a much more critical force, and be more creative and courageous. In this sense, the intelligentsia identity gave strength to Baudouin de Courtenay, the Russian formalists he inspired, and the members of the Prague School. Note too that Osip Brik wrote that

OPOJAZ and its formal methodology was "... a bogeyman for the clergy practicing literature. The brazen attempt to look at poetic sanctities from a scientific point of view provoked loud indignation" (Mayenowa 1970: 42). Some Polish formalists, from Vilnius and Warsaw, tried to act in a similar vein in the late 1930s. Artur Hellich points out, for example, that Franciszek Siedlecki wrote about "the artificiality of the division between theoretical work and artistic creation". He considered it a slogan proclaimed by "bourgeois theoreticians of literature", whom he sarcastically dubbed "priests of Pure Science" (Siedlecki 1936). However, as described above Siedlecki's circle, or more precisely, those of its members who survived the war, were co-opted into the intelligentsia faction led by Stefan Żółkiewski. This was the faction that became part of the dominant sector in the field of power.

Żółkiewski's trajectory perfectly manifests one of the key paradoxes of the development of the Polish social sciences, especially linguistics and literary studies. The ease of entry into the field of power, and additionally, cyclical offers to members of the intelligentsia elite to take up important positions in its dominant sectors, turn out to be corrupting for the development of the scientific field. It can be argued that in the late 1940s Żółkiewski was among the few younger scholars best predisposed to develop a direction that could become not only the trademark of Polish linguistics, but also its main competitive advantage, viz. the synthesis of linguistics with structural sociology. One potential form of such a synthesis was the aforementioned interaction between Marxism and Formalism, which was particularly creative in the USSR in the 1920s. In Poland, similar processes, albeit less pronounced, were cut short first by World War II and then prevented from recommencing by the advent of Stalinism around 1949. However, if Żółkiewski had not held high party posts until 1968, and had been engaged mainly in scientific activity, perhaps not only his creative output, but also the state of the entire field of sciences of language might have looked very different. It is striking that, as Głowiński notes (Głowiński Wołowiec 2018), it was only after being relieved of all his political posts in 1968, that Żółkiewski began to be seen frequenting the IBL library. Shortly thereafter, Żółkiewski, who turned 60 in 1971, wrote several impressive books. The first that should be mentioned is the one devoted to the "literary culture" of interwar

Poland (Żółkiewski 1973). Another ambitious work, which explored the contexts of the development of the literary field in Poland since 1890, was written in the 1980s but only appeared posthumously (Żółkiewski 1995). These studies seem modern even by today's standards. At the time they were written, they were positively trailblazing. They combine structural sociological analysis with modern literary research methodologies. If they were translated into foreign languages, they would probably find an interested readership in the dynamically developing field of historical sociology of literature. In particular, they seem to resonate with Gisèle Sapiro's work on the French literary field (Sapiro 2014). However, in the context of the 1970s and the subsequent polarization of the intellectual field along quite different lines described above, they have attracted little attention. Żółkiewski's standing in the field at the time undoubtedly contributed to this. He was perceived as a retired politician rather than a scholar, and his intellectual interests may have appeared somewhat anachronistic at a time when structuralism was in retreat. It can therefore be said that his political career and, more broadly, his position in the field of power, which he enjoyed until 1968, meant that he failed to realize his intellectual potential in the field of pure science. As I have suggested, this potential resulted not only from his personal abilities, but also from his specific trajectory in the fields of language and literature studies, which predestined him to play the role of a leader of the Polish school of sociology of literature.

Roman Jakobson can also be considered a quintessential intelligentsia member. His social world consisted primarily of the intelligentsia salons, where he met important figures in the field of power from whom he drew countless intellectual inspirations. However, except for a brief moment around 1918, when he found himself in a faction of the intelligentsia associated with the Bolsheviks, who had just become the dominant force in the Russian field of power, he was usually with the dominated factions of the field of power. This, it seems, gave his work dynamism and theoretical momentum. These characteristics of the Russian formalists, politically connected with the left, and thus operating on the margins of official intellectual life in the Russian Empire, also applies to the Prague School. It can be even seen as an extension of the formalists' activity, as several important Russian intellectuals, especially Jakobson and Trubetzkoy, emigrated to Czechoslovakia, where they were able to find conditions conducive to intellectual development in their work with Czechs and others. However, they were all related to the dominated part of the field of power of interwar Czechoslovakia. They benefitted from state support for academic institutions and they enjoyed autonomy, but they were not tied to the dominant factions in the field of power. The Prague school enjoyed a similar status again during the 1960s, when it was going through another period of flourishing. The same can be said of the Tartu-Moscow school, although while it developed dynamically, it was not linked to the dominant forces in the Soviet field of power. Moreover, the caesura of 1968 further clipped the wings of these already dominated circles. The Prague school, in particular, virtually ceased active functioning. Yuri Lotman can also be considered a classical intelligentsia member who used his intelligentsia status as a tool for intellectual and political activity, albeit from a decidedly dominated position in the field of power. For example, during the late Soviet period, he promoted the juxtaposition of a positive, i.e. intelligentsia and sophisticated, Russian identity with a negative Soviet identity, which he defined as primitive, boorish, and politically harmful (Waldstein 2007). This implied an idealization of intelligentsia ideals, but it was also an unequivocal admission of dominated status within the USSR elite. In the Russian field of power, as discussed in early chapters, the liberal intelligentsia only found itself in a position of privilege, although not dominance, for a brief period at the turn of the 1990s. Lotman was thus something of a "consecrated heretic". This is the term Dumont (2020) used to describe Roland Barthes. However, France differs from Poland and Russia in several ways. Crucially, prominent intellectuals in Poland and Russia are all members of the intelligentsia elite. This automatically brings them closer to the field of power and facilitates contacts with other factions of the elite, including politicians and artists. As Dumont (2020) shows, active contacts with these circles helped Barthes achieve intellectual star status in France. In Poland and Russia, this is a structural mechanism to valorize most of the intelligentsia who happen to be scholars.

Similarly, Poland differs significantly from France and Russia. For one thing, Poland occupies a much lower rung in international hierarchies, including the global field of the social sciences. For this reason, very few Polish scientists have distinguished international careers. The most notable exceptions in the field of linguistics are Jan Baudouin de Courtenay and Jerzy Kuryłowicz. Kuryłowicz established strong relations with prominent US linguists, initially by gaining contacts and experience in the interwar period thanks to a Rockefeller Foundation fellowship, and then, from 1957, by visiting leading US universities. It can further be hypothesized that the lack of any wider globally recognized successes in Polish linguistics is mainly due to Poland's peripheral position. The intellectual revival of the 1960s was the exception rather than the rule, and resulted from Poland's intermediary role between the USSR and the West. Dumont (2018) argues that the dynamics of global literary theory in the 1960s was mainly concentrated in a triangle whose vertices were France, the United States and the Soviet Union. This probably explains the successes of both the Russian formalist school and the Prague school of linguistics. These had a common factor in the person of Roman Jakobson—a Russian of Jewish origin who relocated to Prague and later joined the elite of the US academic field. The unique position of Poland as an interface periphery attracted Jakobson to Poland, but it was not enough to have the achievements of Polish linguistics included in the global canons.

Thus, while the conditions for the development of modern linguistics in Poland during the communist period were no less favorable than those in Czechoslovakia or the USSR, no "Polish school" emerged. Ulicka gives several reasons for this spectacular failure. In particular, she calls Polish structuralism "… a cryptonym for an ideological stance that is as much cognitive as it is social, political, and ethical, and which was inherited from the nineteenth-century liberal tradition in its various guises" (Ulicka 2020: 119). She also points out that "Polish structuralism is located in the space of social communication between science and politics. Bear in mind that the ideal of 'pure science' is a recent one in modern science, perhaps only dating back to the first decade of theory building in the 1920s, and even then, mainly among scholars born in the 1890s" (Ulicka 2020: 122). Finally, she hypothesizes that perhaps "… there was no Polish formalism, no structuralism, no Polish phenomenology, no hermeneutics, no psychoanalysis, no sociology of literature, and none of those 'schools' and 'directions' into which the field of literary studies is customarily divided. They were there, but they were intertwined" (Ulicka 2020: 128). These theses are perfectly plausible.

However, not all of the reasons Ulicka gives for Poland's failure and Russia and Czechoslovakia's success are compelling. In particular, her arguments for cultural diversity, or a kind of "spirit of place", being a causal factor are not entirely persuasive. Jakobson promoted a certain "borderland" mythology, or even ideology. I have previously written about this in critical terms, especially in *Ideologies of Eastness in Central and Eastern Europe* (Zarycki 2014). Jakobson argued that Czechoslovakia lay at the crossroads of different cultures, and that the historical peculiarity of its culture was the result of the creative blending of currents from distant sources since the Cyrillo-Methodian era. The interweaving of various, at times even opposing, currents is supposed to explain the appeal of Czech art and ideology during the most creative periods of Czech history. At least this is what Jakobson claimed in a lecture delivered at the IBL in Warsaw in 1958 (Gierowski 2018: 31). The Czech Republic, as a cultural borderland that had combined different traditions for centuries, was supposedly predisposed to create an intellectual space of understanding and cooperation. Gierowski argues that this vision can be seen as a forerunner of more recent popular visions of Central Europe, e.g. the one proposed by Kundera (Gierowski 2018: 33). Ulicka (2007) expands on this idea by describing Prague as the node of the European modernism network that sits at the intersection of the Russian formal school, structuralist linguistics (Roman Jakobson, Petr Bogatyrev), the Czech avant-garde, French surrealism (André Breton, Paul Éluard), and phenomenological philosophy (Edmund Husserl, Jan Patočka). All these cosmopolitan strands have been blended creatively in Bohemia (Ulicka 2007: 272). In a similar vein, Galin Tihanov (2004) links four countries in his analysis of the origins of literary theory, viz. Russia, Hungary, Poland, and Czechoslovakia. His arguments also seem to be strongly culturalist. He identifies the first key factor as "… a strong domestic tradition of philosophy that could impose its authority and thereby prevent the purposeful transformation and modification of established philosophical discourses into tools of literary theory. The intelligentsia in these countries lived after World War I on borrowed philosophical capital, mainly of German-Austrian provenance" (Tihanov 2004:

80). He further argues that "Having been parts of (by-then-defunct) empires before World War I (or having been an empire in its own right, in the case of Russia), each of the four countries was a natural locus of polyglossia and heterotopia (exile), providing -sometimes at the cost, and in the form, of bitter ethnic conflicts - a painfully beneficial environment for theoretical contemplation of literature beyond the presumed naturalness of native tongues and traditions". Tihanov also employs the notion of "regime of relevance" and contends that a country's creativity is a byproduct of its configuration of regimes of relevance (which may overlap), arguing that "In all four countries, there existed a unique blend of alienation from and identification with the type of nation-state formation characteristic of the period - an ambivalence suggestive of the complex interaction between the inherited and new regimes of relevance by which literature was judged and consumed" (Tihanov 2004: 80). This analysis, however, lacks any indication of the key factor in the configuration of the field of power in each country. Crucially, the Polish intelligentsia can obviously have an ambivalent relationship to the state, but regardless of whether the Polish state exists, the intelligentsia has been the dominant force within the Polish field of power since about 1918. Disputes within the Polish intelligentsia are thus mainly disputes between competing variants of the dominant ideology. By contrast, similar intelligentsia ideologies in Czechoslovakia, and especially Russia, were the ideologies of the dominated part of the field of power. All these waves of intellectual activity, especially those of structuralism or formalism, periodically entered into certain synchronizations. They also included the development of Western intellectual fields. However, it is difficult to speak of a clear international homology here. Structuralism was the ideology of much of the liberal intelligentsia in Poland during the 1960s. This group occupied positions homologous to those privileged in the field of power. Some of its representatives could be even seen as being members of the field of power at the time. Prior to 1965, there was something of a balance of power between the liberal and hardline communist factions in the field of power. It could be seen as a configuration creating homologically creative tensions in several intellectual and academic fields. At the same time, the country found itself advantageously placed at an interface periphery between East and

West. This facilitated intellectual exchange. Another important aspect of the intellectual boom of the 1960s were the aspirations of the Soviet bloc to project itself as an alternative model of modernity—sometimes with tangible results. Given the central role of science in the East–West rivalry to become the dominant modernization model, academic debates automatically assumed an additional political dimension. This was not, however, simply a mechanism to legitimize domination, but a genuine belief in science as a major driver of economic and social progress. As long as the Soviet bloc remained confident that it could achieve victory, or at least hold its own, in its confrontation with the West, academia, especially, but by no means solely, the physical sciences, enjoyed many privileges, and theoretical debates were judged for their contribution to scientific progress, as well as their political utility. This led the communist authorities to give more respect to the autonomy of academia, especially after 1956, and allow discussion on Western theoretical concepts, even if they were frowned upon politically. This began to change in the 1970s, when the ruling elites of communist countries realized that it could not challenge Western hegemony in the global production of knowledge.

Thus, once the special conditions that had fostered the "golden 60s" ceased to exist, interest in structuralism waned, and the memory of its Polish school slowly faded with it, even in Poland. An illustration of the thesis of Polish structural otherness may be found in the homology between the structures of the fields of power and academia and the geographic space of cities. Ulicka suggests comparing the geographies of Warsaw and St. Petersburg. In the former case, Oboźna Street separates two crucial institutions: the Department of Polish language (*polnistyka*) at the University of Warsaw, which, during the 1960s and 1970s, was dominated by a group of scholars loyal to the party, a circle best symbolized by the figure of Jan Zygmunt Jakubowski; and Staszic Palace, where the liberal-dominated IBL was located. Similarly, the Neva River in St. Petersburg demarcated official and anti-official Russian literary studies in the 1920s (Ulicka 2020: 92). Ulicka also points to a broader homology with Western countries in that all the currents considered innovative intellectual paradigms, starting with structuralism and formalism, were born "outside the center", both figuratively and geographically (in both countries and major cities), and were always peripheral or "off" in character (Ulicka 2007: 103–104). The same homology which Ulicka recounts in the case of Warsaw is invoked by Głowiński (Głowiński and Wołowiec 2018: 287), but with an additional layer. He sees Oboźna Street as the symbolic border between Europe and Asia, marked as it was by the orientalism that typified the Polish intelligentsia. Obviously, the IBL represented Europe, i.e. progress and civilization, while the Polish Studies Department at University of Warsaw, headed by Jakubowski, represented Asia, i.e. barbarism, tyranny, and backwardness. This metaphor arrogantly but aptly, as it seems, illustrates the power of the intelligentsia in Poland, especially its liberal sector. The camp of loyal linguists symbolized by Jakubowski, even if it enjoyed the support of the communist government, was always in a weaker position symbolically. One possible exception was the relatively brief Stalinist period, but at the time, as discussed earlier, most of the intelligentsia circles that later formed its liberal core actively collaborated with the communist government, so the reading of that configuration may be ambiguous. In fact, the balance of forces between these camps may be also seen as equalized also in the 1960s, but certainly the liberal faction, at least until 1965, cannot be seen as dominated. In any case, as illustrated by Głowiński's statement, the party loyalists were clearly assigned the stigma of inferiority, failure, and moral weakness by the liberals. Even if this view could be considered a "position taking", in the 1960s, it has become a doxa at the end of the 1980s.

4.4 On Possible Lessons from the Polish Case Study

The first part of this book attempts to adapt and develop field of power theory in what can be described, with reference to Wallerstein's world-systems theory, as "peripheral" or "semi-peripheral", and which is exemplified by Poland. Another key concept which is adapted and critically developed in the present book is homology. The mechanism of homology, which is central to Bourdieu's field of power theory, rests on the assumption that the structures of all the fields in a given society are in some degree dependent on the structures of the field of power. This assumption does not disregard the fact that all fields, even those strongly dependent on the field of power, enjoy some degree of autonomy. Bourdieu and his followers have convincingly shown how universal and crucial the mechanism of homology is. As argued in the introductory chapter, most social actions, and especially those that are symbolic in nature, help enforce homology, thereby producing meanings. Importantly, new meanings are only created by partial homology, as full homology is tantamount to producing redundant messages or frames. Academics are involved in the workings of homology by linking (not necessarily in simple or predictable ways) their debates to issues and cleavages defined by the field of power. Some of the Polish linguists and literary scholars discussed in this book illustrate this mechanism quite well. It can also be argued that the degree of homology between the field of power, the field of politics, and specific sub-fields of culture and academia, define not only the autonomy but also the innovative nature of the last as these. In particular, once homology is strong, i.e. when cleavages strongly converge, academic sub-fields merely translate the political debates and cleavages of the elite. When homology decreases, autonomy appears and academic and cultural sub-fields may be seen as more "interesting" and "creative". However, once weakness of homology reaches an "excessive" level, the creative fields can seem "irrelevant" or "detached from reality". The phenomenon of the "Polish golden 1960s" can be construed this way. The creative tension of these years was due to there being just the right degree of homology between the field of power and academic fields. Obviously, the global context played a part. As mentioned above, East–West rivalry created conditions in which the autonomy of academia and the relevance of theoretical debate was widely respected, even by the communist leadership. Once this conviction started to weaken, homology increased in the 1970s to reach even higher levels in the 1990s and beyond. This eventually sapped the creative spirit in most of the social sciences and humanities in Poland.

As argued in the first part of this book, the basic structure of oppositions within the semi-peripheral field of power usually differs from the structure of the field of power in the core states of the world system. This key difference reconfigures the power relations and, consequently, the structures of the main social struggles and meanings in the (semi)periphery. Thus, in Bourdieu's model, the meta-field of power has the economic field at one pole and the field of cultural production at the other with several other fields in between. Similar structures can be also identified in the peripheries. However, due to the limited autonomy of individual fields, the "pro vs. anti-center" axis is the crucial dimension of the peripheral field of power. Consequently, that cleavage, crossing Bourdieu's culture vs. economy (autonomy vs. heteronomy) axis, sets the directions and determines the meanings of the main social struggles in the peripheries. In effect, models of the peripheral field of power, along with other fields (cultural, economic, religious), are divided between the sector nearest to the core of the global, capitalist system (in Wallerstein's sense) and its polar opposite, which keeps its distance and sometimes even manifests its antipathy to it. In other words, we can talk about pro-center and anti-center positions and, consequently, about adopting positions in that field. This pits the Western-oriented, "modern" and "liberal" against the "parochial", "traditional", and "conservative". It should go without saying that this cleavage also appears in some core countries.

Because all fields depend on it, the national field of power is the central generator of social meanings. It achieves this primarily by defining the axes of social and symbolic conflicts that are key to establishing meaningful oppositions for building binary codes of social communication. These functions can be also seen as defining key values and establishing the ritual of the civil culture in society, as demonstrated by Jeffrey Alexander (Alexander 2006). In this way, the field of power defines key social oppositions, hierarchies, classification rules, and the limits of what can be said and what views can be legitimately expressed. These norms are often naturalized in the figure of an ideal citizen (Zarycki et al. 2022).

The two basic standpoint categories, defined by any field of power, can be referred to Bourdieu's notions of orthodoxy and heterodoxy. Additionally, doxa, as used by Bourdieu, refers to naturalized norms of social organization which are practically impossible to challenge, mostly because of their social invisibility, i.e. naturalization. Any attempt to redefine them risks being decried as immoral or irrational, as the fields of power can also define basic moral distinctions, as well as rational norms. It can be expected that it is also in the field of power where the basic mechanisms of governance, as defined by Bob Jessop (2012), are negotiated and enacted. These include establishing a common world-view (social imaginary) for individual action, stabilizing the orientations, expectations, and rules of conduct of the key players, and simplifying models and practices to reduce the perceived complexity of the world. Jessop argues that these mechanisms must be sufficiently variegated to be congruent with real-world processes and to remain relevant to governance objectives and to the development of a capacity for dynamic interactive learning about various causal processes and forms of interdependence, attributions of responsibility and capacity for actions, and to the possibilities of coordination in complex, turbulent environments. The dominant modes of socially legitimate self-reflexivity, which may be seen as a critical technique of governability, are also defined in relation to the norms defined in the fields of power. Particular sub-fields in a given society may be seen as arenas of governing selected social processes through which fundamental norms and meanings, negotiated in the field of power, are imposed on specific realms. These fields include academia and its individual disciplines, which enjoy a specific degree of autonomy and can influence the external world and the field of power, but which can also be seen as tools of the field of power.

Polish linguistics and literary studies is a case in point, given that this field enjoys considerable autonomy and is influenced by strong homological forces exerted by the national field of power. As a result, it has naturalized the doxa of intelligentsia hegemony and reinforced it by legitimizing the intelligentsia-controlled norms of the Polish language and its variants and readings of the national literary canon. At the same time, the field is internally divided along the lines of the main political divide of the field of power, i.e. the cleavage that pits Euro-enthusiastic liberals against Euroskeptic conservatives. The homology is not, however, complete, given that the left-liberal camp in the field of the social sciences and humanities dominates the conservative one. The relationship between the forces in the field of power, as well as in the political field is currently the reverse, with the conservative camp having a clear superiority. However, as has been argued, even if the politically dominant liberals in the field of linguistics and literary studies are Euro-enthusiastic or even cosmopolitan, they are still guardians of the canons of Polish language and literature, and this is defined in very traditional, nation-centered terms. Above all, they are, together with their conservative opponents, the guardians of intelligentsia hegemony, which is legitimized by the "cultured" forms of language use and competence in national literature. This in turn is one of the major components of the Polish cultural capital on which the entire social hierarchy of the country rests.

This demonstrates that specific academic disciplines can be seen as institutionalized spheres for implementing the rules of the field of power. The legitimization these disciplines provide, largely by virtue of their autonomous status, is what enables them to function this way. They play specific roles in producing and naturalizing social hierarchies and classification systems in such realms as legal norms, national identity, and national language norms. In a peripheral nation-state such as Poland, another important function is to define and delimit the boundaries of national autonomy (sovereignty) and identity. Academic disciplines can therefore be seen as tools for defending the autonomy of the national fields of power in a global context. In all such fields, distinctions between "Us-Poles" and significant others, especially other nations, are also systematically delimited and implemented. This defense of the autonomy of specific sub-fields, as well as the national field of power, is not meant to imply that they are immune to external forces. On the contrary, such forces can wield considerable influence and even prove crucial to configuring the peripheral field of power. This strong dependence, however, does not mean the disappearance of autonomy, but only that this autonomy is defined above all within the framework defined by this external dependence.

The state plays a major role in creating a national field of power, and in building its structures, along with those of the other dependent fields. However, this does not mean that this role is always played by the nation-state. It can also be performed by broader state structures, particularly imperial structures, which can create national fields of power in ways that are not entirely intentional. Following George Steinmetz's studies on German colonies in Africa (Steinmetz 2008), on the peripheries of the world system an autonomous proto-field of power that is locally more stable and powerful than that of weak and passing peripheral states or distant empires is likely to be found. In Poland, at the turn of the nineteenth and twentieth centuries, a strata of national intelligentsia (a new social grouping at the time) helped sustain an autonomous proto-field of power in which it assumed a dominant position and which consequently appeared more stable and more influential in Polish society than the state structures governing Polish territory. This perspective could be seen as an innovative approach to the origin and nature of the intelligentsia. Moreover, it is hoped that this view on the emergence of the Polish nation-state, with its own social structures and institutions, independent of the Russian, Austrian, and German empires at the turn of the twentieth century, together with the impact of the Soviet system on the Polish field of power and academia, will contribute to the development of the emerging field of post-colonial and post-imperial sociology, and more generally, to the field of new relational historical sociology.

References

Alexander, Jeffrey C. 2006. *The Civil Sphere*. Oxford: Oxford University Press.

Dumont, Lucile. 2018. Théories sans frontières? Les théories littéraires en France et la construction d'un espace transnational, années 1960-années 1970. *Actes de la recherche en sciences sociales* 2018/4 (224): 46–63.

Dumont, Lucile. 2020. Literary Theorists in and Beyond French Academic Space (1960–1970s). *The Sociological Review* 68 (5): 1108–1123.

Gierowski, Piotr. 2018. *Praski strukturalizm: przekroje, miejsca, tożsamości*. Kraków: Wydawnictwo "scriptum".

Głowiński, Michał and Grzegorz Wołowiec. 2018. *Czas nieprzewidziany: Długa rozprawa bez pana, wójta i plebana*. Warszawa: Wielka Litera.

Hellich, Artur. 2020. Heterologie. Między literaturoznawstwem a literaturą. In *Wiek teorii: Sto lat nowoczesnego literaturoznawstwa polskiego*, ed. D. Ulicka, 369–424. Warszawa: Instytut Badań Literackich PAN.

Janion, Maria and Kazimiera Szczuka. 2012. *Transe, traumy, transgresje*. Warszawa: Wydawnictwo Krytyki Politycznej.

Janus, Elżbieta. 2021. A Page from the History of the University of Vilnius: Selected Biographical Threads. *Vilnius University* Open Series 314–327.

Jessop, Bob. 2012. Social Imaginaries, Structuration, Learning, and 'Collibration': Their Role and Limitations in Governing Complexity. *Zarządzanie Publiczne* 1: 71–84.

Łuczewski, Michał. 2020. Jak nie zostać globalnym intelektualistą? Przypadek Marii Ossowskiej. In *Spotkania z Ossowskim*, ed. A. Sułek, 348–385. Warszawa: Wydawnictwo Naukow Scholar.

Mayenowa, Maria Renata. 1970. Rosyjskie propozycje teoretyczne w zakresie form poetyckich (1916–1930). In *Rosyjska szkoła stylistyki: Wybór tekstów*, ed. M.R. Mayenowa and Z. Saloni, 14–54. Warszawa: Państwowy Instytut Wydawniczy.

Sapiro, Gisèle. 2014 *The French Writers' War, 1940–1953, Politics, History, and Culture*. Durham and London: Duke University Press.

Siedlecki, Franciszek. 1936. Martwa nauka. *Lewar* 3.

Steinmetz, George. 2008. The Colonial State as a Social Field: Ethnographic Capital and Native Policy in the German Overseas Empire Before 1914. *American Sociological Review* 73 (4): 589–612.

Tihanov, Galin. 2004. Why Did Modern Literary Theory Originate in Central and Eastern Europe? (And Why Is It Now Dead?). *Common Knowledge* 10: 61–81.

Ulicka, Danuta. 2007. *Literaturoznawcze dyskursy możliwe: Studia z dziejów nowoczesnej teorii literatury w Europie Środkowo-Wschodniej*. Kraków: Universitas.

Ulicka, Danuta. 2020. Rzut oka na nowoczesne polskie literaturoznawstwo teoretyczne. In *Wiek teorii: Sto lat nowoczesnego literaturoznawstwa polskiego*, ed. D. Ulicka, 9–160. Warszawa: Instytut Badań Literackich PAN.

Waldstein, Maxim K. 2007. Russifying Estonia?: Iurii Lotman and the Politics of Language and Culture in Soviet Estonia. *Kritika: Explorations in Russian and Eurasian History* 8: 561–596.

Waldstein, Maxim. 2010. Theorizing the Second World. *Ab Imperio* 1: 98–117.

Zarycki, Tomasz. 2014. *Ideologies of Eastness in Central and Eastern Europe*. London: Routledge.

Zarycki, Tomasz, Rafal Smoczynski and Tomasz Warczok. 2022. Cultural Citizenship without State: Historical Roots of the Modern Polish Citizenship Model. *Theory and Society* 51 (2): 2022: 269–301

Żółkiewski, Stefan. 1973. *Kultura literacka (1918–1932)*; Wrocław: Zakład Narodowy im. Ossolińskich.

Żółkiewski, Stefan. 1995. *Społeczne konteksty kultury literackiej na ziemiach polskich (1890–1939)*. Warszawa: Wydawnictwo Instytutu Badań Literackich.

Index

T. Zarycki, *The Polish Elite and Language Sciences*,
https://doi.org/10.1007/978-3-031-07345-8

C

K

O

Q

R

U

V

Y

Z

Kolibri 360 GmbH
Ferdinand-Jühlke-Straße 7
99095 Erfurt, Deutschland
produktsicherheit@kolibri360.de